# Neurokinetics

Albert Gjedde • William R. Bauer
Dean F. Wong

# Neurokinetics

## The Dynamics of Neurobiology In Vivo

Springer

Albert Gjedde
Copenhagen University
Department of Neuroscience
and Pharmacology
3 Blegdamsvej
2200 Copenhagen N
Denmark
gjedde@sund.ku.dk

Dean F. Wong
Johns Hopkins Medical Institutions
JHOC Bldg room 3245
601 N. Caroline St.
21287 Baltimore Maryland
USA
dfwong@jhmi.edu

William R. Bauer
2053 Old Washington Road
21157 Westminster Maryland
USA
pet@crofter.org

ISBN 978-1-4419-7408-2 e-ISBN 978-1-4419-7409-9
DOI 10.1007/978-1-4419-7409-9
Springer New York Dordrecht Heidelberg London

Library of Congress Control Number: 2010937988



Printed on acid-free paper

Springer is part of Springer Science+Business Media (www.springer.com)

# Preface

Attempts to understand physiological processes by quantification and interpretation of observations made in vivo have challenged the biological and physical sciences for centuries. From the earliest physiological experiments in living organisms, the joint approaches of biology and physics to the discovery of these processes, from cells to humans, have yielded profound insights and have had a major impact on our understanding of all organ systems and on the modern practice of medicine as a whole. The work of Helmholz (1821–1894) is an example of the early merger of physics and biology that ultimately led to the most recent formulation of a systems biology approach that is no less than the quest for complete quantification of the dynamic processes of entire organisms and organs in health and disease, for example in the shape of the Physiome Project of Bassingthwaighte (2000) and the Blue Brain Project of Markram (2006).

In this compendium, we focus on the dynamics of brain physiology in vivo from the perspective of the methods of tracer kinetics (neurokinetics). Applications of neurokinetics seek to measure the processes that take place in the tissue without disturbing these processes, and subsequently to map these measurements onto images of brain tissue.

Applications of physiological kinetics (including neurokinetics) use "indicators" or markers, ranging from the dyes introduced at the dawn of experimental physiology, via stable (nonradioactive) or unstable (radioactive) isotopes introduced in the 1960s, to the most recent methods of in vivo imaging of optical, magnetic resonance (MR), and magnetic field (MEG) signals for visualization and detection.

The authors dedicate this book to the consolidation of many neurokinetic concepts with roots in the neurophysiology of the mid-20th century with the state-of-the-art imaging and parametric mapping methods of the first decade of the 21st century.

In one of the earliest attempts to quantify the pharmacokinetics of a substance in blood, Widmark (1919) followed the concentration of a single dose of acetone injected into the bloodstream. Widmark and others subsequently examined a number of so-called "model" configurations, including the first account of a one-compartment open model (Widmark and Tandberg 1924) and the later extension to two compartments (Gehlen 1933).

## Models of Living Systems

Apostel (1960) described a model of a living system as an artificial system that "simulates a biological system. The kinetic analysis of the model (which usually describes a dynamic process) tests the validity of the model of the combined kinetic behavior of the elements of each compartment. The model provides the basis for prediction of subsequent behavior. Thus, the model is the mathematical expression of the biological system, and the mathematical analysis is the test of predictions generated by the hypothesis."

Statistical hypothesis testing often is used to judge whether a model is appropriate or not. The model is defined by operational equations that yield a dependent variable for each set of independent variables. The statistical evaluation of the kinetic analysis cannot of itself establish the truth of the model, which is why it is more accurate to describe the validated model as "not yet rejected" and therefore still potentially useful to the solution of a given problem. Likewise, the answers provided by the operational equation are only "consistent" with the experimental or clinical observations. For this reason, it is important to identify those situations in which the model is rejected by the chosen compartmental analysis, e.g., by application of a criterion of information content (Akaike 1974) in which statistical goodness of fit is balanced against the number of parameters fitted.

## *Kinetics and Molecular Biology*

The purpose of kinetic analysis of living matter is to obtain quantitative measures of the rate of molecular reactions. Quantitative approaches were uncommon in biology and medicine prior to the second half of the nineteenth century and only slowly gained ground against traditionally qualitative considerations. The competition between quantitative and qualitative perspectives is felt even today.

The struggle reflects the changing views of disease in the medical sciences in which a disorder originally was thought to represent a major imbalance among qualitatively different matters of nature and life, including the four elements (water, air, fire, and earth) and the four cardinal fluids (blood, phlegm, yellow bile, and black bile).

This imbalance no longer is a valid consideration. The imbalance underlying disease appears to follow minute but specific errors which are now known to create the effects of disease by turning open thermodynamic systems implemented in biochemical and physiological compartments into closed systems that must ultimately fail because entropy rises in closed systems as order is replaced by disorder. Thus, it is a fundamental observation that truly closed systems eventually become incompatible with life.

The concept of imbalance is quantitative, as is the injunction of living matter to respond to exigencies with moderation. Thus, measurement is the modern practice, although it is tied to an increasing understanding of the limits of certainty.

Competing with this understanding is the rise of information technology, according to which the quantitative properties of the components of living systems could be less important, implying that only their structural relations are informative. Thus, there is a current sense that the tide of scientific philosophy is returning in the direction of the holistic and qualitative. Only the practice of meticulous kinetic analysis can correct this misunderstanding.

## *Kinetics and Genomics*

Living matter is distinguished from nonliving matter primarily by its ability to maintain steady-states of incredibly complex molecular compartments far from thermodynamic equilibrium. The information enabling the realization of this enormous potential resides in a remarkably inert and robust molecule called deoxy-ribonucleic acid (DNA). However, the DNA molecule itself does nothing; its entire and completely passive role is to be decoded by a machine or mechanism. Its power to elicit action derives from the ability of other molecules in living tissue to read its instructions at the right time and place.

As the decoder must understand the message of at least the opening sections of the manual (the rest can be learned in due course), in advance of the decoding, the fundamental goal of metabolite and tracer kinetic analysis in biology and medicine is to describe and quantify the processes in their entirety from the conception to the termination of the organism. For example, it is estimated that at the peak of neuronal proliferation during human gestation, as many as 250,000 new brain cells of identical composition are created every minute. Yet, metabolite concentrations everywhere remain inside carefully regulated limits. A snapshot of any one cell would produce an unremarkable image; only the proper tracer kinetic analysis could reveal the astounding dynamics of the metabolite fluxes contributing to this development.

## *Kinetics and Proteomics*

The rate of molecular reactions typically is constrained by proteins. An important measure of health is steady-state, in which proteins maintain the concentrations of metabolites while the molecular fluxes adjust to local and global requirements. Most importantly, the composition of living matter remains constant in steady-state (hence the name) and a momentary glimpse reveals none of the dynamics of the underlying molecular fluxes. The further the steady-state is from a state of equilibrium, the greater is the work required to maintain it, and the greater are the fluxes controlled by the proteins. Only a few processes are near equilibrium and they typically do not interfere with the regulation of the important molecular fluxes of living matter.

When concentrations normally do not change outside tightly controlled limits, past attempts to understand the underlying dynamics by perturbing a system often removed the system from its normal state and sometimes failed to specifically reveal the normal dynamic properties of its kinetics. The introduction of suitably flagged ("labeled") and hence identifiable representatives of the native molecules, called "tracers," accomplishes a minimal perturbation without disturbing the steady-state of the system, provided the quantity of tracer is kept too low to change the system's properties. Methods of doing just that form the core of the tracer kinetic analysis of biological processes.

## *Role of Tracers in the Study of Models*

A physiological/biological process to be studied is often exposed by means of a tracer (not always radioactive), that is a marker of a native molecule relevant to the process that can be detected by an instrument, e.g., radioactive counting or light or magnetic measurement. The tracer must be present in such low mass/quantity that the characteristics of the processes in which the tracer participates do not change (e.g., does not compete with the endogenous processes, in the case of neuroreceptor imaging the tracer does not occupy significant receptor sites to notably compete with endogenous neurotransmitters).

The purpose of this requirement is to rule out the departure from steady-state that would otherwise cause the concentrations of native molecules to change as functions of time. The departure of the native system from steady-state would in turn interfere with the first-order relaxation of tracer compartments, discussed in the text.

Organisms and organs are collections of cells that internalize the tracer in different ways according to the physical and chemical properties of the tracer, and the biochemical and physiological properties of the cells. A physiological model can be formulated as a collection of compartments that represent the different states of the tracer and its metabolites. Strictly speaking, the compartments have no formal relation to the structure of the target organ, except to the extent that the anatomy delineates the processes in which the tracer or its metabolites participate (e.g., a tracer may bind to an active site as a receptor or transport mechanism when its structure fits the receptor or transporter site in the right chemical fashion). For this reason the model may be much simpler than that of the actual native system and still be a valid portrayal of the kinetic behavior of the tracer. In other words, the model is of the tracer, not of the native system. Often, compartments reflect the biochemistry of an organ and refer to quantities of tracer or its metabolites that need not be confined to separate subdivisions of the organ.

Sheppard (1948) defined compartments as quantities of a tracer or its metabolites, the concentrations of which remain the same everywhere, each quantity having a single state that may vary in time but not in space. A quantity is the number of molecules in units of moles (mol), 1 mol holding $6.0225 \cdot 10^{23}$ atoms or particles. Thus, initially, a tracer is neither in a steady-state, nor in equilibrium. However, there are other noncompartmental approaches to physiological quantification that can also be employed (see text).

### *Role of Models as Interpreters of Biological Dynamics*

When researchers interpret processes of physiology and pathophysiology by means of tracers or other marker tools, they examine the results with specific methods that include biochemical measures (e.g., mass spectrometry and radioactivity counting) or external recording in vivo (e.g., positron or single photon emission tomography [PET or SPECT]). There is rather a tendency (often naive) to search for a tracer acting as a "magic bullet" that provides a picture of the entire process by biochemical measures or external imaging of a subject. In the example of external imaging, waiting a specified time after intravenous injection typically occurs in the clinical setting of recording of static images for evaluation of, say, heart or bone in conditions in need of a diagnosis.

However, when attempts are made to understand and quantify a physiological approach with the greatest scientific rigor, evaluation of the full dynamic process prior to a steady-state is necessary, even when mathematical simplifications are later found to be acceptable. This necessity usually includes not only the brain kinetics of the tracer but also the input record which reflects the dynamic history of the tracer itself, circulating from the injection site to the planned target (e.g., the blood volume spaces at the blood–brain barrier interface).

This book is also dedicated to the understanding of the underlying principles of kinetic properties of dynamic biological processes of brain physiology, the so-called "neurokinetics."

## Approaches to Physiological Modeling

Physiological processes are best determined by mathematical descriptions which then are subject to the well established rules of computation of the physical and chemical sciences rather than any qualitative approach that is limited and can lead to erroneous extrapolations beyond the actual empirical data.

### *Compartmental Modeling*

The most common approach to the in vivo quantification of dynamic brain processes (as networks of complex chemical systems) is that of compartmental modeling. This approach divides the physiological processes into definable units. In the case of brain images, it depends at a minimum on records of the tracer input function (usually from plasma or whole-blood samples) and one or more brain compartments . The assumptions and principles are outlined in Chaps. 1 and 2.

## *Non-Compartmental Modeling*

A presentation of noncompartmental models is beyond the scope of this book. However, there are early examples of attempts to quantify physiological properties such as blood flow or blood volume, as in the case of the indicator dilution method reviewed by Zierler (2000).

Distributed models attempt to account for spatial gradients in concentrations (e.g., in blood-tissue exchange) in contrast to compartmental models that depend on concentration averages within each compartment (Kuikka et al. 1991).

In the compartmental models to be discussed in the following chapters, the usual assumption is the existence of homogeneous and fully stirred compartments. There are attempts at modeling that directly address this inhomogeneity with so-called "distributed" models, such as in the case of myocardial blood flow estimation, including fractal analysis, a branch of mathematical analysis (Qian and Bassingthwaighte 2000). Unfortunately, little progress has been made in the field of quantification of dynamic brain processes with noncompartmental models (one attempt was made by Wong and Gjedde in 1996), limited in part by the poorly resolved temporal and spatial sampling of brain images, compared to data obtained from cardiovascular and other systems, perhaps in part due to the more invasive tools used in the study of the latter.

Some distributed models have been proposed for use with PET and external imaging as in the measurement of oxygen consumption (Deussen and Bassingthwaighte 1996), with special attention to small tissue regions (Li et al. 1997). However, the majority of these applications used invasive approaches to imaging, with direct measurement of the tracer concentrations.

## Nomenclature

The nomenclature adopted in this books originated in publications spanning more than 35 years. Other nomenclatures have been presented in more recent attempts to reach consensus that we intend to evaluate in future editions, as discussed in the Glossary section of this book where all terms are explained. In this edition, we chose to retain the nomenclature as originally published while we wait for further refinement of the current consensus reports.

## Acknowledgements

We wish to thank the mentors and teachers who introduced these topics to us in a manner that we could understand, including Ludvik Bass of Australia, Christian Crone of Denmark, and Clifford Patlak, Fred Plum, Louis Sokoloff, and Henry N. Wagner, Jr., of the United States of America. We also thank Arvid Carlsson and Paul Greengard for encouragement and advice, our contemporaries Vin Cunningham, Mike Kuhar, and Sol Snyder for conveying the fundamentals of receptor action, our collaborators Chris Bailey, Per Borghammer, Peter Brust, Paul Cumming, Alain Dagher, Doris Doudet, Alan Evans, Gerhard Gründer, Anker Jón Hansen, Hiroto Kuwabara, Sean Marrett, Anders Rodell, Oliver Rousset, Manou Vafaee, and Yun Zhou for continuous challenges in this discipline, and not the least Rodger Parker who tried to keep us on a most rigorous course for many years. In addition, we wish to thank the funding agencies that over the years enabled the work that is summarized in these pages, as listed in the papers that we cite.

# Contents

# Chapter 1
# Introduction to Compartmental Analysis*

## 1.1 Concept of Compartments

### *1.1.1 Living Systems*

In the context of compartmental analysis, a living organism can be described as an open biological system existing in a *steady-state* far from *thermodynamic equilibrium.* Thermodynamic equilibrium is a state in which no biological processes can occur because there are no potential gradients to drive them; no differences in mechanical potential to drive blood flow, in concentrations to drive diffusion, in chemical potentials to drive metabolism, in electrical potentials to drive ions, and in temperature to drive heat flow. Steady-state and thermodynamic equilibrium share the characteristic that they are invariant in time. Thermodynamic equilibrium is also invariant in space. The steady-state variance of constituent chemicals in space is the focus of compartmental analysis. Spatial variance is assigned to the interfaces between abstract compartments rather than to the living system as a whole. As the compartments by this definition are in thermodynamic equilibrium internally, they are incompatible with life but we choose to ignore this fundamental characteristic.

Compartmental analysis uses the principles of biophysics and mathematics to determine the velocity of exchanges among the compartments (biochemical processes) and the relative size of the individual compartments (biochemical pools) in vivo, using tracer molecules, defined as markers that do not perturb the system.

During a medical study or biological experiment, the tracer and its metabolites assume different states, each of which may be well defined but all of which change and interact as functions of time. Eventually, one or more of these states may reach the steady-state characteristic of the native system, though far from thermodynamic equilibrium. This steady-state can be maintained only in thermodynamically open systems. If energy is no longer provided or expended, potential

*Adapted from Gjedde (1995a) Compartmental analysis. In: *Principles of Nuclear Medicine*, 2nd edition, eds Wagner HNJr, Szabo Z, Buchanan JW. Saunders, Philadelphia, pp. 451–461, with permission from Saunders, Philadelphia.

A. Gjedde et al., *Neurokinetics: The Dynamics of Neurobiology In Vivo*,

DOI 10.1007/978-1-4419-7409-9_1, 

gradients vanish, biological work can no longer occur, thermodynamic equilibrium irreversibly replaces the steady-state, and life ceases to exist. Conversely, when a tracer steady-state of thermodynamic nonequilibrium continues indefinitely, the system turns over energy and, almost by definition, is alive.

When disease intervenes, living systems tend to move toward, and eventually reach, thermodynamic equilibrium, and the individual compartments fuse. This makes tracer kinetic analysis of compartments a useful tool in the characterization of disease. The shift from the healthy steady-state provides both a motive for, and a means of, compartmental analysis in vivo: Only limited information is available in vitro, i.e., postmortem, because the important changes occur prior to death, and the premortem changes define the disease.

### Definition of Compartments as Tracer States

A compartment has a specific mathematical definition. A model is a set of compartments that simulate a biological system. Compartmental analysis tests the validity of the model of the combined kinetic behavior of the elements of each compartment. The model provides the basis for prediction of subsequent behavior. Thus, the model is the mathematical expression of the biological system, and the compartmental analysis is the test of predictions generated by the hypothesis.

As mentioned earlier, Sheppard (1948) defined compartments as quantities of a tracer or its metabolites, the concentrations of which remain the same "everywhere," each quantity having a single state that may vary in time but not in space. A quantity is the number of molecules in units of *mol* ($6.0225 \cdot 10^{23}$). Thus, initially, a tracer is neither in a steady-state nor in thermodynamic equilibrium. Rescigno and Beck (1972a, b) restricted Sheppard's definition of a compartment to that of a tracer state that varies in time only, according to the expression,

$$\frac{\mathrm{d}m}{\mathrm{d}t} = j - k\ m, \tag{1.1}$$

where $m$ is the quantity (*mol*) of tracer that belongs to the compartment (i.e., has the relevant state), $k$ the relaxation ("rate") constant, and $j$ the flux of tracer molecules into the compartment as a function of time. It follows from this definition that the relaxation constant is given by the relationship,

$$k = \frac{j}{m} - \frac{1}{m}\frac{\mathrm{d}m}{\mathrm{d}t}. \tag{1.2}$$

At steady-state ($\mathrm{d}m/\mathrm{d}t = 0$), $k$ is the turnover rate $j/m$, as derived later in (1.12). The definition requires that the escape of tracer from the particular state (the "relaxation" of the state) be a first-order process. It depends on the process responsible for the relaxation whether this requirement is met. One such process (diffusion) will be examined later.[1]

[1] Note the convention that variables are indicated by lower case symbols, except for the relaxation constants.

### 1.1.2 Thermodynamics and Entropy

In most cases, the interfaces between compartments represent diffusion barriers such as cell membranes or chemical reactions involving transporters, receptors, or enzymes. The processes can be spatially well-defined, for example when the diffusion barrier or protein is associated with a cell membrane.

The tracer is subject to the same forces, or potential gradients, that drive the native compartment. Thermodynamically, this "relaxation" results from the production of entropy during the process. When the decay finally ceases, i.e., when the "relaxation" is complete, there is maximum disorder within the system. The potential loss of order enables the system to perform work, measured in units of electrochemical "potential."

The electrochemical potential (of the tracer) is established by the properties of the tracer that enable it to perform work. These include the concentration for a given diffusion coefficient or permeability, the electrical charge for a given electrical field, the volume for a given pressure, the mass for a given gravitational field, and the chemical structure, including the bonds that keep the structure together, for a given chemical environment. Analysis of the speed of dissipation of the electrochemical potential reveals the nature of the system in which the tracer state decays. A closed system will generate entropy and perform work until all parts of the system have the same relative potential of zero, i.e., until thermodynamic equilibrium is achieved. At thermodynamic equilibrium, the potential varies neither in time, nor in space. If the system is open, it may adopt a *steady-state* in which energy is supplied at a rate exactly matched to the dissipation of the system's potential; in that case, the potential varies in space but not in time.

By definition, the tracer behaves in the opposite manner, i.e., it varies with time rather than space. In reality, of course, the tracer varies both with time and in space because it is neither at steady-state nor in thermodynamic equilibrium. To remove the variance in space, the tracer is assigned artificially to pockets or compartments that obey the basic *compartmental assumption* that tracer states are invariant in space. With time, the tracer proceeds *toward*, but never reaches, thermodynamic equilibrium in an open system. Instead, it may attain a *secular* equilibrium in which the relative proportions of two or more quantities approach constants dictated by the rate of escape from the system as a whole. Before the tracer's secular equilibrium turns into thermodynamic equilibrium, the steady-state of the native system is imposed on the tracer which finally becomes part of the system itself.

#### Equilibrium and Steady-State

Assuming there are $N$ compartments, each with mass $m_i(t)$, $1 \leq i \leq N$, the following definitions hold.

*Secular Equilibrium* The masses, $m_i(t)$, are said to be in secular equilibrium if they remain in constant proportion, i.e., $m_i(t)/m_j(t) = constant$ for all $1 \leq i, j \leq N$.

*Transient Equilibrium* The $i$th compartment is said to be in transient equilibrium at time $t$ if $\mathrm{d}m_i(t)/\mathrm{d}t = 0$ at time $t$.

*Steady-State* The system is said to be in steady-state if $\mathrm{d}m_i(t)/\mathrm{d}t = 0$, for all $1 \leq i \leq N$ and all $t$ beyond a certain $t_0$.

*Equilibrium* The term equilibrium, used without a qualifier, means the absence of one or more potential gradients. With a qualifier, the term means absence of a potential gradient of a specific type. For example, chemical equilibrium means the absence of a gradient in chemical potential. Thermodynamic equilibrium refers to the absence of potential gradients of any type.

Work being an integral of force over distance, the absence of a potential gradient implies an inability to do work of the associated type. Thus, the absence of chemical potential implies an inability to do chemical work. As thermodynamic equilibrium is characterized by the absence of all potential gradients, it implies the inability to do work of any type.

## Diffusion

An example of a process that is subject to compartmental analysis is simple diffusion. Diffusion results in the production of entropy such that the system is less orderly at the end of the diffusion than at the onset. Diffusion dissipates the electrochemical potential ($\mu$) which is a function of several factors,

$$\mu = \mu_o + RT \ln c + z\mathcal{F}\phi, \tag{1.3}$$

where $\mu_o$ is the standard chemical potential, $c$ the concentration (activity), $z$ the number of charges per particle, $\mathcal{F}$ Faraday's constant, and $\phi$ its electric potential. Electrochemical equilibrium is present when the ratio defining the *partition coefficient* $\lambda = c_2/c_1 = \alpha_2/\alpha_1$ equals the value of the term $\mathrm{e}^{-(\Delta\mu_o - z\mathcal{F}\Delta\phi)}$ where $c_1$ and $c_2$ are the concentrations and $\alpha_1$ and $\alpha_2$ are the solubilities, relative to water, in two adjoining media.

The speed with which secular equilibrium is reached is an important indicator of the composition of the tissue. The diffusion velocity is the product of the mobility of the particles and the force or potential gradient,

$$-g\frac{\partial \mu}{\partial x} = -g\left(\frac{RT}{c}\frac{\partial c}{\partial x} + z\mathcal{F}\frac{\partial \phi}{\partial x}\right), \tag{1.4}$$

where $g$ is the mobility (inverse frictional resistance) and $\partial\mu/\partial x$ the gradient. When there is no gradient in electrical potential, $(\partial\phi/\partial x = 0)$,

$$j = -cgA\frac{\partial \mu}{\partial x} = -DA\frac{\partial c}{\partial x} \tag{1.5}$$

which is a form of Fick's First Law, where $D$ is the diffusion coefficient, equal to $gRT$, and $A$ the cross-sectional area through which diffusive flux, $j$, occurs.

A mass balance equation, written for a volume extending in a direction perpendicular to $A$ for an incremental length, $\Delta x$, equates the rate of change of mass within the volume to the difference in diffusive flux across its boundaries,

$$j(x + \Delta x, t) - j(x, t) = A\frac{\partial c(x,t)}{\partial t}\Delta x.$$

In the limit, as $\Delta x \to 0$

$$\frac{\partial j(x,t)}{\partial x} = A\frac{\partial c(x,t)}{\partial t}.$$

Substituting Fick's First Law in the last equation yields Fick's Second Law,

$$\frac{\partial c}{\partial t} = -D\frac{\partial^2 c}{\partial x^2}. \tag{1.6}$$

Fick's laws are expressed in differential equations. For their practical application they must be integrated. The particular form of the differentiated equations depends on the conditions of the diffusion, to be discussed later.

### Permeability

Compartmental analysis assumes variation only in time. To make the *compartmental assumption*, it is necessary to reduce the gradient, $\partial c/\partial x$, to a fixed difference covering the width, $\Delta x$, of an interface placed between two compartments. On either side of the interface are concentrations, $c_1(t)$ and $c_2(t)$, which vary only with time. A discrete approximation to (1.5) describes flux across the interface.

$$j(t) = -\mathrm{DA}\frac{c_2(t) - c_1(t)}{\Delta x} = \frac{\mathrm{DA}}{\Delta x}c_1(t) - \frac{\mathrm{DA}}{\Delta x}c_2(t) = j_2(t) - j_1(t), \tag{1.7}$$

where $j_2(t) = c_1(t)\mathrm{DA}/\Delta x$ may be regarded as flux into compartment 2 from compartment 1, and $j_1(t) = c_2(t)\mathrm{DA}/\Delta x$ regarded as flux in the opposite direction. In the many cases in which neither the width nor the composition of the interface is known, the ratio between $D$ and $\Delta x$ is defined as the diffusional *permeability* coefficient, $P_d$. Mass balance equations of the form

$$\frac{\mathrm{d}m_i(t)}{\mathrm{d}t} = j_i(t) - \frac{P_d A}{V_i}V_i c_i(t) = j_i(t) - k_i m_i(t) \tag{1.8}$$

may then be written for each compartment, where $i = 1, 2$, $V_i$ is the volume in which $c_i$ is dissolved, and $k_i$, the relaxation constant, is the ratio $P_d A/V_i$. Thus, $m_i$, the tracer state of compartment $i$, is replenished at the rate of $j_i$ and depleted at the rate of $k_i m_i$.

The concentrations $c_1$ and $c_2$ should be regarded as concentrations in water. In situations covered later in the text, concentrations in plasma and tissue will appear along with their relative solubilities.

### 1.1.3 Fundamental Solution

The solution to (1.1) and (1.8) most simply is obtained by Laplace transformation,

$$p\mathcal{L}(m) = \mathcal{L}(j) - k\mathcal{L}(m) \tag{1.9}$$

according to which, when solved for $\mathcal{L}(m)$,

$$\mathcal{L}(m) = \mathcal{L}(j)\frac{1}{p+k}. \tag{1.10}$$

Equation (1.10) is the transform of a convolution integral,

$$m(T) = j \bigotimes \mathrm{e}^{-kt} = \mathrm{e}^{-kT}\left[ m(0) + \int_o^T j\mathrm{e}^{kt}\ \mathrm{d}t \right]. \tag{1.11}$$

Equation (1.11) is a transcendental equation which can be solved for $k$ only by iteration but may have zero, one, or two solutions. Special cases include $j = 0$ ("wash-out") and $m(0) = 0$. By this definition, wash-out must be monoexponential.

For a monoexponentially declining input function, $j = j(0)\mathrm{e}^{-\beta t}$, where $\beta < k$, a secular equilibrium is eventually reached

$$m = \frac{j}{k-\beta}\left(1 - \mathrm{e}^{-(k-\beta)T}\right) \overset{\mathrm{SE}}{\rightarrow} \frac{j}{k-\beta} \overset{\mathrm{SS}}{\rightarrow} \frac{j}{k}, \tag{1.12}$$

where SE symbolizes the approach to secular equilibrium and SS the approach to steady-state. In this case secular equilibrium means that the relative proportion of $m$ to $j$, i.e., the ratio $m/j$, approaches a constant, $1/(k-\beta)$. This can be seen by substituting $j(0)\mathrm{e}^{-\beta t}$ for $j$ in (1.12) and carrying out the integration. The steady-state value is obtained from (1.11) by setting the time derivative to zero.

A monoexponentially declining function is characteristic of a closed system consisting of two compartments. Steady-state replaces secular equilibrium when $k \gg \beta \gg 1/T$. Once secular equilibrium is established, the compartment essentially has disappeared by absorption into its precursor compartment. This process limits the number of actual tracer states present in any one model.

An alternative method of solution of (1.11) was suggested by Cunningham and Jones (1993), which bridges the gap between compartmental and noncompartmental analysis. This approach has been termed *spectral analysis*. It identifies a *spectrum* of true compartments that together constitute an apparent compartment and thus reveals an underlying compartmental heterogeneity of a tissue. The approach is complicated by the nonunique nature of its solutions.

### 1.1.4 Limitations of Compartmental Analysis

When the number of compartments, and their linkages, are established, the model and its solutions follow automatically but not all models are equally useful.

The more compartments a model has, the less it discriminates between compartments, although it may be of practical value as an operational equation. In addition, the relaxation constants or transfer coefficients of a series of compartments can be distinguished only when their magnitudes are not too different. Slow "relaxations" tend to obscure rapid relaxations as the compartments move toward secular equilibrium. Thus, by analyzing the organ uptake of a tracer as a function of time, only a limited number of compartments and transfer coefficients can be identified.

In transient analysis, the independent and dependent variables are measured as functions of time and the desired coefficients estimated from the fundamental solution by regression analysis, using computerized optimization. The solution expressed in (1.11) is the prototype of an operational equation used for regression analysis in which the input ($j$) and output ($m$) functions are compared to yield the value of the parameter $k$ which "optimizes" the comparison.

Often enough, the results of regression analysis cannot be related to the biological characteristics of the system. Regression analysis is only meaningful when the validity of the model is independently established. It, usually, is impossible to decide the validity of the model and obtain the best estimates of the coefficients at the same time. Both mathematical simulations of the behavior of the model and actual experiments help justify the choice of the model.

## 1.2 Single Tissue Compartment Analysis

A special case of single tissue compartment analysis is quantification of blood flow to a tissue. This procedure involves the whole organ estimate of uptake of tracers with specific properties. The generation or elimination of a substance of this kind in an organ can be calculated by the steady-state perfusion principle of Fick,

$$\Delta j = F c_{\mathrm{a}} - F c_{\mathrm{v}} \tag{1.13}$$

where $\Delta j$ is the quantity of a given material passing into or out of the organ, $F$ the blood flow, $c_{\mathrm{a}}$ the arterial concentration, and $c_{\mathrm{v}}$ is the cerebral venous concentration, assumed to be uniform ("single outlet"). For the transient phase, this "black box" principle can be modified to reflect the behavior of a tracer,

$$\frac{\mathrm{d}m}{\mathrm{d}t} = F c_{\mathrm{a}} - F c_{\mathrm{v}}. \tag{1.14}$$

When the tracer is inert, $m$ is the quantity in the organ. For an inert substance in the steady-state, $\mathrm{d}m/\mathrm{d}t$ must be zero. Thus, when the tracer is used to measure blood flow it is assumed that the tracer is inert and perfectly diffusible, i.e., faces no diffusion barriers between the vascular space and the tissue and that, therefore, tracer delivery to the organ is only limited by the blood flow, $F$. Each volume of the tissue contains a vascular tracer component in addition to that of the parenchyma, of course. In the one-compartment analysis, the tracer in the tissue, including its vascular bed, occupies a single compartment in which the tracer concentration in the aqueous fraction of the vascular volume, as it exits from the tissue, is identical to the concentration in the aqueous phase of the parenchyma. In this case, the tracer in the tissue is distributed in a single compartment with a uniform concentration. The quantity of accumulated labeled material, $m$, in the tissue then can be represented by the product $Vc_{\mathrm{v}}$. For the transient phase, (1.14) may then be written as:

$$\frac{\mathrm{d}m}{\mathrm{d}t} = Fc_{\mathrm{a}} - \frac{F}{V}m, \tag{1.15}$$

where $V$ is the single-compartment tissue-blood partition volume. For tracers that occupy a single state, a single compartment is then established in the tissue with a relaxation constant of $k = F/V$ and $j = Fc_{\mathrm{a}} = kVc_{\mathrm{a}}$. The single compartment is the limiting case of the multicompartment case, derived later. As shown later, it is unlikely that the single-compartment analysis ever represents a complete description of tracer uptake because most tissues include nonexchange vessels that lead to a requirement for additional compartments. The multicompartment analysis completed below indicates that $F$ determined according to (1.15) estimates blood flow exactly only when the volume of vessels (arterial, capillary, and venous) is negligible compared to the volume of the tissue as illustrated in Fig. 1.1.

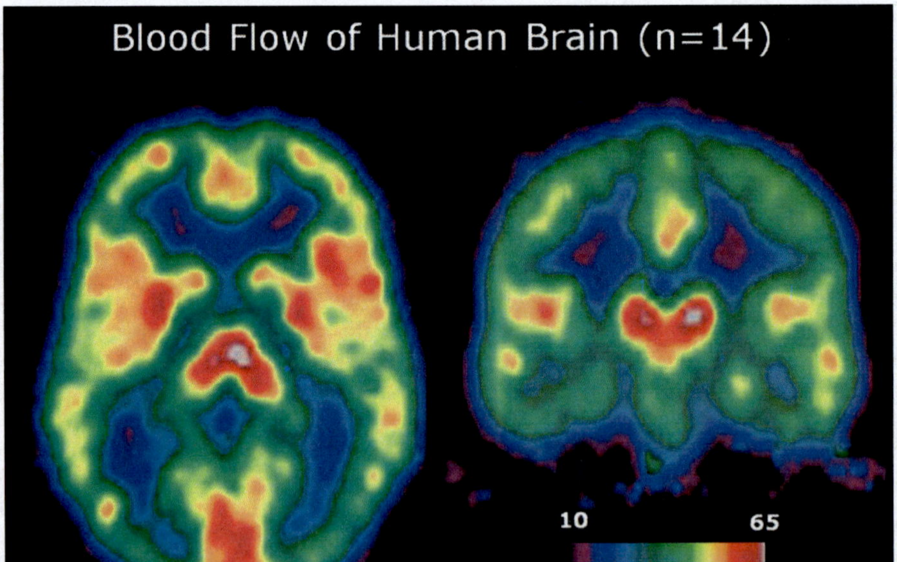


**Fig. 1.1** The case of the single compartment: Average cerebral blood flow rates of 14 healthy young adult volunteers in units of ml $\mathrm{hg}^{-1}$ $\mathrm{min}^{-1}$, determined by positron emission tomography of brain uptake of [$^{15}$O]water after i.v. administration, by solution of (1.15), according to the method of Ohta et al. (1996). Images prepared by Christopher Bailey, PhD, Aarhus PET Center

## 1.3 Two Tissue Compartment Analysis

Special cases of the two tissue compartment analysis include the quantification of oxygen consumption by a tissue, when tracers occupy separate vascular and tissue compartments. Exchange vessels are vascular channels in which the tracer molecules exchange with the surrounding tissue. In many organs of the body, the capillary endothelial cells have leaky "tight" junctions which do not impede the escape of small polar solutes from the circulation. In brain, the capillary endothelial cells in the tissue have particularly tight (so-called *tight* tight) junctions that do form such a barrier. Thus, in brain, the concentration difference of newly administered polar tracer solutes between the two sides of the endothelium initially is so great that the endothelium may be the only significant barrier to the distribution of such tracers in brain. For these tracers, the brain has two kinetic compartments, the vascular space and an extravascular space, separated by a blood–brain barrier. The two states of the tracer cannot be detected separately in vivo because the interface between the compartments is inside the brain tissue. For nonbrain tissue, the leaky junctions of the capillary endothelium give access to the first significant barrier to small polar molecules, namely the tissue cell membranes.

Endothelial permeability is measurable as an index of tracer clearance ($K$) from the circulation. The clearance does not reflect the permeability surface area product directly because the tracer concentration falls from the arterial to the venous end of the capillary. For this reason, the vascular space in the tissue may appear not to be a true compartment. However, it can be shown that the tracer in the extravascular space (i.e., in the "extravascular" state) obeys (1.1) and hence functions kinetically as a compartment.

It is the purpose of two-compartment transient analysis to estimate the tracer's rate of unidirectional clearance ($K_1$) from the vascular compartment to the extravascular compartment, as an indirect measure of the permeability of the blood-tissue barrier to the tracer, and the tracer's volume of partition between the circulation and the extravascular tissue space ($V_e$). Neither can be determined exactly but as an approximation only.

### 1.3.1 Compartmental Assumptions

In this analysis, each compartment is assumed to hold homogenously distributed and fully mixed tracer contents. When local properties ($F$, $P_d A$, $V_d$) are all uniform within a region of interest, tracer transport can be described by a pair of mass-balance equations,

$$V_c \frac{\partial c_c}{\partial t} = FL \frac{\partial c_c}{\partial x} - P_d A \left( \frac{c_c}{\alpha_1} - c_e \right) \tag{1.16}$$

and

$$V_\mathrm{d}\frac{\partial c_\mathrm{e}}{\partial t} = FL\frac{\partial c_\mathrm{e}}{\partial x} + P_d A\left(\frac{c_\mathrm{c}}{\alpha_1} - c_\mathrm{e}\right), \tag{1.17}$$

where $F$ is blood flow, $V_\mathrm{c}$ the volume of distribution in the capillary, $L$ the length of the exchange vessels, $c_\mathrm{c}$ the blood (or plasma) concentration of the tracer, $c_\mathrm{e}$ the tissue concentration, $\alpha_1$ the solubility in plasma, and $V_\mathrm{d}$ is the volume of distribution of extravascular tracer with the concentration $c_\mathrm{e}$.

If $c_\mathrm{e}$ is uniform from the entrance to the exit of the exchange vessel (capillary) by instantaneous axial diffusion, such that the incremental addition to $c_\mathrm{e}$ during a single passage is negligible, and if discontinuities exist at both ends of the capillary, such that the concentration is $c_\mathrm{a}$ at the inlet, $\bar{c}_\mathrm{c}$ in the capillary, and $c_\mathrm{o}$ at the outlet, then these compartmental assumptions lead to the ordinary differential equations:

$$V_\mathrm{c}\frac{\mathrm{d}\bar{c}_\mathrm{c}}{\mathrm{d}t} = F(c_\mathrm{a} - c_\mathrm{o}) - P_d A\left(\frac{\bar{c}_\mathrm{c}}{\alpha_1} - c_\mathrm{e}\right) \tag{1.18}$$

and

$$V_\mathrm{d}\frac{\mathrm{d}c_\mathrm{e}}{\mathrm{d}t} = P_d A\left(\frac{\bar{c}_\mathrm{c}}{\alpha_1} - c_\mathrm{e}\right). \tag{1.19}$$

The ratio $V_\mathrm{d}/\alpha_1$ is the partition volume $V_\mathrm{e}$. Equations (1.18) and (1.19) are not directly soluble because $c_\mathrm{o}$ is unknown.

**Capillary Compartment** To solve (1.18) and (1.19), the following strategy must be adopted. To determine $\bar{c}_\mathrm{c}$, let $c_\mathrm{c}(\tau)$ be the capillary tracer concentration in a small volume of plasma, $\Delta V_\mathrm{c}$, at the time $\tau$ after its entry into the capillary. This small volume is not stationary, but moving with plasma at precisely the same rate of flow. As a result, there is no net flow through the volume, and tracer concentration within it varies only by exchange of tracer with tissue. For simplicity, set $\Delta P = P_d \Delta A/\alpha_1$, $P = P_d A/\alpha_1$, $\Delta p = P_d \Delta A/V_\mathrm{d}$, and $p = P_d A/V_\mathrm{d}$. Thus, $\Delta P/\Delta p = P/p = V_\mathrm{d}/\alpha_1 = V_\mathrm{e}$ and

$$\Delta V_\mathrm{c}\frac{\mathrm{d}c_\mathrm{c}(\tau)}{\mathrm{d}\tau} = \Delta p\, m_\mathrm{e} - \Delta P c_\mathrm{c}(\tau) \tag{1.20}$$

where $m_\mathrm{e}$ and $c_\mathrm{a}$ are the tracer variables assumed not to change appreciably during the time of one capillary transit. This time, $\bar{\tau}$, equals $V_\mathrm{c}/F$. Recalling that $\Delta P/\Delta V_\mathrm{c} = P/V_\mathrm{c}$, and that $c_\mathrm{c}(0) = c_\mathrm{a}$, integration from $\tau = 0$ to $\tau = u$ yields:

$$c_\mathrm{c}(u) = c_\mathrm{a}\mathrm{e}^{-Pu/[F\bar{\tau}]} + m_\mathrm{e}\frac{p}{P}\left(1 - \mathrm{e}^{-Pu/[F\bar{\tau}]}\right). \tag{1.21}$$

As the mean capillary tracer content is the weighted integration of the capillary concentration along the capillary's length, a second integration yields:

$$\bar{c}_\mathrm{c} = \frac{1}{\bar{\tau}} \int_0^{\bar{\tau}} c_\mathrm{c}(u)\mathrm{d}u = \frac{F}{P}\left(1-\mathrm{e}^{-P/F}\right)c_\mathrm{a} + \frac{m_\mathrm{e}}{V_\mathrm{e}}\left[1-\frac{F}{P}\left(1-\mathrm{e}^{-P/F}\right)\right], \quad (1.22)$$

where the term $1-\mathrm{e}^{-P/F}$ is defined as the extraction fraction, $E_o$, for unidirectional transfer of tracer into the tissue (Crone 1963).

The concentration $\bar{c}_\mathrm{c}$ varies both with $c_\mathrm{a}$ and with the concentration of tracer in the tissue interstitial fluid (precursor pool), $c_\mathrm{e}$, which in turn varies with $c_\mathrm{a}$. Generally speaking, therefore, $\bar{c}_\mathrm{c}$ relative to $c_\mathrm{a}$ is lowest initially when $c_\mathrm{e} \sim 0$ and highest at the steady-state level $c_\mathrm{a}$.

It is now possible to introduce the transfer coefficients $K_1$ and $k_2$. The symbol $K_1$ represents a clearance,

$$K_1 = F\left(1-\mathrm{e}^{-P/F}\right) = E_o F. \quad (1.23)$$

The symbol $k_2$ represents a fractional clearance,

$$k_2 = F\left(1-\mathrm{e}^{-P/F}\right)/V_\mathrm{e} = E_o F/V_\mathrm{e} = K_1/V_\mathrm{e}. \quad (1.24)$$

Equation (1.23) was derived twice, first by Bohr (1909) and then by Crone (1963), to describe the relationship between clearance ($K_1$), perfusion ($F$), and permeability-surface area product ($P_d A$) of the blood-tissue barrier.

From the average capillary concentration expressed in (1.22), the quantity of tracer in the capillary compartment can now be calculated

$$m_c = V_\mathrm{c}\bar{c}_\mathrm{c} = V_o c_\mathrm{a} + (V_\mathrm{c} - V_o)\frac{m_\mathrm{e}}{V_\mathrm{e}}, \quad (1.25)$$

where $V_o$ is the initial volume of distribution, $V_\mathrm{c}K_1/P$. This description of the contents of the capillary indicates that the tracer in the capillary has two states, one equal to the arterial concentration in the volume $V_o$, and one present at the "plasma-equivalent" tissue concentration (which is lower than the capillary outflow concentration, as shown later) in the volume $V_\mathrm{c} - V_o$.

### Tissue Compartment

The net rate of tracer transfer across the endothelium equals the difference between the unidirectional rates of transfer. Thus, the net transport of tracer across the endothelium is given by (1.19), as modified by (1.22),

$$\frac{\mathrm{d}m_\mathrm{e}}{\mathrm{d}t} = F\left(1-\mathrm{e}^{-P/F}\right)c_\mathrm{a} - F\left(1-\mathrm{e}^{-P/F}\right)\frac{m_\mathrm{e}}{V_\mathrm{e}} = K_1 c_\mathrm{a} - k_2 m_\mathrm{e} \quad (1.26)$$

in which $m_e$ is the amount of exchangeable inert tracer accumulated until the time $t$. The tissue compartment is revealed as a compartment with $k_2$ as the relaxation constant. For $j = K_1 c_a$, and $m_e(0) = 0$, (1.11) provides the solution to this equation for the transient phase,

$$m_e = K_1 \int_o^T c_a e^{-k_2(T-t)}\, dt \overset{\text{SE}}{\rightarrow} c_a \frac{k_2 V_e}{k_2 - \beta} \overset{\text{SS}}{\rightarrow} V_e c_a. \tag{1.27}$$

This fundamental equation, developed by Kety (1951, 1960a, b) and Johnson and Wilson (1966) for inert substances, shows that the extravascular compartment fulfills the definition of a real compartment. It describes the transient phase of exchange en route to secular equilibrium, as expressed in (1.11). The secular equilibrium, as always, requires a monoexponentially declining input function, $c_a(0)e^{-\beta t}$. It describes a constant ratio between $m_e$ and $c_a$ during the monoexponential decline of $c_a$.

## 1.3.2 Combined Compartments

For a two-compartment system consisting of the compartments $m_c$ and $m_e$, the total amount of tracer in the region of interest is the sum of $m_e$ and (1.27),

$$m = m_c + m_e = V_o c_a + (V - V_o)\frac{m_e}{V_e}, \tag{1.28}$$

where $V$ is defined as the sum $V_c + V_e$.[2] Inserting (1.27) into (1.28), we obtain the total tracer quantity in the tissue,

$$m = V_o c_a + (V - V_o)\, k_2 \int_o^T c_a e^{-k_2(T-t)}\, dt \tag{1.29}$$

which is the solution to the equation,

$$\frac{dm}{dt} = V_o \frac{dc_a}{dt} + k_2(V c_a - m), \tag{1.30}$$

where $V_o$ is the initial volume of distribution in the tissue. During the transient phase, the three parameters, $V_o$, $V$, and $k_2$ can be determined by nonlinear regression, on the basis of at least three separate observations. The ratio $(V - V_o)/V_e$ is close to unity: The capillary volume is about 0.5% of the tissue volume, and $V_o$ is but a fraction of $V_c$. Depending on the tracer solubility in tissue, $V_e$ is close to

[2] $V_o$, $V_c$, or $V_e$, are true, physical volumes because $V_o = V_c K_1/P$, $V_c = V_c^w \alpha_1$ where $V_c^w$ is the capillary whole-blood or plasma water volume, and $V_e = V_d/\alpha_1 = V_d^w \alpha_2/\alpha_1)$ where $V_d^w$ is the tissue water volume.

95% of the tissue volume. For low values of $P$ relative to $K_1 V_e$, $V_o$ approaches the capillary volume $V_c$, and $V - V_o \rightarrow V_e$. For very large values of $P$, $V_o$ approaches zero, and $V \rightarrow V_e + V_c$.

At infinity, the secular equilibrium established for a monoexponentially declining function, $c_a(0)e^{-\beta t}$, where $\beta < k_2$, follows from (1.29),

$$m \overset{\text{SE}}{\rightarrow} V_o c_a + (V - V_o)\left(\frac{k_2}{k_2 - \beta}\right) c_a \overset{\text{SS}}{\rightarrow} V c_a \tag{1.31}$$

such that the steady-state volume of distribution is $V$ $(= V_c + V_e)$.

Multilinear regression analysis is occasionally preferable to nonlinear regression. By integration of (1.30) for $m_e(0) = c_a(0) = 0$, a multilinear equation in measurable variables is obtained,

$$m = V_o c_a + k_2 V \int_o^T c_a \, dt - k_2 \int_o^T m \, dt. \tag{1.32}$$

Estimates of $V$, $V_o$, and $k_2$ can be made by regression of the normalized solution of the equation earlier. For an example of the use of (1.32), see Fig. 5.3. The linearity of the equation can be further evaluated by normalization to eliminate one of the independent variables. In this case, the normalization refers to the process of dividing the quantity of tracer in the tissue by the arterial concentrations of tracer obtained in individual experiments,

$$\frac{m}{c_a} = V_o + k_2 V \left(\frac{\int_o^T c_a \, dt - \int_o^T \frac{m}{V} \, dt}{c_a}\right), \tag{1.33}$$

where the $m/c_a$ ratio is the tissue volume of distribution. When $V$ is sufficiently large, and $T$ small, the equation shrinks to a formula derived in abstract by Fenstermacher et al. (1979) and in full publication by Gjedde (1981a, b) for blood–brain barrier permeability and extended to irreversible metabolism by Gjedde (1982),

$$\frac{m}{c_a} = V_o + k_2 \, V \, \Theta, \tag{1.34}$$

where $\Theta$ is the normalized integral $\int_o^T c_a dt / c_a$ with a dimension of time which represents a virtual time variable. The fraction $m/c_a$ is a linear function of $\Theta$ with a slope of $k_2 V$. An example of the use of this slope-intercept plot to determine capillary permeability in brain is shown in Fig. 1.2.

### 1.3.3 Arteries and Veins

For tissues that include arteries, arterioles, venules, and veins (and most tissues do), $V_o$ and $V$ take on additional meaning, and the difference between $V$, the total

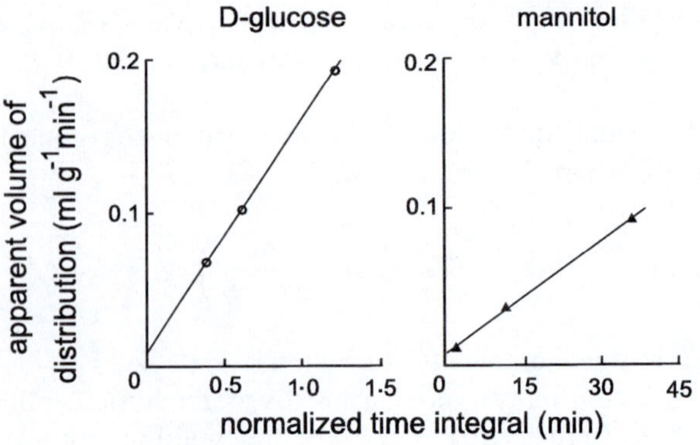


**Fig. 1.2** First published slope-intercept or Gjedde–Patlak plot of capillary permeability (Gjedde 1981b; Patlak et al. 1983). *Left panel* illustrates D-glucose permeability, *right panel* mannitol permeability. The abscissa is normalized time-integral of tracer concentration in plasma (min). The ordinate is apparent volume of distribution in brain (ml $g^{-1}$ $min^{-1}$). Note different scale of the two abscissae, indicating orders of magnitude difference between the two slopes. From Gjedde (1981b)

volume, and $V_e$, the tissue "exchangeable" volume, becomes even larger. In the case of arterial and venous volumes, represented by $V_a$ and $V_v$, respectively, in (1.28) and (1.31), the following augmentations occur,

$$V = V_a + V_c + V_e + V_v \tag{1.35}$$

and

$$V_o = V_a + V_c(K_1/P) + V_v(1 - E_o). \tag{1.36}$$

For significant capillary and venous volumes, the discrepancies between $k_2V$, $k_2(V - V_o)$, and $K_1 = k_2V_e$ cannot be ignored. When $P_dA$ is very large, as it is for a blood flow tracer, $V_o$ reduces to $V_a$.

## 1.4 Three Tissue Compartment Analysis

The three tissue compartment model is a description of three tissue compartments that include vascular, extravascular, and an additional biochemically defined pool, separated from the extravascular compartment by a barrier represented by proteins such as transporters, receptors, or enzymes.

This model is an extension of the two-compartment model which greatly complicates the solution of the tracer kinetic equations, unless simplifying assumptions are introduced. The most fundamental simplifying assumption holds that only tracer concentrations are present. The introduction of a third compartment establishes a third tracer state, $m_m$, defined as follows:

$$\frac{dm_m}{dt} = j_3 - k_4\, m_m \tag{1.37}$$

in which $k_4$ is the relaxation constant. The definitions of the influx, $j_3$, and the relaxation constant, $k_4$, depend entirely on the process responsible for the establishment of a third tracer state, be it metabolism or binding to receptors.

### *1.4.1 Compartmental Assumptions*

Biological "trapping" of the tracer often occurs because of recognition by, and binding to, a receptor prior to enzymatic action, or import by a membrane protein, according to classical Michaelis–Menten kinetics. The binding is often so rapid that a secular equilibrium is reached in time for the enzymatic action or the transport to be the dominant process of relaxation, rather than the dissociation from the receptors. The processes of association and dissociation, therefore, may not be "measurable" by compartmental analysis. Nonetheless, the binding compartment can be described later as a prelude to the discussion of the slower processes of catalysis or transport.

#### Receptor or Metabolite Compartment

Simple binding can be described by the equation derived from the formalism of Michaelis and Menten (1913) and later applications to tracer kinetics (Mintun et al. 1984; Gjedde et al. 1986; Wong et al. 1986a, b). Binding is considered the result of the three opposing processes of association, dissociation, and possible transport or catalysis,

$$\begin{aligned}\frac{\mathrm{d}m_b}{\mathrm{d}t} &= k_{\mathrm{on}}(B_{\mathrm{max}} - m_b)\frac{m_{\mathrm{e}}}{V_{\mathrm{d}}} - (k_{\mathrm{off}} + k_{\mathrm{cat}})m_b \\ &= \frac{k_{\mathrm{on}}}{V_{\mathrm{d}}}B_{\mathrm{max}}m_{\mathrm{e}} - (k_{\mathrm{off}} + k_{\mathrm{cat}})\left(1 + \frac{m_{\mathrm{e}}}{V_{\mathrm{d}}K_d^{\mathrm{w}}}\right)m_b,\end{aligned} \tag{1.38}$$

where $k_{\mathrm{on}}$ is the bimolecular association constant, $B_{\mathrm{max}}$ the number of available binding sites, $m_b$ the quantity of bound tracer molecules, $k_{\mathrm{off}}$ the dissociation constant, and $k_{\mathrm{cat}}$ the catalytic (i.e., turn-over) rate of the binding protein. This equation shows that the bound tracer represents a real compartment only when the precursor compartment, $m_{\mathrm{e}}$, is negligible or constant, relative to $K_d^{\mathrm{w}}$. This equation defines $K_d^{\mathrm{w}}$, the Michaelis–Menten half-saturation concentration, as the $(k_{\mathrm{off}} + k_{\mathrm{cat}})/k_{\mathrm{on}}$ ratio. The binding defines a tracer state only when $m_{\mathrm{e}}$ is approximately constant, or when $m_{\mathrm{e}}$ is always negligible relative to $V_{\mathrm{d}}K_d^{\mathrm{w}}$, in other words when the tracer concentration is too low to occupy a measurable fraction of the binding sites. The solution to (1.38) is (1.11) for negligible $m_{\mathrm{e}}$ and $m_b(0) = 0$,

$$m_b = \frac{k_{\mathrm{on}}}{V_{\mathrm{d}}}B_{\mathrm{max}}\int_o^T m_{\mathrm{e}}\mathrm{e}^{-(k_{\mathrm{off}}+k_{\mathrm{cat}})(T-t)}\,\mathrm{d}t \overset{\mathrm{ss}}{\rightarrow} \left(\frac{B_{\mathrm{max}}}{V_{\mathrm{d}}K_d^{\mathrm{w}}}\right)m_{\mathrm{e}} = \left(\frac{V_b}{V_{\mathrm{e}}}\right)m_{\mathrm{e}}, \tag{1.39}$$

where $V_b$ is a virtual volume equal to the $B_{\text{max}}/(\alpha_1 K_d^{\text{w}})$ ratio. In theory, (1.39) provides the means to determine the dissociation rate from the binding site. The rates of association and dissociation are often so rapid that the transient phase of approach toward secular equilibrium is too short lasting to be measured. Instead, continued monoexponential decline of the precursor pool at the rate of $\beta$, i.e., relaxation only of the state occupied by $m_{\text{e}}$, leads to secular equilibrium. This derivation is consistent with the Michaelis–Menten solution to the binding equation when the magnitude of $\beta$ is negligible compared to dissociation and catalysis. As it is a common experience that the association and dissociation rates are rapid, the monoexponential washout of the precursor pool indeed often may be negligible. Simple solubility is a special case of binding which can be described as a separate state, rather than as the solubility discussed earlier.

The actual quantity of bound tracer depends strongly on the affinity ($K_d^{\text{w}}$), relative to $B_{\text{max}}$. In general, the larger the magnitude of $k_{\text{cat}}$, the lower the quantity $m_b$. As expressed by the rate constant $k_{\text{cat}}$, the binding of the tracer to a receptor may be the prelude to an enzymatic reaction or facilitated transport through a membrane. At the steady-state of binding expressed by (1.40), the flux imposed by these processes in (1.37) is given by:

$$j_3 = k_{\text{cat}} m_b \overset{\text{ss}}{\rightarrow} k_{\text{cat}} \frac{B_{\text{max}}}{V_{\text{d}} K_d^{\text{w}}} m_{\text{e}} = \frac{k_{\text{on}}}{V_{\text{d}}} \left( \frac{k_{\text{cat}} B_{\text{max}}}{k_{\text{off}} + k_{\text{cat}}} \right) m_{\text{e}} \equiv k_3 m_{\text{e}}. \tag{1.40}$$

The rate-limiting step in the two-step process consisting of binding and catalysis (facilitation) can now be evaluated by considering the relative magnitudes of $k_{\text{off}}$ and $k_{\text{cat}}$. The relaxation constant $k_3$ is defined by this equation as the steady-state (combined forward) rate constant of the binding.

### Precursor Compartment

The introduction of a binding, metabolism, or transport compartment that reaches secular equilibrium affects the description of the compartment $m_{\text{e}}$ that directly exchanges contents with the circulation. The description depends on the reversibility of the tracer accumulation in the compartment to which $m_{\text{e}}$ is precursor.

**Reversible Metabolism or Transport** In the ordinarily reversible case defined in (1.37), the compartment equation describing the precursor pool expands to an expression which depends on the properties of the metabolite compartment,

$$\frac{\text{d}m_{\text{e}}}{\text{d}t} = K_1 c_{\text{a}} + k_4 m_{\text{m}} - (k_2 + k_3) m_{\text{e}} \overset{\text{ss}}{\rightarrow} K_1 c_{\text{a}} - k_2 m_{\text{e}}. \tag{1.41}$$

The steady-state solution is equal to the limiting case of $k_3 = 0$. The solution to (1.37) and (1.41) is,

$$m_e = K_1 \left[ \left( \frac{q_2 - k_4}{q_2 - q_1} \right) \int_o^T c_a e^{-q_2(T-t)} \, dt - \left( \frac{q_1 - k_4}{q_2 - q_1} \right) \int_o^T c_a e^{-q_1(T-t)} \, dt \right] \overset{\text{ss}}{\rightarrow} V_e c_a, \quad (1.42)$$

where $q_1$ and $q_2$ are composite constants,

$$2q_1 = \left( k_2 + k_3 + k_4 - \sqrt{(k_2 + k_3 + k_4)^2 - 4k_2k_4} \right) \quad (1.43)$$

and

$$2q_2 = \left( k_2 + k_3 + k_4 + \sqrt{(k_2 + k_3 + k_4)^2 - 4k_2k_4} \right). \quad (1.44)$$

For an example of the use of (1.44), see Fig. 7.4.

**Irreversible Metabolism or Transport** In the case of an irreversible metabolism or transport reaction ($k_4 = 0$), the composite relaxation constants $q_1$ and $q_2$ assume the values $q_1 = 0$ and $q_2 = k_2 + k_3$. This causes the solution to (1.42) to be greatly simplified,

$$m_e = K_1 \int_o^T c_a e^{-(k_2+k_3)(T-t)} \, dt \overset{\text{ss}}{\rightarrow} V_f c_a, \quad (1.45)$$

where $V_f$ equals the steady-state volume of distribution $K_1/(k_2 + k_3)$. This description of the third compartment refers to no particular process of trapping. In actual practice, the tracer may be trapped by a number of different mechanisms, reversible or irreversible. In the final analysis, trapping represents merely an expansion of the distribution space, as will be explained later.

### Transport or Metabolism Compartment

The interpretation of the process symbolized by the rate constant $k_3$ depends on the steady-state solution to the binding equation as expressed earlier in (1.38) and is a function of the relationship between the two relaxation constants $k_{off}$ and $k_{cat}$.

For $k_{off} \ll k_{cat}$, the rate-limiting step is the association with the recognition site. For this case, $k_3$ equals $k_{on}B_{max}/V_d$ as described in (1.38). It is surprising to realize that the highest binding affinities render the accumulation of the tracer in the third compartment a function of the rate of association of the tracer to the recognition site. Figure 1.3 illustrates the binding of the dopamine $D_{2-4}$ receptor ligand $N$-methylspiperone to its receptor. This binding is of such high affinity that it continues irreversibly for the duration of the measurement.

For $k_{off} \gg k_{cat}$, the rate-limiting step is the catalysis by the protein. In this case, $k_3$ equals $k_{cat}B_{max}/(V_d K_d^w)$, according to which the accumulation in the third

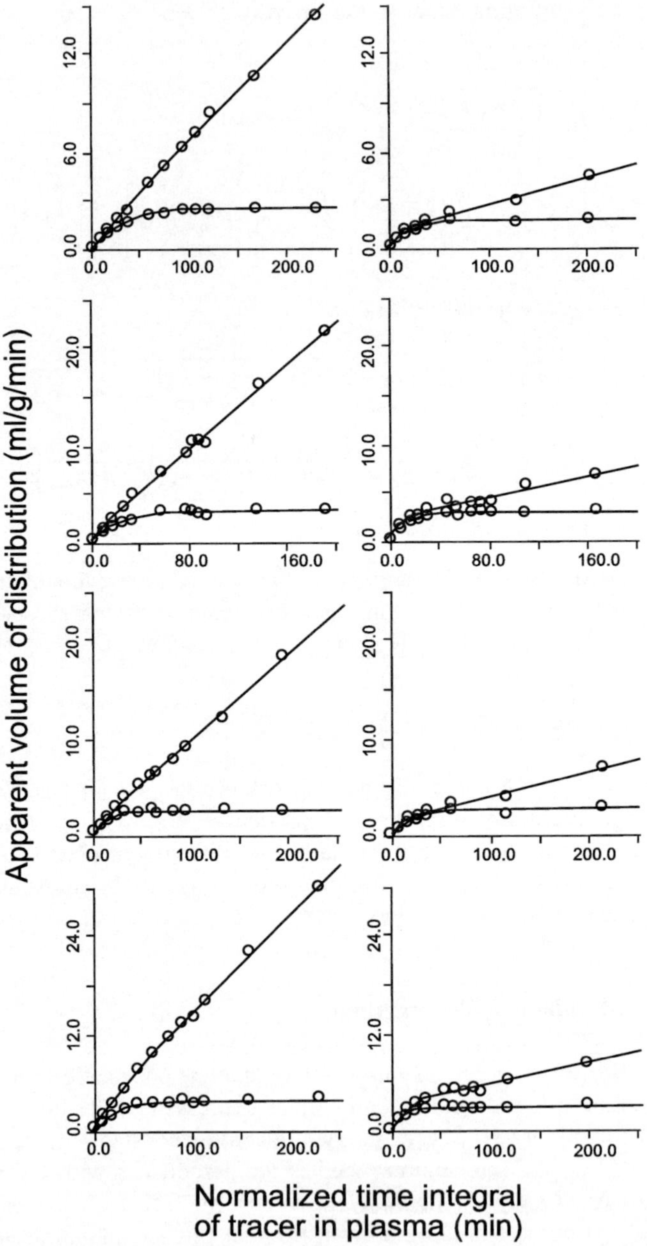


**Fig. 1.3** First published slope-intercept or Gjedde–Patlak plot of receptor binding in vivo: $N$-methylspiperone (NMSP) binding to caudate nucleus (*upper curves*) and cerebellum (*lower curves*), plotted as volume of distribution vs. normalized integral in four normal human volunteers. *Left panel*: Before haloperidol blockade. *Right Panel*: After 90% blockade of NMSP binding sites in neostriatum by haloperidol, 2.5 ng ml$^{-1}$ in plasma. Abscissae and ordinates as in Fig. 1.2. Note reduction of slope due to haloperidol blockade of $B_{\max}$ and hence reduction of $k_3$. From Wong et al. (1986b)

compartment is a function of the rate of catalysis (activity) of the enzyme or transporter. This relationship defines the maximal velocity of an enzymatic reaction, or transport, $J_{\max}$, such that $k_3$ equals $J_{\max}/(V_d K_d^w)$. It is customary in this case to refer to the Michaelis half-saturation constant as $K_m$ rather than $K_d$, although the two constants are exactly equivalent.

The meaning of $k_4$ depends on the process of relaxation of the third compartment. For a reversible enzymatic reaction, $k_4$ equals $J'_{\max}/(V_d K'^w_d)$ where $J'_{\max}$ is the maximal velocity of the reverse reaction, and $K'^w_d$ its Michaelis–Menten half-saturation constant.

**Reversible Metabolism or Transport** Most transporter and enzyme processes are reversible to a lesser or greater extent. With the substitutions defining $k_3$ and $k_4$, the metabolite compartment can be described by (1.37),

$$\frac{\mathrm{d}m_\mathrm{m}}{\mathrm{d}t} = k_3 m_\mathrm{e} - k_4 m_\mathrm{m} = \left(\frac{J_{\max}}{V_\mathrm{e} K_d}\right) m_\mathrm{e} - \left(\frac{J'_{\max}}{V_\mathrm{e} K'_d}\right) m_\mathrm{m}. \tag{1.46}$$

According to (1.11) and (1.37), the solution to, and steady-state reached by, (1.46) is

$$m_\mathrm{m} = k_3 \int_o^T m_\mathrm{e} \mathrm{e}^{-k_4(T-t)}\, \mathrm{d}t \overset{\mathrm{SS}}{\rightarrow} \frac{k_3}{k_4} m_\mathrm{e} = p_\mathrm{B}\, m_\mathrm{e}. \tag{1.47}$$

The $k_3/k_4$ ratio defines an expansion of the precursor pool volume $V_\mathrm{e}$, represented by the factor $p_\mathrm{B}$, also known as the binding potential or receptor availability in the case of receptors. In the case of a reversible enzymatic reaction, $p_\mathrm{B}$ is the ratio between the Michaelis–Menten constants of the two directions of the reaction. The solution to (1.37) and (1.46) is

$$m_\mathrm{m} = \frac{K_1 k_3}{q_2 - q_1}\left(\int_o^T c_\mathrm{a} \mathrm{e}^{-q_1(T-t)}\, \mathrm{d}t - \int_o^T c_\mathrm{a} \mathrm{e}^{-q_2(T-t)}\, \mathrm{d}t\right) \overset{\mathrm{SS}}{\rightarrow} \frac{K_1 k_3}{k_2 k_4} c_\mathrm{a} = p_\mathrm{B} V_\mathrm{e} c_\mathrm{a}, \tag{1.48}$$

where the composite relaxation constants $q_1$ and $q_2$ are those defined in (1.43) and (1.44).

**Irreversible Metabolism or Transport** In the case of an irreversible process, a complete secular equilibrium is not reached by all compartments. Instead, the composite relaxation constants assume the values 0 and $k_2 + k_3$, respectively, causing the metabolism compartment to continue to expand according to the formula:

$$m_\mathrm{m} = k_3 \int_o^T m_\mathrm{e}\, \mathrm{d}t = \frac{K_1 k_3}{k_2 + k_3}\left[\int_o^T c_\mathrm{a}\, \mathrm{d}t - \int_o^T c_\mathrm{a} \mathrm{e}^{-(k_2+k_3)(T-t)}\, \mathrm{d}t\right]. \tag{1.49}$$

This expression describes irreversible trapping as a compartment with two components, one accumulating, and one equilibrating.

### 1.4.2 Combined Compartments

With metabolism or transport, both depending on binding to a recognition site, the total quantity of tracer in the tissue is the sum of the contents of individual compartments, $m = m_c + m_e + m_b + m_m$. Although this sum has four components, the assumption is routinely made that $m_b$ and $m_m$ never co-exist such that a measurable state of binding ($m_b$) precludes a measurable state of metabolism ($m_m$), and *vice versa*, although in reality this need not be so.

**Binding** In the case of pure binding with no subsequent transport or metabolism, the total quantity of tracer in the tissue approaches a steady-state dictated by the binding capacity and affinity,

$$m = m_c + m_e + m_b = V_o c_a + \rho m_e \overset{\text{ss}}{\rightarrow} (V + V_b)\, c_a, \tag{1.50}$$

where $\rho$ is the volume ratio $(V_c + V_e + V_b - V_o)/V_e$. This result identifies the binding essentially as a virtual expansion of the distribution space $V$. For enzymes and transporters, the bound quantity is considered negligible. The binding capacity and affinity are often inversely related such that low capacity is associated with high affinity. For the highest capacity – lowest affinity combinations the designation nonspecific binding is often used because steady-state is reached almost instantly ($k_3 \gg \beta \gg 1/T$).

**Reversible Metabolism or Transport** In the case of reversible metabolism or transport, the sum of $m_e$ and $m_b$ is given by (1.42) and (1.49),

$$m = m_c + m_e + m_b + m_m = V_o c_a + \rho K_1 \left[ \left( \frac{q_2 - \left(\frac{k_3}{\Omega} + k_4\right)}{q_2 - q_1} \right) \right.$$
$$\left. \times \int_o^T c_a e^{-q_2(T-t)} dt - \left( \frac{q_1 - \left(\frac{k_3}{\rho} + k_4\right)}{q_2 - q_1} \right) \int_o^T c_a e^{-q_1(T-t)} dt \right], \tag{1.51}$$

where the composite relaxation constants $q_1$ and $q_2$ are those defined in (1.43) and (1.44). This equation was derived in principle by Phelps et al. (1979). Eventually, it leads to a steady-state,

$$m \overset{\text{ss}}{\rightarrow} (V + V_b + p_B\, V_e)\, c_a. \tag{1.52}$$

**Irreversible Transport or Metabolism** Irreversible transport or metabolism of the tracer occurs when $k_4$ is zero, of course. In this case, adding the three main compartments (and the negligible binding), an equation for the transient phase of the tracer's approach toward limited secular equilibrium is obtained,

$$m = V_o c_a + \frac{K_1 k_3}{k_2 + k_3} \int_o^T c_a \, dt + \rho K_1 \left(1 - \frac{k_3/\rho}{k_2 + k_3}\right) \int_o^T c_a e^{-(k_2+k_3)(T-t)} \, dt \tag{1.53}$$

which for negligible $B_{\max}$ and a $(V - V_o)/V_e$ ratio close to unity (i.e., $\rho \cong 1$) reduces to,

$$m \cong V_o c_a + k_3 V_f \int_o^T c_a \, dt + k_2 V_f \int_o^T c_a e^{-(k_2+k_3)(T-t)} \, dt, \tag{1.54}$$

where $V_f$, the precursor pool volume, equals $K_1/(k_2 + k_3)$. This equation was first derived in principle by Sokoloff et al. (1977). The equation reaches a secular equilibrium which in shape is identical to that expressed in (1.34),

$$\frac{m}{c_a} \overset{ss}{\rightarrow} \left[ V_o + \frac{k_2}{k_2 + k_3} V_f \right] + k_3 V_f \Theta = V_g + K\Theta \tag{1.55}$$

as described by Gjedde (1982). In this equation, it is convenient to redefine the $y$-intercept and the slope, such that $V_g$ is the combined representation of the precursor pool volumes $V_o + k_2 V_f/(k_2 + k_3)$ and $K$ the net clearance of the tracer, equal to $k_3 V_f$. An example of the use of (1.55) to describe the metabolism of deoxyglucose in rat brain is given in Fig. 1.4. The third compartment is established by the irreversible phosphorylation of deoxyglucose to deoxyglucose-6-phosphate discussed in greater detail in a later chapter.

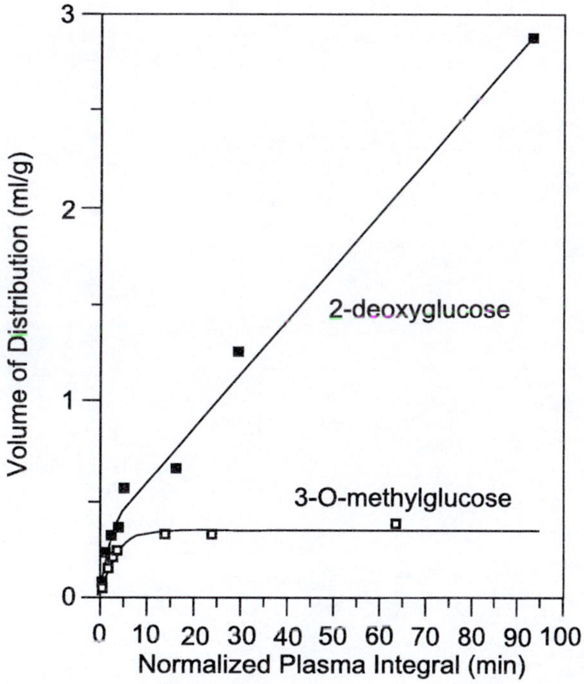


**Fig. 1.4** Redrawn from first published Gjedde–Patlak plot of irreversible tracer metabolism: Slope-intercept plot of irreversible metabolism of 2-deoxyglucose and absent metabolism of 3-$O$-methyl-glucose in rat brain. Abscissae and ordinates as in Figs. 1.2 and 1.3. Note difference between initial slopes (signifying transport across blood–brain barrier) and steady-state slopes (signifying net metabolism). Redrawn from Gjedde (1982)

This reduction in complexity between (1.53) and (1.55) is typical of the general approach to compartmental analysis where the number of relevant compartments must be reduced to the absolute minimum by reasonable assumptions. A compartmental model is a description of the behavior of the tracer, not of nature. It is used to provide knowledge about the natural system that can be used to solve specific problems.

# Chapter 2
# Fundamentals of Compartmental Kinetics*

## 2.1 Definition of Relaxation Constants

The key to tracer kinetic analysis is the concept of compartment, a group of atoms or molecules which behave in such an identically predictable manner that the introduction of a few additional but labeled atoms or molecules does not change the behavior significantly. Compartments may be large or small but they are fundamental abstractions, regardless of their size. As such they can be said to defy the very concept they were created to represent, because they require that the contents are at equilibrium and hence allow no interactions among members.

By being relegated to the interfaces between compartments, the kinetic processes studied by tracer kinetic analysis are discontinuous and hence fundamentally at variance with the real nature of kinetic processes, which must be continuous ("distributed"). The mathematically abstract compartments can of course be widely dispersed and physically intermixed with other compartments, but the definition does not allow the members of individual compartments to interact.

Non-linear modeling of distributed processes is possible in theory, provided the measurements have the necessary power of resolution, but this is so rarely the case that distributed models often arise as assemblies of commensurately diminished compartments and as such can be regarded as extensions of the linear compartments considered here. However, the resulting nonlinear kinetics will not be examined in this text. The compartment is proof of Niels Bohr's dictum that the measurement must invalidate the measured because it ignores the quantum nature of the distribution. Its saving grace is its usefulness to the practical problems of biology.

---

* Adapted from Gjedde (2003) Modelling metabolite and tracer kinetics, in *Molecular Nuclear Medicine*, eds L. E. Feinendegen, W. W. Shreeve, W. C. Eckelman, Y. W. Bahk and H. N. Wagner Jr., Springer-Verlag, Berlin Heidelberg, Chap. 7, pp. 121–169, with permission from Springer-Verlag, Berlin Heidelberg.

A. Gjedde et al., *Neurokinetics: The Dynamics of Neurobiology In Vivo*,
DOI 10.1007/978-1-4419-7409-9_2, 

### 2.1.1 Single Compartment

A group of molecules, however confined or identified, form a single kinetic compartment when their rate of escape from the compartment (also called "decay" or "relaxation") is constant and proportional to the number of remaining molecules. In this sense, the compartment is an abstraction never found in real life. However, in many cases, although not in all, it is an excellent approximation when the rate of exchange among the members is much faster than the rate of relaxation. This condition is believed to be fulfilled when the compartment is well-stirred. Thus, if the stirring is not vigorous enough, the compartmentation is said to fail, and the compartment collapses into smaller compartments or noncompartmental distributions. When the product of a chemical reaction forms a single kinetic compartment, the simplest precursor–product relationship obeys a linear differential equation of the form (see Chap. 1),

$$\frac{\mathrm{d}m}{\mathrm{d}t} = j - k\, m, \tag{2.1}$$

where $j$ is the time-variable precursor influx function and $k$ is the relaxation or rate constant. How, or to which location, the product escapes is immaterial to the definition. In this respect, the compartment is a one-dimensional construct with time as the only dimension of change. The relaxation constant is the solution to (2.1),

$$k = \frac{j}{m} - \frac{1}{m}\left(\frac{\mathrm{d}m}{\mathrm{d}t}\right), \tag{2.2}$$

where $m$ is the mass of molecules. The equation shows that the relaxation constant $k$ is the ratio between the influx and the compartment content at steady-state when $\mathrm{d}m/\mathrm{d}t = 0$ (Fig. 2.1).

It is important to emphasize that the relaxation creates the compartment. Neither membranes nor anatomical subdivisions contribute to the compartmentation, except to the extent that they establish or allow the relaxation. Nor is the relaxation necessarily the result of a single mechanism. Frequently, the relaxation occurs because multiple so-called first-order mechanisms combine linearly to form a single process.

The mechanisms of passive transport across membranes, enzymatic steps of metabolic pathways, and radioactive decay, all generate compartments when they decay at a constant rate ($k$), such that the loss of members per unit time is

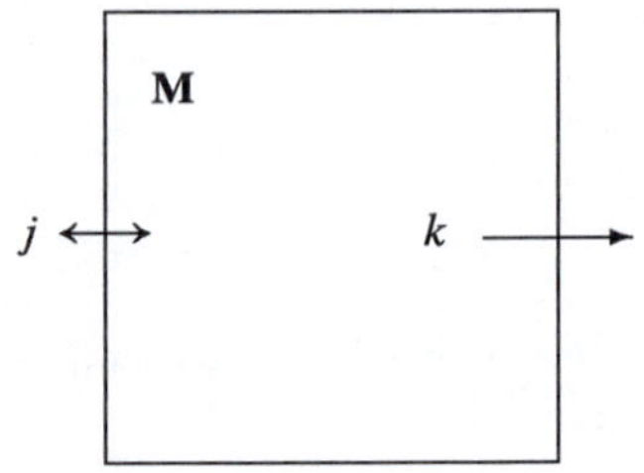


**Fig. 2.1** Model of single compartment **M** occupied by the quantity $m$. Note that net fluxes are represented by *bidirectional arrows*, with rate constants and clearances represented by *unidirectional arrows*

proportional to the number of remaining members. Often, but not always, the number of remaining members is proportional to the concentration of the members in an aqueous medium or other solvent. The proportionality depends on several physical and chemical factors, the simplest being the volume of the solvent, but it is not always possible to judge whether they have remained constant to the satisfaction of the required proportionality between number and concentration of members.

The single isolated compartment does not exist in reality but may arise in approximation when the supply of substance is absent, instantaneous, infrequent, or unpredictable, i.e., when the function $j$ is zero, a Dirac delta function, or when it is given only numerically.

For this compartment the transient solution to (2.1) is the convolution integral (see Chap. 1),

$$m = \mathrm{e}^{-kT}\left[m(0) + \int_o^T j\,\mathrm{e}^{kt}\,\mathrm{d}t\right], \tag{2.3}$$

where $\mathrm{e}^{-kt}$ is the *impulse response function*, which accounts for the monoexponentially changing loss of the original member molecules, as well as for the monoexponentially changing disappearance of new members for whom a correction must be made for the fact that they arrive at different times and hence have different risks of being expelled before a certain time. The steady-state solution to the contents of the compartment is,

$$M = \frac{J}{k} \tag{2.4}$$

in keeping with the definition of a compartment, leading to the steady-state solution to the contents of the compartment,

$$J = k\,M, \tag{2.5}$$

where variables at steady-state are given upper-case symbols.

### 2.1.2 Two Compartments

When two or more compartments occupy positions in series, the products have precursors which themselves form single compartments. Applying the definition, (2.1) becomes,

$$\frac{\mathrm{d}m_2}{\mathrm{d}t} = k_1\,m_1 - k_2\,m_2, \tag{2.6}$$

where $m_1$ refers to the precursor compartment and $m_2$ to the product compartment. Note that $m_1$ and $m_2$ represent numbers of molecules, usually in units of *mol* when normalized against Avogadro's number ($6 \cdot 10^{23}$). When clearance (which has unit of flow) is the precursor's mechanism of decay, the quantity of molecules can be converted to concentration in the precursor solvent with a known volume ($V_1$),

$$\frac{dm_2}{dt} = V_1\, k_1\, \frac{m_1}{V_1} - k_2\, m_2, \tag{2.7}$$

yielding,

$$\frac{dm_2}{dt} = K_1\, c_1 - k_2\, m_2 \tag{2.8}$$

and

$$\frac{dm_1}{dt} = V_1\, \frac{dc_1}{dt}, \tag{2.9}$$

where $K_1$ symbolizes the product $V_1 k_1$ and is known as a clearance with unit of flow and $c_1$ is the concentration of the precursor. The clearance is the flux from compartment 1 to compartment 2 relative to the concentration. The total flux into the two compartments (and all other compartments fed by this closed system) is,

$$j_o = \sum_{i=1}^{2} \left[\frac{dm_i}{dt}\right] = V_1\, \frac{dc_1}{dt} + K_1\, c_1 - k_2\, m_2. \tag{2.10}$$

The distinction between quantity and concentration of substrate is the reason for the distinction between the rate constant $k_1$ and the clearance $K_1$. The use of concentrations instead of masses requires knowledge of the relative solvent volumes of the two compartments. For example, associating the rate constant $k_1$ with the concentration $c_1$ creates an inconsistency unless $m_1$ and $m_2$ occupy the same volume. Thus, to convert the masses included in (2.6) to concentrations, all terms are divided by the same volume, e.g., $V_2$, which means that the apparent concentration $c_1$ is really the concentration of $m_1$ in the solvent of $\mathbf{M}_2$, i.e., $m_1/V_2$, i.e. (Fig. 2.2),

$$\frac{dc_2}{dt} = \frac{K_1}{V_2}\, c_1 - k_2\, c_2 \tag{2.11}$$

in which the first coefficient is different from the mass balance relationship, indicating that masses and concentrations are not exchangeable, such that (2.11) and (2.12) are incompatible,

$$\frac{dm_2}{dt} = \frac{K_1}{V_1}\, m_1 - k_2\, m_2. \tag{2.12}$$

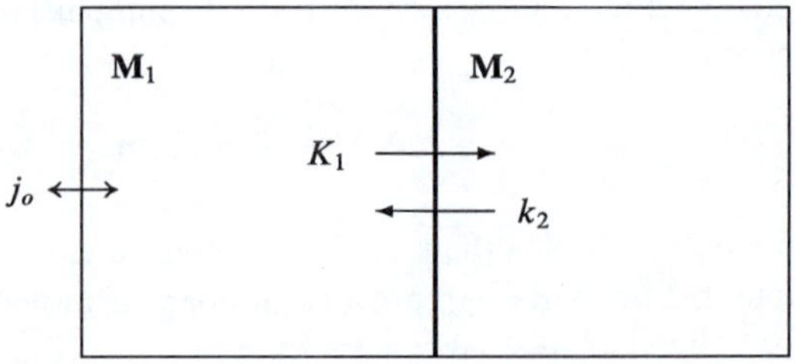


**Fig. 2.2** Model of two compartments exchanging contents consisting of the product ($m_2$) of a reaction and $m_1$ is the precursor with the concentration $c_1$

The transiently nonlinear solution to the differential (2.6) of this relationship is again the convolution integral,

$$m_2 = \mathrm{e}^{-k_2 T}\left[m_2(0) + k_1 \int_o^T m_1\, \mathrm{e}^{k_2 t}\, \mathrm{d}t\right], \tag{2.13}$$

where $k_1$ is the rate constant of the reaction which converts the precursor to the product and $k_2$ is the rate constant for the reaction which reconverts the product to the precursor. With the substitution of $c_1$ for $m_1$, the convolution integral of (2.8) is,

$$m_2 = \mathrm{e}^{-k_2 T}\left[m_2(0) + K_1 \int_o^T c_1\, \mathrm{e}^{k_2 t}\, \mathrm{d}t\right] \tag{2.14}$$

which leads to the combined solution for $m_1$ and $m_2$ when $m_2(0) = 0$,

$$\sum_{i=1}^{2} m_i = V_1\, c_1 + K_1 \int_o^T c_1\, \mathrm{e}^{-k_2(T-t)}\, \mathrm{d}t \tag{2.15}$$

and to the combined solution for $n-1$ recipient compartments supplied by the same delivery compartment,

$$\sum_{i=1}^{n} m_i = V_1 c_1 + \sum_{h=1}^{n-1} K_{1_h} \int_o^T c_1\, \mathrm{e}^{-k_{2_h}(T-t)} \mathrm{d}t. \tag{2.16}$$

A transient multilinear expression for the net flux across the interface is obtained by integration of (2.8),

$$j_1 = K_1\, c_1 - k_2 m_2 = K_1 c_1 - K_1\, k_2 \int_o^T c_1\, \mathrm{d}t + (k_2)^2 \int_o^T m_2\, \mathrm{d}t, \tag{2.17}$$

where $j_1$ is the flux between compartments $\mathbf{M}_1$ and $\mathbf{M}_2$, which yields the total accumulation in the two compartments at the time $T$. By simple integration when $m_1(0) = m_2(0) = 0$,

$$m_2 = K_1 \int_o^T c_1\, \mathrm{d}t - k_2 \int_o^T m_2\, \mathrm{d}t \tag{2.18}$$

and, using $m_1 = V_1 c_1$, the total content as a function of the concentration in compartment 1 ($\mathbf{M}_1$) may be written as:

$$\sum_{i=1}^{2} m_i = V_1\, c_1 + (K_1 + k_2 V_1) \int_o^T c_1\, \mathrm{d}t - k_2 \int_o^T \sum_{i=1}^{2} m_i\, \mathrm{d}t. \tag{2.19}$$

For $k_2 = 0$, (2.19) becomes,

$$\sum_{i=1}^{2} m_i = V_1\, c_1 + K_1 \int_o^T c_1\, \mathrm{d}t \tag{2.20}$$

which, when divided by $c_1$ becomes the equation underlying the Gjedde–Patlak plot (Gjedde 1981a),

$$\sum_{i=1}^{2} v_i = V_1 + K_1 \int_o^T \frac{c_1}{c_1(T)}\, \mathrm{d}t, \tag{2.21}$$

where $v_i(T)$ is an expanding "virtual" volume and $\int_o^T c_1 \mathrm{d}t / c_1$ is an expanding "virtual" time, useful to physiologists (Sarna et al. 1977). For $k_2 > 0$, the steady-state solution to (2.6) is,

$$M_2 = \frac{k_1}{k_2}\, M_1$$

while the steady-state form of (2.15), obtained by setting derivatives to zero in (2.7) through (2.10), is,

$$\sum_{i=1}^{2} M_i = \left[ V_1 + \left( \frac{K_1}{k_2} \right) \right] C_1, \tag{2.22}$$

where $V_1$ is the distribution volume of the molecules in compartment $\mathbf{M}_1$ and $K_1/k_2$ defines an additional volume $V_e$ as a "partition volume." The size of the partition volume is dictated by the magnitudes of $K_1$ and $k_2$ and is different from the distribution volume of the molecules in compartment $\mathbf{M}_2$ ($V_2$). The total volume of distribution is, therefore, $V_1 + (K_1/k_2)$.

At steady-state, by definition, the flux between the compartments is nil,

$$J_1 = K_1 C_1 - k_2 M_2 = 0 \tag{2.23}$$

showing that this "system" of two compartments does nothing more than regulate the magnitude of compartment $\mathbf{M}_2$: The steady-state size of $\mathbf{M}_2$ follows passively from the unidirectional fluxes across the interface, as shown by (2.12).

### *2.1.3 Two Compartments with Sink*

To regulate a steady-state flux without changing the size of compartment $\mathbf{M}_2$, a diversion must be established for the efflux. The receptacles of the efflux from a compartment are known as sinks. A compartment may have multiple sinks and still fulfill the criteria of compartmental behavior, as shown in the model underlying the linked differential equations (Fig. 2.3),

$$\frac{\mathrm{d}m_1}{\mathrm{d}t} = V_1 \frac{\mathrm{d}c_1}{\mathrm{d}t} \tag{2.24}$$

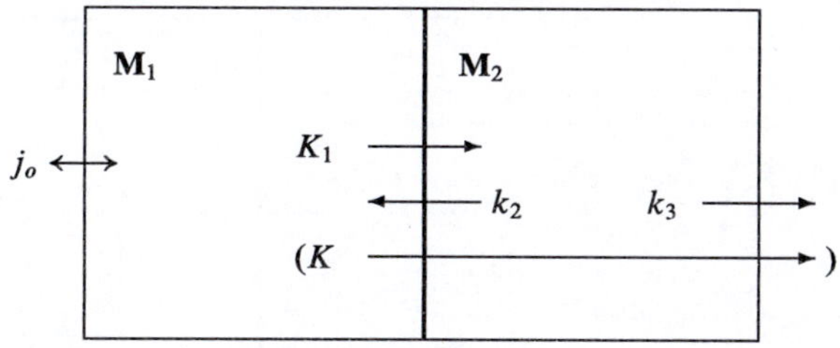


**Fig. 2.3** Model of precursor and product compartments with sink. At steady-state, the sink defines a single steady-state clearance, $K$, equal to $K_1 k_3/(k_2 + k_3)$, which reaches $K_1$ when $k_3 \gg k_2$

and

$$\frac{dm_2}{dt} = K_1\, c_1 - (k_2 + k_3)\, m_2, \tag{2.25}$$

where $K_1$ as the product $V_1 k_1$ is the clearance with unit of flow and $c_1$ is the concentration of the precursor in the volume $V_1$. The total flux into the system of compartments, and all other compartments fed by it, is given by an equation identical to (2.10),

$$j_o = V_1 \frac{dc_1}{dt} + K_1\, c_1 - k_2\, m_2. \tag{2.26}$$

The content of $\mathbf{M}_2$ is given by the transiently nonlinear solution,

$$m_2 = e^{-(k_2+k_3)T} \left[ m_2(0) + K_1 \int_o^T c_1\, e^{(k_2+k_3)t}\, dt \right] \tag{2.27}$$

which yields the combined solution for $m_2(0) = 0$,

$$\sum_{i=1}^{2} m_i = V_1\, c_1 + K_1 \int_o^T c_1\, e^{-(k_2+k_3)(T-t)}\, dt. \tag{2.28}$$

The flux into the sink at any time is given by the transiently multilinear expression,

$$j_2 = k_3\, m_2 = K_1 k_3 \int_o^T c_1\, dt - k_3\,(k_2 + k_3) \int_o^T m_2\, dt. \tag{2.29}$$

Using the integral forms of (2.24) and (2.25), and assuming initial conditions are zero, an integral equation for the total, $m_1 + m_2$, may be written as follows.

$$\sum_{i=1}^{2} m_i = V_1\, c_1 + [K_1 + V_1\,(k_2 + k_3)] \int_o^T c_1\, dt - (k_2 + k_3) \int_o^T \sum_{i=1}^{2} m_i\, dt. \tag{2.30}$$

The steady-state solution to (2.25) may be written as:

$$M_2 = \left( \frac{K_1}{k_2 + k_3} \right) C_1 \tag{2.31}$$

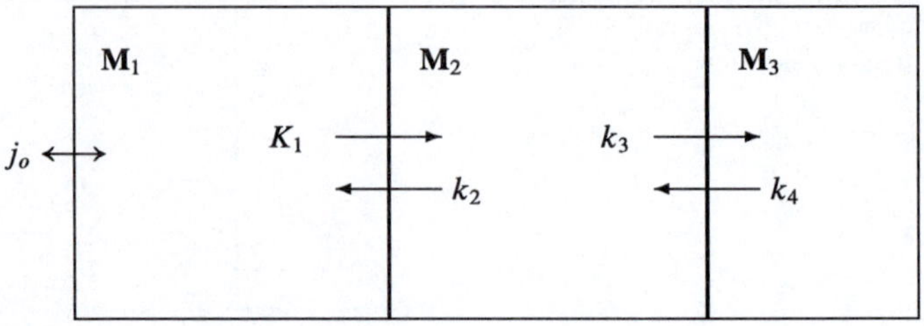


**Fig. 2.4** Closed system of three compartments

which yields the steady-state flux through the system, when the content of $\mathbf{M_2}$ is inserted into (2.29),

$$J_o = k_3 M_2 = K C_1, \tag{2.32}$$

where $K$ is the "net" clearance of the precursor from compartment $\mathbf{M_1}$ to compartment $\mathbf{M_2}$ (Fig. 2.4),

$$K = K_1 \left( \frac{k_3}{k_2 + k_3} \right) = \frac{K_1}{1 + \frac{k_2}{k_3}}. \tag{2.33}$$

### 2.1.4 *Three Compartments*

The previous system was open with a throughput. Closed compartments under normal circumstances require a return outlet as shown in the model above (Fig. 2.4), where

$$\frac{\mathrm{d}m_1}{\mathrm{d}t} = V_1 \frac{\mathrm{d}c_1}{\mathrm{d}t}, \tag{2.34}$$

$$\frac{\mathrm{d}m_2}{\mathrm{d}t} = K_1\, c_1 + k_4\, m_3 - (k_2 + k_3)\, m_2 \tag{2.35}$$

and

$$\frac{\mathrm{d}m_3}{\mathrm{d}t} = k_3\, m_2 - k_4\, m_3, \tag{2.36}$$

where the total flux into the three compartments is given by an equation similar to (2.10) and (2.26),

$$j_o = \sum_{i=1}^{3} \left[ \frac{\mathrm{d}m_i}{\mathrm{d}t} \right] = V_1 \frac{\mathrm{d}c_1}{\mathrm{d}t} + K_1\, c_1 - k_2\, m_2. \tag{2.37}$$

The linked differentials (2.34–2.36) command particular attention because they underlie major applications of tracer kinetic analysis. They have been solved both for $k_4 = 0$ and $k_4 > 0$. The solutions appear complex but are easily obtained by Laplace transformation into second-order polynomials.

**Transiently Non-Linear Solution to Reversible Accumulation** For $k_4 > 0$, the transiently nonlinear solution to the coupled equations (2.34–2.36) is obtained in two steps. The general solution to the differential (2.36) is the ordinary convolution integral,

$$m_3 = \mathrm{e}^{-k_4 T}\left[m_3(0) + k_3 \int_o^T m_2\, \mathrm{e}^{k_4 t}\, \mathrm{d}t\right], \tag{2.38}$$

where $k_3$ is the rate constant of the reaction which converts the precursor in $\mathbf{M}_2$ to the product in $\mathbf{M}_3$ and $k_4$ is the rate constant for the reaction which reconverts this product to its precursor. When (2.38) is inserted, the solution to (2.35) for $m_2(0) = m_3(0) = 0$ is,

$$m_2 = K_1\left[\left(\frac{q_2 - k_4}{q_2 - q_1}\right)\int_o^T c_1 \mathrm{e}^{-q_2(T-t)}\mathrm{d}t - \left(\frac{q_1 - k_4}{q_2 - q_1}\right)\int_o^T c_1 \mathrm{e}^{-q_1(T-t)}\mathrm{d}t\right], \tag{2.39}$$

where

$$q_1 = \frac{k_2 + k_3 + k_4 - \sqrt{(k_2 + k_3 + k_4)^2 - 4k_2k_4}}{2} \tag{2.40}$$

and

$$q_2 = \frac{k_2 + k_3 + k_4 + \sqrt{(k_2 + k_3 + k_4)^2 - 4k_2k_4}}{2} \tag{2.41}$$

which in turn changes the form of (2.38) to

$$m_3 = \left[\frac{K_1 k_3}{q_2 - q_1}\right]\left(\int_o^T c_1\, \mathrm{e}^{-q_1(T-t)}\mathrm{d}t - \int_o^T c_1\, \mathrm{e}^{-q_2(T-t)}\mathrm{d}t\right) \tag{2.42}$$

such that the sum of the contents of $\mathbf{M}_1$, $\mathbf{M}_2$, and $\mathbf{M}_3$ now is,

$$\begin{aligned}\sum_{i=1}^{3} m_i &= V_1 c_1 + K_1\left(\frac{q_2 - (k_3 + k_4)}{q_2 - q_1}\right)\int_o^T c_1\, \mathrm{e}^{-q_2(T-t)}\mathrm{d}t \\ &+ K_1\left(\frac{(k_3 + k_4) - q_1}{q_2 - q_1}\right)\int_o^T c_1\, \mathrm{e}^{-q_1(T-t)}\mathrm{d}t\end{aligned} \tag{2.43}$$

which expresses the general property that,

$$\sum_{i=1}^{3} m_i = V_1 c_1 + \sum_{h=1}^{2} K_{1_h} \int_o^T c_1\, \mathrm{e}^{-k_{2_h}(T-t)}\mathrm{d}t,$$

where $K_{1_1} = K_1\,[q_2 - (k_3 + k_4)]/[q_2 - q_1]$, $K_{1_2} = K_1\,[(k_3 + k_4) - q_1]/[q_2 - q_1]$, $k_{2_1} = q_2$, and $k_{2_2} = q_1$.

**Transiently Non-Linear Solution to Irreversible Accumulation** For $k_4 = 0$, the nonlinear solution of (2.43) reduces to the one given by Sokoloff et al. (1977), for $m_2(0) = m_3(0) = 0$,

$$\sum_{i=1}^{3} m_i = V_1\, c_1 + \frac{K_1 k_2}{k_2 + k_3} \int_o^T c_1\, \mathrm{e}^{-(k_2+k_3)(T-t)}\, dt + \frac{K_1 k_3}{k_2 + k_3} \int_o^T c_1\, \mathrm{d}t \quad (2.44)$$

for which it should be kept in mind that closed systems without backflux have no obvious biological role, although they may exist under experimental circumstances for certain tracers. The solution reduces to,

$$\sum_{i=1}^{3} m_i = V_1 c_1 + K \left[ \int_o^T c_1\, \mathrm{d}t + \left(\frac{k_2}{k_3}\right) \int_o^T c_1 \mathrm{e}^{-(k_2+k_3)(T-t)}\, \mathrm{d}t \right], \quad (2.45)$$

where the “net” clearance is,

$$K = K_1 \left( \frac{k_3}{k_2 + k_3} \right) = \frac{K_1}{1 + \frac{k_2}{k_3}} \quad (2.46)$$

as defined in (2.33). Gjedde (1982) showed that this equation, when normalized against the concentration $c_1$ at time $T$, defines a continuously increasing apparent volume of distribution in the manner of the Gjedde–Patlak solution to (2.19) (Gjedde 1982),

$$\sum_{i=1}^{3} v_i = \left[ V_1 + \left( \frac{K_1 k_2}{k_2 + k_3} \right) \int_o^T \left[ \frac{c_1(t)}{c_1(T)} \right] \mathrm{e}^{-(k_2+k_3)(T-t)}\, \mathrm{d}t \right] + K \int_o^T \frac{c_1(t)}{c_1(T)}\, \mathrm{d}t \quad (2.47)$$

if compartments $\mathbf{M}_1$ and $\mathbf{M}_2$ reach a steady balance (“secular” equilibrium), depending on the magnitudes of $k_2$ and $k_3$.

**Transiently Linear Solution to Reversible Accumulation** For $k_4 > 0$, the transient multilinear solution to (2.34–2.36) was given by Evans (1987) for $m_2(0) = m_3(0) = 0$,

$$\sum_{i=1}^{3} m_i = a_1 c_1 + a_2 \int_o^T c_1\, \mathrm{d}t + a_3 \int_o^T \int_o^u c_1\, \mathrm{d}t\, \mathrm{d}u$$
$$+ a_4 \int_o^T \sum_{i=1}^{3} m_i\, \mathrm{d}t + a_5 \int_o^T \int_o^u \sum_{i=1}^{3} m_i\, \mathrm{d}t\, \mathrm{d}u, \quad (2.48)$$

where,

$$a_1 = V_1$$
$$a_2 = K_1 + V_1(k_2 + k_3 + k_4)$$
$$a_3 = K_1(k_3 + k_4) + k_2 k_4 V_1$$

$$a_4 = -(k_2 + k_3 + k_4)$$
$$a_5 = -k_2 k_4.$$

**Transiently Linear Solution to Irreversible Accumulation** For $k_4 = 0$, the multilinear solution to (2.34–2.36) was first given by Blomqvist (1984). It is obtained here by setting $k_4 = 0$ in (2.48) for $m_2(0) = m_3(0) = 0$,

$$\sum_{i=1}^{3} m_i = V_1 c_1 + [K_1 + V_1(k_2 + k_3)] \int_o^T c_1 \, dt + K_1 k_3 \int_o^T \int_o^u c_1 \, dt \, du - (k_2 + k_3) \int_o^T \sum_{i=1}^{3} m_i \, dt \tag{2.49}$$

**Steady-State Solution to Reversible Accumulation** When $k_4 > 0$, the steady-state flux ($J_o$) through the system is zero because of the absent sink. The transient solutions describe the time-course of the approach of each of the compartments toward steady-state. The net steady-state flux of zero is associated with the following distinct sizes of each compartment, depending on the magnitudes of the rate constants,

$$M_3 = \left(\frac{k_3}{k_4}\right) M_2 = \left(\frac{K_1 k_3}{k_2 k_4}\right) C_1 \tag{2.50}$$

and

$$M_2 = \left(\frac{K_1}{k_2}\right) C_1 \tag{2.51}$$

yielding the sum of the three compartments,

$$\sum_{i=1}^{3} M_i = \left[V_1 + \frac{K_1}{k_2}\left(1 + \frac{k_3}{k_4}\right)\right] C_1, \tag{2.52}$$

where $M$ is the mass of molecules. The equation shows that in this closed system of three compartments, the content of one or more compartments is maintained at a specific level dictated by the magnitude of the relaxation constants. It is important to note that only the relative magnitudes of the constants matter. The steady-state volume of distribution is $\sum_{i=1}^{3} M_i / C_1$ according to (2.52) (Fig. 2.5).

**Pseudo-Steady-State Solution to Irreversible Accumulation** When $k_4 = 0$, the content of the last of the compartments ($\mathbf{M}_3$) continues to increase according to (2.47) and, thus, never reaches true steady-state. When $T \gg k_2 + k_3$, the "virtual" volume of distribution of (2.47) approaches a line of slope $K$ and ordinate intercept $V_1 + K_1 k_2 / (k_2 + k_3)^2$ with time. The pseudo-steady-state solution to (2.47), also known as the Gjedde–Patlak plot is (Gjedde 1982; Patlak et al. 1983),

$$\sum_{i=1}^{3} v_i(T) \cong \left[V_1 + \frac{K_1 k_2}{(k_2 + k_3)^2}\right] + K \int_o^T \frac{c_1(t)}{C_1(T)} \, dt. \tag{2.53}$$

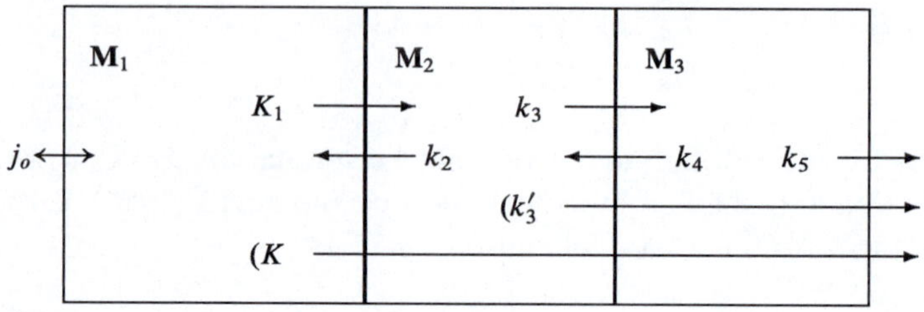


**Fig. 2.5** Model of multiple compartment with multiple sinks. Steady-state $k_3'$ equals $k_3 k_5 / (k_4 + k_5)$ and reaches $k_3$ when $k_5 \gg k_4$. Steady-state $K$ equals $K_1 k_3' / (k_2 + k_3')$ and reaches $K_1$ when $k_3' \gg k_2$ at near-equilibrium

### 2.1.5 *Three Compartments with Sink*

An open system with a throughput can be created by adding another rate constant to compartment **M**$_3$ as shown in Fig. 2.5,

$$\frac{\mathrm{d}m_1}{\mathrm{d}t} = V_1 \frac{\mathrm{d}c_1}{\mathrm{d}t}, \tag{2.54}$$

$$\frac{\mathrm{d}m_2}{\mathrm{d}t} = K_1\, c_1 + k_4\, m_3 - (k_2 + k_3)\, m_2 \tag{2.55}$$

and

$$\frac{\mathrm{d}m_3}{\mathrm{d}t} = k_3\, m_2 - (k_4 + k_5)\, m_3 \tag{2.56}$$

with the total flux into the compartments (and all other compartments fed by this open system), equal to

$$j_o = V_1 \frac{\mathrm{d}c_1}{\mathrm{d}t} + K_1\, c_1 - k_2\, m_2. \tag{2.57}$$

The time course of the sum of the magnitudes of all three compartments is a complex solution to (54–56). An easier approach is the numerical solution obtained by simply adding the analytical solutions for the individual compartment, but the number of coefficients makes the solution less useful, and it will not be given here. More important is the net flux through the system at steady-state.

#### Generalized Steady-State Solution

The steady-state magnitudes of the compartments are

$$M_3 = \left[\frac{k_3}{k_4 + k_5}\right] M_2 = \left(\frac{K_1}{k_2 + k_3'}\right) \left[\frac{k_3}{k_4 + k_5}\right] C_1 \tag{2.58}$$

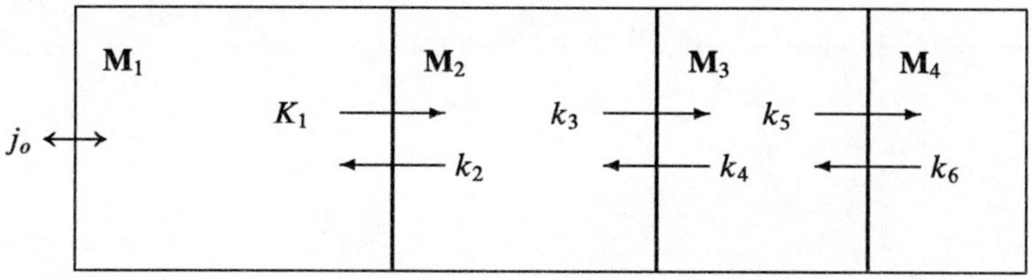


**Fig. 2.6** Model of closed system of multiple exchanging compartments

and, of course,

$$M_2 = \left(\frac{K_1}{k_2 + k_3'}\right) C_1 \tag{2.59}$$

such that

$$\sum_{i=1}^{3} M_i = \left(V_1 + \left(\frac{K_1}{k_2 + k_3'}\right)\left[1 + \left(\frac{k_3}{k_4 + k_5}\right)\right]\right) C_1, \tag{2.60}$$

where the apostrophe ("prime") of the apparent relaxation constant $k_3'$ refers to its definition as a composite of the rate constants of several compartmental interfaces affecting compartment $\mathbf{M}_3$ (Fig. 2.6),

$$k_3' = k_3 \left(\frac{k_5}{k_4 + k_5}\right). \tag{2.61}$$

The steady-state flux through the system is

$$J_o = k_5\, M_3 = K\, C_1, \tag{2.62}$$

where

$$K = \frac{K_1}{1 + \frac{k_2}{k_3}\left(1 + \frac{k_4}{k_5}\right)} = \frac{K_1}{1 + \frac{k_2}{k_3'}} \tag{2.63}$$

which can now be generalized to any number of nested compartments with a terminal sink,

$$K = \frac{K_1}{1 + \frac{k_2}{k_3}\left(1 + \frac{k_4}{k_5}\left[1 + \cdots + \left(1 + \frac{k_{2n-2}}{k_{2n-1}}\right)\right]\right)}, \tag{2.64}$$

where $n$ is the number of compartments ending with a sink described by the rate constant $k_{2n-1}$. The corresponding size of the terminal compartment is given by:

$$M_n = \left(\frac{K}{k_{2n-1}}\right) C_1 \tag{2.65}$$

and the combined magnitude of all compartments is

$$\sum_{i=1}^{n} M_i = \left[ V_1 + K \left( \frac{1}{k'_3} + \frac{1}{k'_5} + \frac{1}{k'_7} + \cdots + \frac{1}{k'_{2i-1}} + \cdots + \frac{1}{k_{2n-1}} \right) \right] C_1, \tag{2.66}$$

where

$$k'_{2i-1} = \frac{k_{2i-1}}{1 + \frac{k_{2i}}{k_{2i+1}} \left[ 1 + \frac{k_{2i+2}}{k_{2i+3}} \left( 1 + \cdots + \left[ 1 + \frac{k_{2n-2}}{k_{2n-1}} \right] \right) \right]}. \tag{2.67}$$

### 2.1.6 *Four or More Compartments*

The serial system shown in Fig. 2.6 without a throughput can be created by adding additional compartments (**M**$_4$-**M**$_n$), such that the system consists of the delivery compartment,

$$\frac{\mathrm{d}m_1}{\mathrm{d}t} = V_1 \frac{\mathrm{d}c_1}{\mathrm{d}t}, \tag{2.68}$$

the precursor compartment,

$$\frac{\mathrm{d}m_2}{\mathrm{d}t} = K_1 \, c_1 + k_4 \, m_3 - (k_2 + k_3) \, m_2, \tag{2.69}$$

and multiple product compartments in series, including

$$\frac{\mathrm{d}m_3}{\mathrm{d}t} = k_3 \, m_2 + k_6 \, m_4 - (k_4 + k_5) \, m_3 \tag{2.70}$$

and, for $4 < i < n$,

$$\frac{\mathrm{d}m_i}{\mathrm{d}t} = k_{2i-3} m_{i-1} + k_{2i} m_{i+1} - (k_{2i-1} + k_{2i-2}) m_i \tag{2.71}$$

and ending with the $n$th compartment,

$$\frac{\mathrm{d}m_n}{\mathrm{d}t} = k_{2n-3} \, m_{n-1} - k_{2n-2} \, m_n, \tag{2.72}$$

where the total flux into the four or more compartments generalizes to,

$$j_o = \sum_{i=1}^{n} \left[ \frac{\mathrm{d}m_i}{\mathrm{d}t} \right] = V_1 \frac{\mathrm{d}c_1}{\mathrm{d}t} + K_1 \, c_1 - k_2 \, m_2 \tag{2.73}$$

as the flux across interior boundaries cancels out in the sum.

### Convolution of Impulse Response Function as Generalized Transient Solution

The impulse response function is the function that defines the changes of the compartmental contents with time when convolved with the input or "forcing" function, which defines the contents of the delivery compartment, be it the vascular bed or another (reference) tissue. In the case of compartments defined by linear first-order differential equations, the ratio between the integrals of the tissue curve and the forcing function, extrapolated to infinity, yields the ratio between the impulse response function coefficients $K_{1_h}$ and $k_{2_h}$. Linear systems theory shows that the generalized transiently nonlinear solution to the linked differential equations (2.68–2.72), for $\sum_{i=1}^{n} m_i(0) = 0$, is the convolution of the impulse response function given in (2.16) with the vascular "forcing" function (Gunn et al. 2001),

$$\sum_{i=1}^{n} m_i = V_1 c_1 + \sum_{h=1}^{n-1} K_{1_h} \int_o^T c_1 \, \mathrm{e}^{-k_{2_h}(T-t)} \, \mathrm{d}t, \tag{2.74}$$

where the coefficients $K_{1_h}$ define the total clearance,

$$K_1 = \sum_{h=1}^{n-1} K_{1_h} \tag{2.75}$$

and the relaxation constants $k_{2_h}$ define the steady-state volume of distribution,

$$V = \sum_{i=1}^{n} V_i = V_1 + \sum_{h=1}^{n-1} \frac{K_{1_h}}{k_{2_h}} = V_1 + \int_o^\infty r_{\mathrm{I}}(t) \, \mathrm{d}t, \tag{2.76}$$

where $V$ is the total steady-state volume of distribution, and the impulse response function $r_{\mathrm{I}}$ is defined as:

$$r_{\mathrm{I}}(t) = \sum_{h=1}^{n-1} K_{1_h} \, \mathrm{e}^{-k_{2_h} t}. \tag{2.77}$$

The relaxation constants $k_{2_h}$, individually as well as collectively, have no simple biological meaning (see (2.43)). They are, in a sense, *descriptive* rather than *determinant* of the system. The relationship between the parameters of the impulse response function and the relaxation constants of each compartment must be worked out for each constellation of compartments. For the current example of at least four compartments in series, the relationship is worked out later.

## Generalized Steady-State Solutions

**Reversible Accumulation** For $k_{2n-2} > 0$, the steady-state flux through the closed system is zero ($J_o = 0$), but the steady-state contents of the compartments are,

$$M_4 = \left(\frac{k_5}{k_6}\right) M_3 = \left(\frac{K_1 k_3 k_5}{k_2 k_4 k_6}\right) C_1,$$

$$M_3 = \left(\frac{K_1 k_3}{k_2 k_4}\right) C_1,$$

and

$$M_2 = \left(\frac{K_1}{k_2}\right) C_1$$

such that the sum of the magnitudes of the four compartments results in a chain of nested coefficient ratios,

$$\sum_{i=1}^{4} M_i = \left[V_1 + \frac{K_1}{k_2}\left(1 + \frac{k_3}{k_4}\left[1 + \frac{k_5}{k_6}\right]\right)\right] C_1 \tag{2.78}$$

which can be generalized to a closed system of $n$ compartments in series,

$$\sum_{i=1}^{n} M_i = \left[V_1 + \frac{K_1}{k_2}\left(1 + \frac{k_3}{k_4}\left[1 + \frac{k_5}{k_6}\left(1 + \cdots + \left[1 + \frac{k_{2n-3}}{k_{2n-2}}\right]\right)\right]\right)\right] C_1, \tag{2.79}$$

where the last, or $n$th compartment, relative to the first compartment, has the size,

$$M_n = \left(\frac{K_1 k_3 k_5 ... k_{2n-3}}{k_2 k_4 k_6 ... k_{2n-2}}\right) C_1,$$

where $n$ is the number of compartments. The generalization shows that a closed system maintains the steady-state content of one or more compartments at a level regulated by the relaxation constants of all preceding compartments in the series. For example, the last compartment may represent a neurotransmitter in the synaptic cleft.

The impulse response function yields the total content of the compartments in terms of its own relaxation constants according to (2.73),

$$\sum_{i=1}^{n} M_i = \left[V_1 + \sum_{h=1}^{n-1} \frac{K_{1_h}}{k_{2_h}}\right] C_1 \tag{2.80}$$

which yields the relationship between the individual compartmental relaxation constants and the powers of the exponentials of the impulse response function by the substitution of (2.79).

**Irreversible Accumulation** For $k_{2n-2} = 0$, the last compartment never reaches a steady-state. The compartment continues to expand while the preceding compartments eventually reach secular equilibria. The steady-state solution to this process is again the equation underlying the Gjedde–Patlak plot (see (2.53)),

$$\sum_{i=1}^{4} v_i(T) = V_1 + K\left(\frac{1}{k_3'} + \frac{1}{k_5}\right) + K\int_o^T \frac{c_1(t)}{C_1(T)}\mathrm{d}t, \tag{2.81}$$

where the apparent relaxation constant $k_3'$ is,

$$k_3' = k_3\left(\frac{k_5}{k_4 + k_5}\right) \tag{2.82}$$

and

$$K = \frac{K_1}{1 + \frac{k_2}{k_3}\left(1 + \frac{k_4}{k_5}\right)} = \frac{K_1}{1 + \frac{k_2}{k_3'}} \tag{2.83}$$

which can be generalized to $n$ compartments,

$$\sum_{i=1}^{n} v_i(T) = \left[V_1 + K\left(\frac{1}{k_3'} + \frac{1}{k_5'} + \frac{1}{k_7'} + \cdots + \frac{1}{k_{2i-1}'} + \cdots + \frac{1}{k_{2n-3}}\right)\right]$$
$$+K\int_o^T \frac{c_1(t)}{C_1(T)}\mathrm{d}t, \tag{2.84}$$

where

$$k_{2i-1}' = \frac{k_{2i-1}}{1 + \frac{k_{2i-2}}{k_{2i-1}}\left[1 + \frac{k_{2i}}{k_{2i+1}}\left(1 + \cdots + \left[1 + \frac{k_{2n-4}}{k_{2n-3}}\right]\right)\right]} \tag{2.85}$$

and

$$K = \frac{K_1}{1 + \frac{k_2}{k_3}\left(1 + \frac{k_4}{k_5}\left[1 + \cdots + \left(1 + \frac{k_{2n-4}}{k_{2n-3}}\right)\right]\right)}. \tag{2.86}$$

The impulse response function defines the net clearance $K$ in terms of the "unidirectional" clearance of the irreversibly accumulating compartment, $K_{1_{n-1}} = K$.

### 2.1.7 *Multiple Compartments in Series and in Parallel*

With a few exceptions (see Gunn et al. 2001), the impulse response function defines any constellation of linear compartmental systems, which can be described by first-order linear differential equations (Gunn et al. 2002), regardless of the structure.

Thus, also compartments in parallel follow the general principles underlying the steady-state solutions. The impulse-response function describes the constellation but its biological significance must be extracted from an analysis of the relationship between the relaxation constants of each compartment.

## Generalized Steady-State Solution of Closed System of Branching Compartments

In a closed ("reversible") system as shown in Fig. 2.7, the steady-state flux through the system is of course zero ($J_o = 0$), but the solution to the steady-state magnitude of the combined compartments, generalized for any closed system of several parallel branches extending $n$ steps from the origin, shows that parallel compartments are additive, while serial compartments are nested,

$$\sum_{i=1}^{n} M_i = \left[ V_1 + \frac{K_1}{k_2} \left( 1 + \frac{k_{3_1}}{k_{4_1}} \left[ 1 + \frac{k_{5_{1_1}}}{k_{6_{1_1}}} + \cdots \right] \right.\right.$$
$$\left.\left. + \frac{k_{3_2}}{k_{4_2}} \left[ 1 + \frac{k_{5_2}}{k_{6_2}} \left( 1 + \cdots + \left[ 1 + \frac{k_{(2n-3)_2}}{k_{(2n-2)_2}} \right] \right) \right] \right) \right] C_1, \quad (2.87)$$

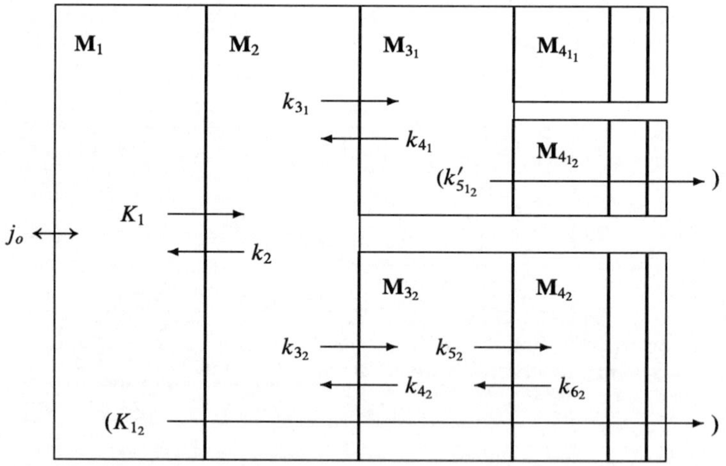


**Fig. 2.7** Multiple compartments in series and parallel. Compartments in parallel are additive; compartments in series are nested. Branches $1_2$ and 2 are open ("irreversible"), branch $1_1$ is closed ("reversible"). Net clearances for open system are shown in brackets. Note that *apostrophes* ("primes") refer to composite rate constants or clearances which traverse several membranes

where the additional subscripts refer to the branch in question. The last compartment of the $n_{2nd}$ branch, relative to the first compartment, has the size

$$M_{n_2} = \left( \frac{K_1\, k_{3_2}\, k_{5_2} \ldots k_{(2n-3)_2}}{k_2\, k_{4_2}\, k_{6_2} \ldots k_{(2n-2)_2}} \right) C_1, \tag{2.88}$$

where $n$ is the number of steps removed from the source. The generalization again shows that the closed system maintains the steady-state content of one or more compartments at a level that is regulated by the rate constants of the involved mechanisms. For example, the last compartment may represent a neurotransmitter and its level in the synaptic cleft.

### Generalized Steady-State Solution of Open System of Branching Compartments

For an open system of $n$ compartments in one branch ("2"), the net flux out of the branch is dictated by the net clearance to the sink of the branch,

$$K_{1_2} = \frac{K_1}{1 + \frac{k_2}{k_{3_2}}\left(1 + \frac{k_{4_2}}{k_{5_2}}\left[1 + \cdots + \left(1 + \frac{k_{(2n-2)_2}}{k_{(2n-1)_2}}\right)\right]\right)}, \tag{2.89}$$

where $K = K_{1_2}$ is the net clearance by branch 2 and $n$ is the number of compartments leading to the sink governed by the rate constant $k_{2n-1_2}$. The corresponding magnitude of the $n$th compartment in this branch (open system) is given by:

$$M_{n_2} = \left( \frac{K_{1_2}}{k_{(2n-1)_2}} \right) C_1 \tag{2.90}$$

and the combined magnitude of all compartments (excluding the sink) in this open branch is

$$\sum_{i=1}^{n} M_{i_2} = \left[ V_1 + K'_{1_2} \left( \frac{1}{k'_{3_2}} + \frac{1}{k'_{5_2}} + \frac{1}{k'_{7_2}} + \cdots + \frac{1}{k'_{(2i-1)_2}} + \cdots + \frac{1}{k_{(2n-1)_2}} \right) \right] C_1, \tag{2.91}$$

where

$$k'_{(2i-1)_2} = \frac{k_{(2i-1)_2}}{1 + \frac{k_{(2i)_2}}{k_{(2i+1)_2}}\left[1 + \frac{k_{(2i+2)_2}}{k_{(2i+3)_2}}\left(1 + \cdots + \left[1 + \frac{k_{(2n-2)_2}}{k_{(2n-1)_2}}\right]\right)\right]}, \tag{2.92}$$

where the apostrophe refers to the composite nature of the rate constant, which traverses multiple membranes.

## 2.2 Interpretation of Relaxation Constants

The biological reality is that concentrations vary in space but not normally in time. The compartmental kinetics must be translated from the space-invariant and time-variant model systems to the space-variant but time-invariant real pathways. Note that time-invariant (steady-state) variables, as earlier, are given as upper-case symbols.

In many cases, the relaxation constant $k$ represents one of just a few processes, including flow, diffusion, membrane permeability, enzyme reaction, facilitated diffusion across membranes, or receptor binding. Depending on the processes known to establish the compartments occupied by a metabolite or by tracer molecules, relaxation constants can be estimated and interpreted in specific ways. In the case of tracer molecules, the molecules trace a specific set of processes, often because the tracer has been designed for the purpose of revealing the kinetics of just these processes.

### 2.2.1 *Flow*

When the flow of a solvent links two compartments, the relaxation of the downstream compartment $\mathbf{M}_2$ is due to the washout of the molecules dissolved in the effluent. This is an example of an open vascular system. The products $k_1 V_1$ and $k_2 V_2$ both equal the flow rate $F$, and $k_2$ equals the flow through the system per unit volume of the second compartment. Thus, the process establishes a proper compartment when $F$ is constant,

$$k_2 = F/V_2 \tag{2.93}$$

as shown in Fig. 2.8.

The direct use of this model is limited to the dissolution of inert flow tracers in the bloodstream, which allow the estimation of flow rate, volume of distribution, and mean transit time ("bolus tracking") as in Ostergaard et al. (1998).

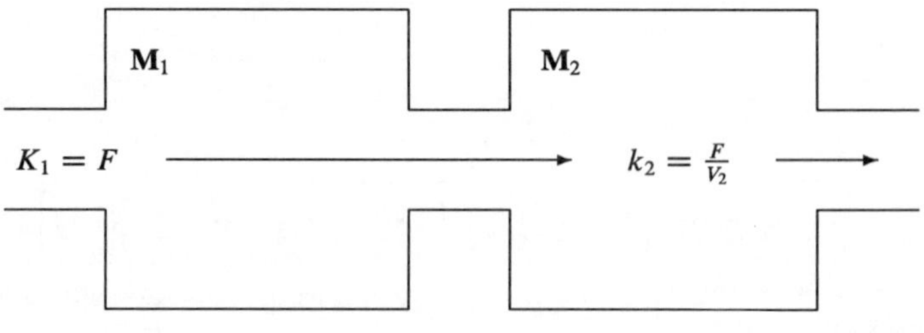


**Fig. 2.8** Convection model of precursor and product compartments perfused by the solvent

### 2.2.2 *Passive Diffusion*

Diffusion is a fundamental process of elimination of concentration differences and release of entropy leading to a flux in the direction of a concentration gradient (see Chap. 1). In the unidirectional case, according to Fick's First Law, the one-dimensional flux is given by (Gjedde 1995a),

$$J = -\mathrm{DA}\,\frac{\mathrm{d}C}{\mathrm{d}x}, \tag{2.94}$$

where $D$ is the diffusion coefficient, $A$ the cross-sectional area of the volume through which the unidirectional diffusion occurs, and $\mathrm{d}C/dx$ is the *gradient* ($G$), equal to the slope of the concentration (or activity) curve or plane at the coordinate $x$. In a closed system, the concentration gradient is eventually dissipated until $J = 0$ when $\mathrm{d}C/dx = 0$. The system is at *equilibrium* when net flux is no longer taking place and $\mathrm{d}C/\mathrm{d}t$ reaches zero. If the diffusing substance undergoes no change other than the diffusion itself, the temporal rate of elimination of the concentration gradient is proportional to the spatially negative rate of change of the gradient,

$$\frac{\partial c}{\partial t} = -D\,\frac{\mathrm{d}G}{\mathrm{d}x}, \tag{2.95}$$

where $G$ is the gradient, such that,

$$\frac{\partial c}{\partial t} = -D\,\frac{\partial^2 c}{\partial x^2}, \tag{2.96}$$

where the diffusion coefficient is the proportionality factor. Thus, when the substance is uniformly distributed in the volume, diffusion ceases. In this sense, passive diffusion and compartmentation are mutually incompatible concepts, as compartments have no internal concentration gradients because the gradients are placed at two-dimensional interfaces between the compartments.

The purpose of the treatment below is to establish the extent to which varying concentration profiles within compartments are consistent with compartmental kinetics, i.e., the extent to which a diffusion process can be given a pseudo-compartmental description which satisfies the criteria of compartment kinetics. The task becomes one of the determination of the concentration profiles and total contents of substance which different structural arrangements give rise to, while maintaining the steady-state of the concentration of the diffusing substances.

#### Steady-State One-Dimensional Diffusion

In the simplest steady-state condition the rate of change of the diffusing substance with respect to time is nil ($\partial c_1/\partial t = 0$) but the rate of change of the diffusing substance with respect to distance (one-dimensional space) is significant

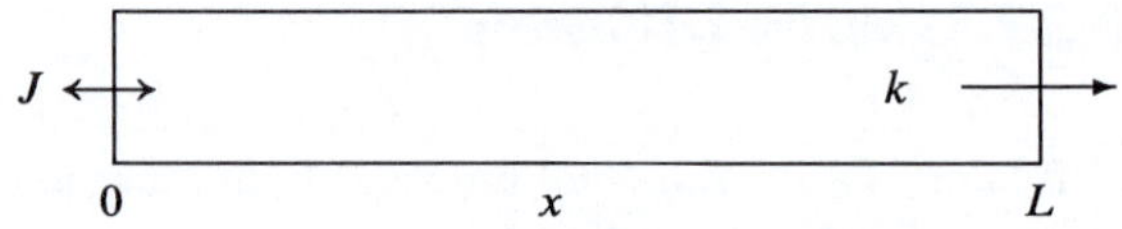


**Fig. 2.9** Single compartment with simple passive one-dimensional diffusion due to Brownian motion

($\partial C_1/\partial x \neq 0$). This condition can be evaluated by application of the diffusion equation to the case of one-dimensional diffusion in one direction ($x$) through a volume ("box") in which no concentration gradients exist in the orthogonal directions $y$ and $z$, and substance is lost or gained only at the ends of the box as shown in Fig. 2.9.

When the concentration of the substance is kept constant, the spatial rate of change of the gradient is nil,

$$\frac{d^2C}{dx^2} = 0 \tag{2.97}$$

and the result is a constant (or uniform) gradient,

$$G(x) = \frac{dC}{dx} = \alpha \tag{2.98}$$

with a linearly stationary concentration profile,

$$C(x) = \alpha\, x + C(0) \tag{2.99}$$

with a slope of the magnitude $\alpha$,

$$\alpha = \frac{C(x) - C(0)}{x} \tag{2.100}$$

and a steady-state flux for a box of length $x = L$ of,

$$J_o = -\mathrm{DA}\,\frac{C(L) - C(0)}{L} = -\mathrm{DA}\,\frac{\Delta C}{L} \tag{2.101}$$

as well as a steady-state content of the "box" of,

$$M = L\,A\,\frac{C(L) + C(0)}{2} = L\,A\,\overline{C}, \tag{2.102}$$

where $\overline{C}$ is the average steady-state concentration of the compartment. Then, the effective relaxation constant of the efflux is

$$k = \frac{J}{M} = -\left(\frac{D\,\Delta C}{L^2\,\overline{C}}\right), \tag{2.103}$$

where $L^2/(2D) = \tau_D$ is the characteristic time constant of diffusion along the distance $L$ obtained from the equation derived by Einstein (1908), such that

$$k = -\left(\frac{\Delta C}{2\,\tau_D\,\overline{C}}\right), \tag{2.104}$$

where the relaxation constant of the substance undergoing passive diffusion is the inverse of the time constant of diffusion if the concentration $C(0)$ is kept constant by quantitative removal to the external medium. From the opposite perspective, the product of $k$ and $\tau_D$ can be said to determine the difference between the concentrations at the two extremes of the diffusion path,

$$\Delta C = -2\,k\,\tau_D\,\overline{C}, \tag{2.105}$$

where $\tau_D$ is proportional to the square of the length of the diffusion path. Thus, a large concentration difference is favored by long diffusion paths and, conversely, a small or negligible concentration difference is favored by very short diffusion paths, once the relaxation constant of the compartment is given.

### Steady-State Passive Diffusion with Concentration-Dependent Loss

Unless the external concentrations are kept constant, equilibrium will eventually ensue. Equilibrium is anathema to the biological *steady-state* in which $\mathrm{d}C_1/\mathrm{d}t = 0$ because the gradient is preserved by energy-requiring processes which maintain the flux by establishing constant concentrations at both ends of the diffusion path. Alternatively, substance may be removed continuously from the diffusion path by a concentration dependent process occurring in the directions orthogonal to the direction of passive diffusion ($x$) as shown in Fig. 2.10. If the volume is sufficiently narrow in the orthogonal directions (infinitely long narrow tube with a cross-sectional area of $A$), and the product of the rate constant of removal and the time constant of diffusion is very small, no concentration gradients exist in the orthogonal directions.

The spatial rate of change of the gradient is now a function of the rate of removal of the substance along the length of the tube,

$$\frac{\mathrm{d}^2C}{\mathrm{d}x^2} = \left(\frac{k}{D}\right) C(x), \tag{2.106}$$

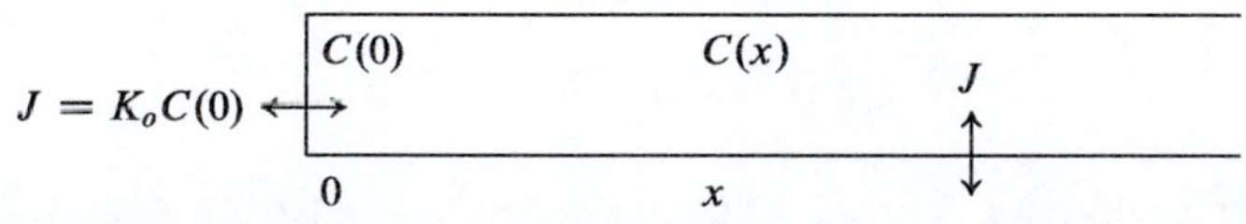


**Fig. 2.10** Single compartment with simple passive one-dimensional diffusion through infinitely long tube with concentration-dependent loss

where $x$ is the distance along the length of the tube and $k$ is the relaxation constant of a concentration-dependent loss in the orthogonal $y$ and $z$ directions. At steady-state, the gradient is the monoexponential function satisfying (2.106),

$$G(x) = \frac{dC}{dx} = -C(0)\, e^{-x\sqrt{k/D}}\, \sqrt{k/D} \tag{2.107}$$

with the concentration profile,

$$C(x) = C(0)\, e^{-x\sqrt{k/D}} \tag{2.108}$$

as well as the total content of the infinitely long tube,

$$M = A \int_o^\infty C(x)\, dx = A\, C(0)\, \sqrt{D/k}, \tag{2.109}$$

where $M$ is the steady-state content of the infinitely long tube in which the concentration declines to zero. The constant $\sqrt{D/k}$ is a characteristic length constant, $l_D$, which depends only on the system. The constant $\sqrt{k/D}$ is therefore the inverse length constant, and the constant $\sqrt{Dk}$ is the ratio between the length constant and the time constant of diffusion, $l_D/\tau_D$. The entire loss of substance therefore occurs in the orthogonal directions, i.e., through the wall of the tube. The efflux through the wall of this virtual compartment is given by:

$$J = k\, A \int_o^\infty C(x)\, dx = A\, C(0)\, \sqrt{Dk} \tag{2.110}$$

which confirms that $k$ is indeed the steady-state relaxation constant of the compartment,

$$k = \frac{J}{M} = D \left( \frac{A\, C(0)}{M} \right)^2, \tag{2.111}$$

where the term $M/[A\; C(0)]$ reveals that the characteristic length constant $l_D$ in reality is the diffusion distance effectively reached in the tube. As the compartment is in steady-state and no substance is lost or gained at the end of the tube, the flux into the origin of the tube must be of the same magnitude as the flux through the wall. The influx is determined by the value of the gradient at $x = 0$,

$$J = -D\, A\, G(0) = A\, C(0)\, \sqrt{Dk} \tag{2.112}$$

such that (2.108) becomes also,

$$C(x) = C(0)\, e^{-x\, kA\, C(0)/J} \tag{2.113}$$

in which the spatial rate constant of the monoexponential decline is a function of the rate constant of loss or gain through the wall of the tube and the apparent clearance of substance by diffusion into the tube. We now define a clearance as the volume of the precursor compartment which per unit time is cleared of the substance by the diffusion into the infinitely long tube,

$$C(x) = C(0)\,\mathrm{e}^{-x\,kA/K_o}, \tag{2.114}$$

where $K_o$ is a characteristic clearance of substance into the tube, equal to $A\sqrt{Dk}$ or $A\,l_\mathrm{D}/\tau_\mathrm{D}$. As the ratio $J/C(0)$, this clearance depends only on the properties of the system that include the cross-sectional area and the length and time constants of the tube (Fig. 2.10).

### Steady-State Diffusion with Concentration-Dependent Loss and Backflux

Substance can also be added to the diffusion path by a concentration-independent transfer in the directions orthogonal to the direction of passive diffusion. Such a process could be a second compartment continuously feeding the diffusion path, as shown in Fig. 2.11.

The spatial rate of change of the gradient is now a function of the net removal or addition of substance along the length of the tube,

$$\frac{\mathrm{d}^2C_1}{\mathrm{d}x^2} = \left(\frac{k}{D}\right)[C_1(x) - C_2], \tag{2.115}$$

where $x$ is the distance along the length of the tube, $k$ the relaxation constant of the concentration-dependent loss, and $kC_2$ is the constant rate of gain by flux into the diffusion path, when the concentration $C_2$ is kept constant by the flux $J_2$. The general solution to (2.115) is given by:

$$C_1(x) = C_2 + \alpha\mathrm{e}^{-(L-x)\sqrt{k/D}} + \beta\mathrm{e}^{-(L-x)\sqrt{k/D}}$$

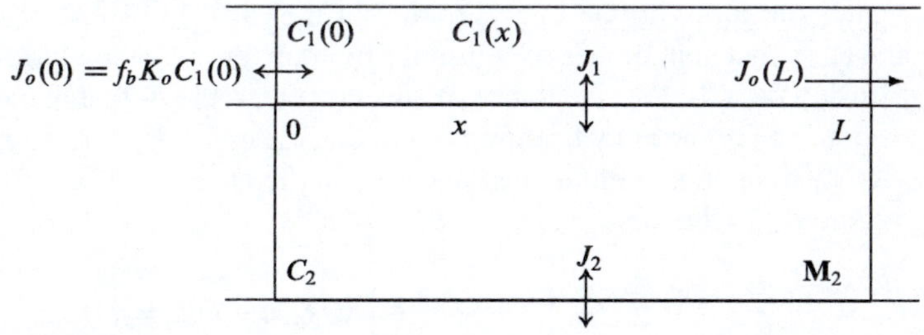


**Fig. 2.11** System of two compartments with simple passive one-dimensional diffusion through tube of finite length with concentration dependent loss to, and concentration independent gain from, the surrounding mantle compartment

as can easily be verified by substitution. The constants of integration, $\alpha$ and $\beta$, are determined by boundary conditions. The constants may be written in terms of the boundary values, $C_1(0)$ and $C_1(L)$, as follows:

$$\alpha = \frac{\mathrm{e}^{L\sqrt{k/D}}[C_1(0) - C_2] - [C_1(L) - C_2]}{\mathrm{e}^{L\sqrt{k/D}} - \mathrm{e}^{-L\sqrt{k/D}}}$$

and

$$\beta = \frac{\mathrm{e}^{L\sqrt{k/D}}[C_1(L) - C_2] - [C_1(0) - C_2]}{\mathrm{e}^{L\sqrt{k/D}} - \mathrm{e}^{-L\sqrt{k/D}}}.$$

In cases such as that of non-synaptic transmission, where diffusion is over relatively long distances, $C_1(x)$ tends to decay exponentially, as if $\beta \approx 0$. This will be the case if $L\sqrt{k/D}$ is relatively large and the ratio $[C_1(L) - C_2]/[C_1(0) - C_2]$ is relatively small. Making that assumption,

$$C_1(x) \approx C_2 + [C_1(0) - C_2]\mathrm{e}^{-x\sqrt{k/D}},$$

which is essentially (2.117), below. At steady-state, the gradient is given by the appropriate boundary conditions,

$$G(x) = \frac{\mathrm{d}C_1}{\mathrm{d}x} = \sqrt{k/D}\ [C_2 - C_1(0)]\ \mathrm{e}^{-x\sqrt{k/D}}, \tag{2.116}$$

where $\sqrt{k/D}$ is the inverse length constant $(1/l)$. The equation yields the concentration profile when integrated,

$$C_1(x) = C_2 + (C_1(0) - C_2)\,\mathrm{e}^{-x\sqrt{k/D}} \tag{2.117}$$

that approaches $C_2$ rather than the zero of the previous condition. The backflux reduces the gradient of the substance in the tube and, hence, reduces the clearance of substance into the origin of the tube. The influx of substance at the origin of the tube can be evaluated from the gradient at $x = 0$,

$$J_o(0) = -D\ A\ G(0) = A\ \sqrt{Dk}\ (C_1(0) - C_2) \tag{2.118}$$

which relates the influx to the concentration of the diffusing substance at the entrance and to the net gain or loss of substance by transfer in the direction orthogonal to the diffusion path. Let the clearance of substance into the diffusion path from the external source be reduced by backflux to the real clearance by diffusion, $K_o'$, from the previously defined maximum clearance $K_o$ in the absence of backflux from the compartment $\mathbf{M}_2$ to the tube,

$$K_o' \equiv \frac{J_o(0)}{C_1(0)} = A\sqrt{Dk}\left(1 - \frac{C_2}{C_1(0)}\right) = f_b\ A\sqrt{Dk} = f_b\ K_o, \tag{2.119}$$

where the fraction $f_b$, equal to $1 - [C_2/C_1(0)]$, depends on the concentration difference at the entrance to the tube. The term $f_b$ is a key factor in the remaining treatment. The fraction $f_b$ accounts for the reduction of the clearance ("backflux") by the bidirectional transfer, such that the influx is,

$$J_o(0) = f_b\, K_o\, C_1(0) = K_o'\, C_1(0) \tag{2.120}$$

for which the concentration profile along the diffusion path is,

$$C_1(x) = C_1(0)\left[1 - f_b\left(1 - e^{-xf_bkA/K_o'}\right)\right] \tag{2.121}$$

and the flux of substance at $x = L$ likewise becomes,

$$J_o(L) + J_1(L) = -DA\, G(L) = A\sqrt{Dk}\ (C_1(0) - C_2)\ e^{-L\sqrt{k/D}}$$

of which the efflux is,

$$J_o(L) = K_o'\, C_1(L) = K_o'\, C_1(0)\left[1 - f_b\left(1 - e^{-f_bkLA/K_o'}\right)\right], \tag{2.122}$$

where the product $LA$ is the volume of the tubular diffusion path. The net flux in the directions orthogonal to the diffusion path can now be calculated as the difference,

$$J_1 = J_o(0) - J_o(L) = f_b\, K_o'\, C_1(0)\ \left(1 - e^{-f_bkLA/K_o'}\right) \tag{2.123}$$

which is zero when $f_b$ is zero.[1]

### 2.2.3 *Properties of Delivery Compartment*

#### Diffusion-Limited Membrane Permeability

Membrane permeability is a special case of passive diffusion without loss or gain in which the dimensions are fixed but the concentrations on the two sides of the membrane may vary. In this treatment, the membrane is the lining of the narrow tube discussed earlier (Fig. 2.12).

---

[1] The solution is not defined for $k$ equal to zero, because the gradient expressed in (2.116) then has the simpler solution,

$$G(x) = \frac{dC}{dx} = \frac{I}{D}\,x + H_1, \tag{2.124}$$

where $C_2 \neq I/k$, which yields the concentration profile,

$$C(x) = \frac{I}{2D}\,x^2 + H_1x + H_2, \tag{2.125}$$

where $H_1$ and $H_2$ depend on the conditions. This formula can be applied to the concentration independent consumption of oxygen diffusing unidirectionally through a tissue.

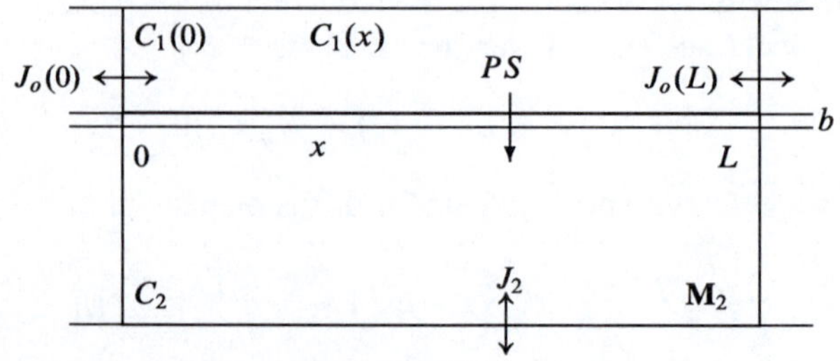


**Fig. 2.12** Two compartments separated by membrane with diffusion-limited permeability through wall of thickness $b$ of tube of finite length $L$

The diffusion flux is described by (2.101), modified to include the concentrations of the diffusing substance in the medium surrounding the membrane, rather than in the membrane itself,

$$dJ = -D'dS\,\frac{C_1(x,b) - C_1(x,0)}{b} = -D'dS\,\frac{\Delta C_1}{b}, \tag{2.126}$$

where $b$ is the width of the membrane. The ratio $D'/b$ is defined as the permeability coefficient $P$, the prime referring to the possibly different solubilities of the diffusing substance in the membrane and the surrounding media, and $S$ is defined as the surface area of the membrane to distinguish that area from $A$, the cross-sectional area of the tube. Thus,

$$dJ = -PdS\,\Delta C_1, \tag{2.127}$$

where $PdS$ (as a clearance) has unit of flow. Let a tube of length $L$ and inner volume $V_1 = AL$ be lined by a membrane of thickness $b$ and surface area $S$ for which $C_1(x,b) = C_2$ for all $x$. The relaxation constant of the efflux through the membrane is then given by (2.93),

$$k = P\,dS/dV_1 = PS/V_1 \tag{2.128}$$

such that (2.121) together yield the concentration profile along the length of the tube,

$$C_1(x) = C_1(0)\left[1 - f_b\left(1 - e^{-f_b\,(x/L)(PS/K_o')}\right)\right], \tag{2.129}$$

where $LA = V_1$, the volume of the entire length of the tube. The relation expresses the monoexponential decline of the concentration along the length of the tube when diffusion in the tube is accompanied by permeation of the membrane lining the tube. For $x = L$, the concentration declines to the expression,

$$C_1(L) = C_1(0)\left[1 - f_b\left(1 - e^{-f_b\,PS/K_o'}\right)\right], \tag{2.130}$$

where, as earlier, $K'_o = f_b\, A\sqrt{Dk} = f_b\, \sqrt{DAPS/L}$, such that,

$$C_1(L) = C_1(0)\left[1 - f_b\left(1 - e^{-\sqrt{PSL/(DA)}}\right)\right] = C_1(0)\left[1 - f_b\left(1 - e^{-L\sqrt{2P/(Dr)}}\right)\right], \tag{2.131}$$

where $P$ is the permeability of the membrane, $D$ the diffusion coefficient of the substance diffusing through the tube, and $r$ is the radius of the tube. It turns out that this expression can be generalized also to the case of convection through the tube in which the clearance into the proximal end of the tube, and hence the diffusion, is replaced by convection.

When the tube is of finite length, the relaxation constant of the compartment formed by the tube no longer equals just $PS/V_1$. The content of the compartment is,

$$\begin{aligned} M_1(L) &= A\int_o^L C_1(0)\left(1 - f_b\left[1 - e^{-\frac{xf_bPS}{LK'_o}}\right]\right)dx \\ &= V_1 C_1(0)\left[1 - f_b\left(1 - \frac{1 - e^{-f_bPS/K'_o}}{f_bPS/K'_o}\right)\right] \end{aligned} \tag{2.132}$$

where $M_1(L)$ is the total mass of the substance in the tube of finite length $L$ in which the concentration declines to $C_1(L)$. The decay of the compartment, therefore, occurs both through the wall and at the end of the tube. The net flux across the wall of the tube is given by:

$$J_1 = \frac{PSA}{V_1}\int_o^L (C_1(x) - C_2)\ dx = f_b\, K'_o\, C_1(0)\, \left(1 - e^{-f_bPS/K'_o}\right) \tag{2.133}$$

while the efflux from the end of the tube is given by:

$$J_o(L) = K'_o\, C_1(0)\, \left[1 - f_b\, \left(1 - e^{-f_bPS/K'_o}\right)\right] \tag{2.134}$$

such that the total flux, equal to the flux at the entry, is,

$$J_o(0) = J_1 + J_o(L) = K'_o\, C_1(0) \tag{2.135}$$

which yields the true relaxation constant of the entire contents of the compartment formed by the tube of finite length,

$$\begin{aligned} k_o = \frac{J_o(0)}{M_1(L)} &= \left[\frac{K'_o}{V_1}\right]\left(\frac{PS}{PS(1 - f_b) + K'_o\,(1 - e^{-f_bPS/K'_o})}\right) \\ &= \frac{PS/V_1}{1 - e^{-PS/K_o}}\ \text{for}\, f_b = 1 \end{aligned} \tag{2.136}$$

which approaches either $K'_o/V_1$ when the $PS$ product is small relative to the magnitude of the clearance (observe the equivalence with (2.93)) or $PS/V_1$ when the $PS$

product is large relative to $K_o'$ and $f_b$ is unity ($K_o' = K_o$, no gain by influx from compartment $\mathbf{M}_2$). The effective relaxation constant of the loss to compartment $\mathbf{M}_2$ is,

$$k_1 = \frac{J_1}{M_1} \tag{2.137}$$

such that

$$k_1 = \left[\frac{K_o'}{V_1}\right]\left(\frac{1 - e^{-f_b PS/K_o'}}{1 - f_b\left(1 - \frac{1 - e^{-f_b PS/K_o'}}{f_b PS/K_o'}\right)}\right) = \left[\frac{K_o'}{V_o}\right]\left(1 - e^{-f_b PS/K_o'}\right), \tag{2.138}$$

where $V_o$ is an apparent volume defined as $M_1/C_1(0)$, which expresses the relaxation of the diffusion path in terms of a virtual compartment with a uniform concentration equal to $C_1(0)$. This convention converts the diffusion path to a virtual compartment, which conforms to the fundamental definition.

### Flow-Limited Membrane Permeability

Flow-limited membrane permeability is a special case of flow-dependent exchange between two compartments when solvent flows through the precursor compartment, and a fixed fraction of the solvent is cleared of the solute by permeation of the solute into the second compartment. The two compartments are separated by a membrane which is impermeable to the solvent but not to the solute (semipermeable membrane). A typical example of this system is the capillary with its surrounding tissue mantle.

The exchange of substance between the capillary and the tissue is assumed to be nonenergy-requiring (passive) and, thus, to proceed only in the direction of the solute concentration difference between the two compartments. The system is considered to be closed if the solute in the product compartment decays exclusively by return to the precursor compartment.

As described earlier, the contents of the capillary bed do not reside in a true compartment, because the relaxation constant for the flux through the wall, by means of which a fraction of the solute ($E = J_1/J_o(0)$) enters the exchange compartment, equals $k_1 = K_o'\left(1 - e^{-f_b PS/K_o'}\right)/V_o$ rather than $PS/V_1$. Nonetheless, as derived earlier (2.138), it is possible to reduce the capillary to the status of a compartment by showing when it decays as required of a compartment.

The clearance of a fraction of the contents of the delivery compartment to the exchange compartment is a function both of the permeability-surface area product of the membrane through which the transfer occurs and the flow of the solvent in the tube lined by this membrane, as derived by Crone (1963). Crone derived the clearance on the condition that no more than a negligible amount of capillary solute actually accumulate in the tissue compartment (i.e., when $C_1(x, b) = 0$ for all $x$), for example because the tissue volume of distribution is very large, or because the solute is quantitatively consumed in the tissue. In this condition, $f_b = 1$, where $1-f_b$ is the concentration ratio between tissue and capillary at the site of arterial entry.

The clearance equations can be derived for the bidirectional case, in which $f_b$ is less than unity. The flow of a solute through the delivery compartment, combined with the transaxial loss of solute to the surrounding mantle, can be represented by an apparent diffusion coefficient. Let the delivery compartment be a narrow tube of length $L$, cross-sectional area $A$, volume $V_1$, wall permeability-surface area $PS$, and flow $F$. The flow term $F$ replaces the term for the clearance as reduced by backflux ($K_o'$). The apparent diffusion coefficient of the solute is then the solution to (2.119) where $F$ replaces $K_o'$,

$$D_{\text{app}} = \left(\frac{F}{f_b A}\right)^2 \frac{V_1}{PS}, \tag{2.139}$$

where, again, $f_b$ is the bidirectionality fraction which reduces the effective clearance by the backflux. The fundamental flux equation for the loss of solute at any distance along the capillary then is given by an expression, which is formally analogous to the diffusion equation,

$$J_1 = -D_{\text{app}} A \frac{\mathrm{d}C_1}{\mathrm{d}x} = -\left(\frac{F}{f_b}\right)^2 \left[\frac{V_1}{A\,PS}\right] \frac{\mathrm{d}C_1}{\mathrm{d}x} \tag{2.140}$$

where the spatial rate of change of the solute gradient in the delivery compartment at steady-state is given by,

$$\frac{\mathrm{d}^2 C_1}{\mathrm{d}x^2} = \left(\frac{f_b PS}{\mathrm{FL}}\right)^2 (C_1(x) - C_2) = \left(\frac{f_b PS}{\mathrm{FL}}\right)^2 (C_1(x) - [1 - f_b]\, C_1(0)) \tag{2.141}$$

such that the gradient is,

$$\frac{\mathrm{d}C_1}{\mathrm{d}x} = \left(\frac{f_b PS}{\mathrm{FL}}\right) \left[(f_b - 1)\, C_1(0) - f_b\, C_1(0)\, \mathrm{e}^{-x f_b PS/(\mathrm{FL})}\right] \tag{2.142}$$

and the concentration profile of the solute in the delivery compartment as before is,

$$C_1(x) = C_1(0) \left(1 - f_b \left[1 - \mathrm{e}^{-x f_b PS/(\mathrm{FL})}\right]\right). \tag{2.143}$$

**Capillary Model of Bidirectional Flux**

In the living organism, the delivery compartment usually (but not always) is a capillary. In those cases, replace $C_1(0)$ by $C_a$, the symbol of the arterial concentration, $x$ by $L$, and $C_1(L)$ by $C_v$, the symbol of the venous concentration. The fundamental equations of the capillary exchange then emerge from (2.143) as:

$$C_v = C_a \left[1 - f_b \left(1 - \mathrm{e}^{-f_b PS/F}\right)\right] \tag{2.144}$$

in which the net flux across the capillary wall is given by (2.123) and (2.133) for $C_1(0) = C_a$ and $K'_o = F$,

$$J_1 = f_b \, C_a \, F \left(1 - e^{-f_b PS/F}\right) \tag{2.145}$$

and

$$J_o(0) = F \, C_a \tag{2.146}$$

whence the extraction fraction associated with this unidirectional transfer arises as the ratio between the flux into the capillary and the efflux through the wall,

$$E = 1 - \frac{C_v}{C_a} = \frac{J_1}{J_o(0)} = f_b \left(1 - e^{-f_b PS/F}\right) \tag{2.147}$$

which defines the steady-state extraction fraction. Only in the case of zero tissue concentration, i.e., $f_b = 1$, does this equation reduce to the relationship derived by Crone (1963),

$$E_o = 1 - e^{-PS/F} \tag{2.148}$$

which thus is valid only for substances of sufficiently low permeability. The total actual content of the capillary compartment is given by (2.132),

$$M_1 = V_1 \, C_a \left[1 - f_b \left(1 - \frac{1 - e^{-f_b PS/F}}{f_b PS/F}\right)\right] = V_o C_a \tag{2.149}$$

in which $V_o$ defines a virtual volume of the delivery compartment with the uniform concentration $C_a$. The virtual delivery compartment represents the redistribution of $M_1$ in a volume, $V_o$, so small that the concentration uniformly is $C_1(0)$. The redistribution shows that $M_1$ fulfills the criteria of a compartment with a volume of $V_o$

$$V_o = V_1 \left[1 - f_b \left(1 - \frac{1 - e^{-f_b PS/F}}{f_b PS/F}\right)\right], \tag{2.150}$$

where the real average concentration in the physical volume $V_1$ is,

$$\overline{C}_1 = C_a \left[1 - f_b \left(1 - \frac{1 - e^{-f_b PS/F}}{f_b PS/F}\right)\right] \tag{2.151}$$

such that the "virtual" and the real compartments have the same contents but different apparent volumes.

In the case of bidirectional transfer, $f_b$ is less than unity. The simplest way to describe the transfer in terms of compartmental kinetics is to split the net flux into efflux and influx, as shown by (2.135) and (2.145), such that

$$J_1 = C_a \, F \, E_o, \tag{2.152}$$

where $K_1$ is an apparent clearance, also known as the Capillary Diffusion Capacity, in the direction of transfer from delivery compartment to exchange compartment. It is also the ratio between a fictive unidirectional flux through the wall of the capillary and the arterial concentration, obtained by setting $f_b = 1$,

$$K_1 = F\left(1 - \mathrm{e}^{-PS/F}\right) = F\,E_o \tag{2.153}$$

such that the same fictive unidirectional flux is given by:

$$\overrightarrow{J_1} = f_b K_1\, C_\mathrm{a}, \tag{2.154}$$

where the relaxation constant $k_1$ for this fictive unidirectional flux is,

$$k_1 = \frac{\overrightarrow{J_1}}{M_1} = F\left(1 - \mathrm{e}^{-PS/F}\right)/\,V_o = K_1/\,V_o. \tag{2.155}$$

The fictive unidirectional flux in the opposite direction is then,

$$\overleftarrow{J_1} = J_1 - \overrightarrow{J_1} = (1 - f_b)\; C_\mathrm{a}\; K_1 \tag{2.156}$$

with the relaxation constant given by the fractional clearance,

$$k_2 = \frac{\overleftarrow{J_1}}{M_2} = \frac{F}{V_2}\left(1 - \mathrm{e}^{-PS/F}\right) = K_1/\,V_2, \tag{2.157}$$

where $M_2$ equals $(1 - f_b)\; C_\mathrm{a}\; V_2$, indicating that the compartment exists when the flow, volume, and permeability-surface area terms are at steady-state, as shown in the model of Fig. 2.13.

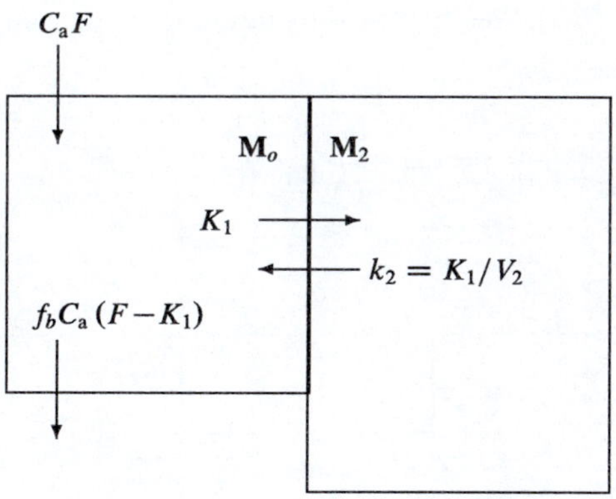


**Fig. 2.13** Compartmental model of capillary and tissue compartments in which the tissue compartment clears precursor from solvent flowing through the capillary compartment by means of semi-permeable membrane. $\mathbf{M_0}$ is a virtual compartment with the smaller (virtual) volume $V_o$, uniform concentration $C_\mathrm{a} = C_1(0)$, and a substance content of $M_1$. The fraction $1 - f_b$ is the ratio between concentrations of solute in tissue and capillary compartments

Very large values of the *PS* term relative to $F$ turn the clearance compartment into a flow-limited compartment in which the magnitude of the flow determines the influx to the compartment. On the other hand, very low values of the *PS* product relative to $F$ turn the compartment into a diffusion-limited compartment in which the magnitude of the permeability determines the influx to the product compartment. The ratio between the parameters $K_1$ and $K_1/V_2$ is the volume of the solvent $V_2$ in the product compartment.

Thus, depending on the magnitude of the *PS* product, the rate constant is an index either of the flow through the system, relative to its volume, or of the permeability-surface area product of the interface, also relative to the volume of the system.

### 2.2.4 Protein–Ligand Interaction

In their simplest form, the three processes of enzyme reaction, facilitated diffusion across membranes, and receptor binding are variations on the same underlying theme. The theme received its steady-state formulation by Briggs and Haldane (1925), as an extension of the original equilibrium solution of Michaelis and Menten (1913). The key mechanism is the binding of a ligand to sites on the protein. The interacting compartment $\mathbf{M}_2$ is normally either open when it acts as a catalyst or transporter, or closed when it predominantly acts as a receptor (Fig. 2.14).

#### Competitive Interaction

The protein releases the bound ligand ($m_2$) at a constant rate to the same or a different compartment, either as the intact precursor ($m_1$) or as a transformed or translocated product ($m_3$), thus freeing the protein for new occupation, according to the general formula,

$$\frac{\mathrm{d}m_2}{\mathrm{d}t} = (j_1 + j_2) - (k_2 + k_3)\,m_2. \tag{2.158}$$

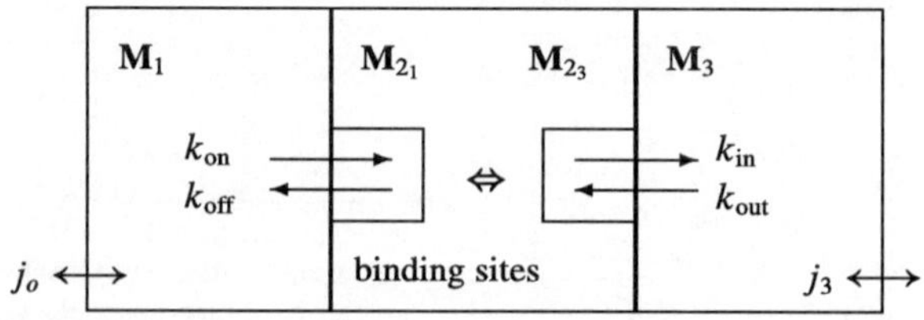


**Fig. 2.14** Model of protein–ligand interaction. Unbound ligands ($m_1$ and $m_3$) in respective compartments ($\mathbf{M}_1$ and $\mathbf{M}_3$) interact with sites on protein ($\mathbf{M}_{2_1}$ and $\mathbf{M}_{2_3}$). Maximum number of available sites is $B_{max}$

The size of the protein compartment is a function of the magnitude of these coefficients, some of which combine to define the interaction of the protein with two or more competitors for the same sites,

$$\frac{dm_2}{dt} = \left[\frac{k_{on}\,(B_{max} - m_2)}{V_1}\right] m_1 + \left[\frac{k_{out}\,(B_{max} - m_2)}{V_3}\right] m_3 - (k_{off} + k_{in})\, m_2, \tag{2.159}$$

where $m_1$ and $m_3$ are the masses of competitors in compartments $\mathbf{M}_1$ and $\mathbf{M}_3$, and $m_2$ is the sum of the competitors attached to the binding sites. Equation (2.159) is usually rearranged to allow an evaluation of whether $\mathbf{M}_2$ has the characteristics of a compartment.

Combination of the variables of (2.159) from this perspective yields an equation with $m_2$ as the independent variable,

$$\begin{aligned}\frac{dm_2}{dt} &= \left[\frac{k_{on}\,B_{max}}{V_1}\right] m_1 + \left[\frac{k_{out}\,B_{max}}{V_3}\right] m_3 - (k_{off} + k_{in})\, m_2 \\ &\quad -(k_{off} + k_{in}) \left(\left[\frac{k_{on}}{k_{off} + k_{in}}\right]\left[\frac{m_1\, m_2}{V_1}\right] + \left[\frac{k_{out}}{k_{off} + k_{in}}\right]\left[\frac{m_3\, m_2}{V_3}\right]\right)\end{aligned} \tag{2.160}$$

which shows that the bound ligand constitutes a simple compartment only when $m_1$ and $m_3$ remain constant (or effectively nil). The terms $(k_{off} + k_{in})/k_{on}$ and $(k_{off} + k_{in})/k_{out}$ define the Michaelis half-saturation concentrations $K_{M_1}$ and $K_{M_3}$, and the terms $k_{off}/k_{on}$ and $k_{in}/k_{out}$ the dissociation constants $K_{d_1}$ and $K_{d_3}$, e.g.,

$$K_{M_1} = \frac{k_{off} + k_{in}}{k_{on}} = K_{d_1} + \frac{k_{in}}{k_{on}} \tag{2.161}$$

which in turn shows that $K_M$ is always greater than $K_d$ when there is significant translocation of substrate.

**Competitive Inhibition** For the steady-state of $dm_2/dt = 0$, (2.160) yields the famous Michaelis–Menten equation for competition between two ligands (Dixon 1953),

$$M_2 = B_{max}\left[\frac{C_1}{K_{M_1}\left(1 + \frac{C_3}{K_{M_3}}\right) + C_1} + \frac{C_3}{K_{M_3}\left(1 + \frac{C_1}{K_{M_1}}\right) + C_3}\right], \tag{2.162}$$

where $C_1 = M_1/V_1$ and $C_3 = M_3/V_3$. The ratio $M_2/B_{max}$ is the combined occupancies ($\sigma$) of the ligands,

$$\sum \sigma = \frac{\frac{C_1}{K_{M_1}} + \frac{C_3}{K_{M_3}}}{1 + \frac{C_1}{K_{M_1}} + \frac{C_3}{K_{M_3}}} = \frac{\chi_1 + \chi_3}{1 + \chi_1 + \chi_3} = \frac{\sum \chi}{1 + \sum \chi}, \tag{2.163}$$

where $\chi_1$ is the ratio $C_1/K_{M_1}$, the normalized or relative concentration of the ligand in the precursor compartment, and $\chi_3$ is the ratio $C_3/K_{M_3}$. The occupancy of ligand $m_1$ is then,

$$\sigma_1 = \frac{\chi_1}{1+\sum\chi} \tag{2.164}$$

which expresses the effect of the competition between the two ligands when the occupation by one ligand causes the occupancy of other ligands to decline.

The bound ligand relative to the unbound ligand at steady-state can be expressed as a binding potential, obtained from (2.162), as in Gjedde et al. (1986),

$$p_{B_1}(\chi_1) = \frac{M_{2_1}}{M_1} = \frac{\sigma_1\ B_{max}}{M_1} = \frac{B_{max}}{V_1\ K_{M_1}\ (1+\sum\chi)} = \frac{p_{B_1}(0)}{1+\sum\chi}, \tag{2.165}$$

where $p_B$ is a binding potential, the change of which reflects any change of $B_{max}$, $C_1$, or $C_2$. This equation linearizes to the Eadie–Hofstee equation (Eadie 1952),

$$M_{2_1} = B_{max} - [V_1\ K_{M_1}\ (1+\chi_3)]\ p_{B_1}(\chi_1) \tag{2.166}$$

according to which competitive interaction increases the magnitude of the Michaelis constant in proportion to the magnitude of competitor concentrations, $\chi_3$, provided they are constant.

Depending on its function, the protein exerts its function when it unites with the occupant, either by reacting with other molecules or by translocating the product to another compartment. If the translocated product is identical to the precursor, the process is transport with a maximum velocity of $k_{in}\ B_{max}$, symbolized by $T_{max}$. If the product is different from the precursor, the protein is an enzyme catalyst with the maximum reaction rate $k_{in}\ B_{max}$, symbolized by $V_{max}$.

Many processes retain transitional features, combining transport and catalysis, or binding and transport. The significance of the transitional processes depends on the relationship between the rate constants $k_{off}$ and $k_{in}$. Little or no transport or catalysis takes places when $k_{in}$ is equal or close to zero. Little or no binding occurs, on the other hand, when $k_{off}$ is equal or close to zero. Thus, whether binding takes precedence over translocation or transformation depends on the relative rates of product release.

In principle, protein-ligand interaction does not fulfill the basic requirement of compartmental kinetics because the rate of relaxation depends on the concentration of the decaying substance. However, under special circumstances of the relaxation, depending on the magnitudes and time courses of the ligand concentrations, it is possible to define relaxation constants which reflect these circumstances. The following definitions are valid only when the ligand concentrations are effectively constant or nil, relative to their Michaelis constants, or when the total ligand occupancy of the receptors is constant. The distinction between the definitions does not apply to true steady-states, in which neither concentrations nor occupancies vary with time.

The common definitions of relaxation constants are based on the association constants,

$$k_1 = \left(1 - \sum \sigma\right) \frac{k_{\text{on}} B_{\max}}{V_1} \quad \text{and} \quad k_4 = \left(1 - \sum \sigma\right) \frac{k_{\text{out}} B_{\max}}{V_3} \tag{2.167}$$

are valid only when $\sum \sigma$ is constant or effectively nil and the concentration of any ligand varies as a function of time; in those cases, the dissociation constants are,

$$k_2 = k_{\text{off}} \quad \text{and} \quad k_3 = k_{\text{in}}. \tag{2.168}$$

while, *alternatively*, the less common definitions based on the dissociation constants,

$$k_2 = (1 + \chi_1)\, k_{\text{off}} \quad \text{and} \quad k_3 = (1 + \chi_3)\, k_{\text{in}} \tag{2.169}$$

are valid only when the concentrations of all ligands are constant and $\sum \sigma$ varies; in that case, the definitions based on the association constants are,

$$k_1 = \frac{k_{\text{on}} B_{\max}}{V_1} \quad \text{and} \quad k_4 = \frac{k_{\text{out}} B_{\max}}{V_3} \tag{2.170}$$

which cover the common test tube cases of build-up of bound substance from a constantly maintained medium but do not easily apply to living matter, in which changes of concentrations lead to changes of binding, rather than the reverse. It is important to keep in mind that the equations can be extended for any number of ligands in any combinations, some of which may be of constant concentration while others are effectively nil. At normal steady state, all compartments have constant magnitudes. In that situation only the ratios between the relaxation constants have meaning, and neither set of definitions is valid.

**Competitive Activation** In competitive activation, the affinity of the binding sites for a ligand rises when another ligand binds to the same sites, i.e., the dissociation rates $k_{\text{off}}$ and $k_{\text{in}}$ decline. In the simplest way, the action is explained by the occupancy of the secondary ligand preventing the dissociation of the primary ligand for as long as the secondary ligand is bound. The occupancy of the ligand is then:

$$\sigma_1' = \frac{C_1}{K_{\text{M}_1}\,(1 - \sigma_3) + C_1} = \frac{C_1}{K'_{\text{M}_1} + C_1} = \frac{\chi_1\,(1 + \chi_3)}{1 + \chi_1\,(1 + \chi_3)} = \frac{\chi_1'}{1 + \chi_1'},$$

where $K'_{\text{M}_1} = K_{\text{M}_1}/(1 + \chi_3)$ and $\chi_1' = \chi_1\,(1 + \chi_3)$. The action of the secondary ligand changes the mass of bound ligand relative to the unbound ligand at steady-state expressed as the binding potential

$$p_{\text{B}_1}(\chi_1) = \frac{M_{21}}{M_1} = \frac{\sigma_1'\, B_{\max}}{M_1} = \frac{B_{\max}}{V_1\, K_{\text{M}_1}\,(1 + \chi_1 - \sigma_3)} = \frac{p_{\text{B}_1}(0)}{1 + \chi_1 - \sigma_3} \tag{2.171}$$

which predicts an increase of the binding potential in the presence of the secondary ligand. This equation also linearizes to the Eadie–Hofstee equation (Eadie 1952),

$$M_{2_1} = B_{\max} - [V_1\ K_{M_1}\ (1 - \sigma_3)]\ p_{B_1}(\chi_1) \tag{2.172}$$

according to which the competitive activation reduces the magnitude of the Michaelis constant in proportion to the magnitude of activator concentration, $\chi_3$, provided they are constant.

### Non-Competitive Interaction

In noncompetitive interaction, the occupation of the sites by the noncompetitively interacting ligand is not influenced by the changing concentrations of other ligands, either because their concentrations are too low relative to their Michaelis constants ("tracers") or because the noncompetitively interacting ligand permanently renders particular sites unavailable to occupation by other ligands, i.e., lowers the association rate for other ligands so much that their concentrations are now insignificant relative to their noncompetitively elevated Michaelis constants. There are two kinds of noncompetitive interaction, inhibition and activation.

**Non-Competitive Inhibition** In noncompetitive inhibition, a fraction of the protein sites are permanently occupied by the inhibitor, regardless of the concentration of other ligands, or they are rendered permanently unavailable for occupation by any ligand. If the permanently unavailable sites are $\mathbf{M}_{2_3}$, the steady-state binding potential, obtained from (2.162) is,

$$p_{B_1}(\chi_1) = \frac{M_{2_1}}{M_1} = \frac{\sigma_1\,(1-\varsigma_3)\,B_{\max}}{M_1} = \frac{(1-\varsigma_3)\,B_{\max}}{V_1\ K_{M_1}\ (1+\chi_1)} = \frac{1-\varsigma_3}{1+\chi_1}\ p_{B_1}(0), \tag{2.173}$$

where $\varsigma_3$ is now the occupancy of the noncompetitively interacting ligand, which is independent of the concentrations of other ligands. This equation also linearizes to the Eadie–Hofstee equation,

$$M_{2_1} = (1-\varsigma_3)\ B_{\max} - V_1\ K_{M_1}\ p_{B_1}(\chi_1) \tag{2.174}$$

which confirms that noncompetitive inhibition reduces the number of available binding sites, but leaves the affinity of the available sites intact.

As earlier, the corresponding definitions of relaxation constants for mixed competitive and noncompetitive interaction, i.e.,

$$k_1 = (1-\sigma_1)\,\frac{k_{\text{on}}\,(1-\varsigma_3)\,B_{\max}}{V_1} \quad \text{and} \quad k_4 \cong 0 \tag{2.175}$$

are valid only when $\sigma_1$ is constant or effectively nil and the concentration of any ligand varies as a function of time; in those cases, the dissociation constants are,

$$k_2 = k_{\text{off}} \quad \text{and} \quad k_3 = k_{\text{in}}. \tag{2.176}$$

*Alternatively*, the less common definitions based on the dissociation constants, i.e.,

$$k_2 = (1 + \chi_1)\ k_{\text{off}} \quad \text{and} \quad k_3 = (1 + \chi_1)\ k_{\text{in}}$$

are valid only when the concentrations of all ligands are constant and $\sigma_1$ varies; in that case, the definitions based on the association constants are,

$$k_1 = \frac{k_{\text{on}}\ (1 - \varsigma_3)\ B_{\text{max}}}{V_1} \quad \text{and} \quad k_4 \cong 0$$

which, as earlier, can be extended for any number of ligands in any combinations, some of which may be of constant concentration, while others are effectively nil.

**Non-Competitive Activation** In noncompetitive activation, the interaction is coupled to the presence of an activator which gives the ligand access to the binding sites. For example, transport of some solutes against a concentration gradient is directly or indirectly coupled to the breakdown of ATP. In these cases, the binding potential is given by:

$$p_{\text{B}_1}(\chi_1) = \frac{\sigma_1\ \varsigma_3\ B_{\text{max}}}{M_1} = \frac{\varsigma_3\ B_{\text{max}}}{V_1\ K_{\text{M}_1}\ (1 + \chi_1)} = \frac{\varsigma_3\ p_{\text{B}_1}(0)}{1 + \chi_1}, \tag{2.177}$$

where $\varsigma_3$ is now the occupancy of the agonist. The interaction of G-proteins with G-protein-linked receptors is of this kind, being activated by the primary receptor ligand, e.g., dopamine. Likewise, the Eadie–Hofstee plot becomes,

$$M_{2_1} = \varsigma_3 B_{\text{max}} - [V_1\ K_{\text{M}_1}]\ p_{\text{B}_1}(\chi_1) \tag{2.178}$$

and the relaxation constant is

$$k_1 = \frac{(1 - \sigma_1)\ k_{\text{on}}\ \varsigma_3\ B_{\text{max}}}{V_1}$$

## 2.2.5 *Receptor Binding*

The protein is said to function primarily as a receptor for a single ligand type when $k_{\text{off}} \gg k_{\text{in}}$, and $K_{\text{M}_1} \cong K_{d_1}$ (see (2.161)). Thus, when the translocation of the ligand ("internalization") is minimal, only association and dissociation occur. This is the situation originally conceived by Michaelis and Menten (1913) who assumed that the entity later termed the Michaelis constant would equal the dissociation constant $k_{\text{off}}/k_{\text{on}}$ of the protein–ligand complex, because of near-equilibrium between $\mathbf{M}_1$ and $\mathbf{M}_2$. Subsequently, it was shown that this kind of near-equilibrium exists mostly in the cases of receptor binding. The interaction between a single ligand and the receptor protein is a simple exchange between two compartments (adapted from Fig. 2.2) as shown in Fig. 2.15:

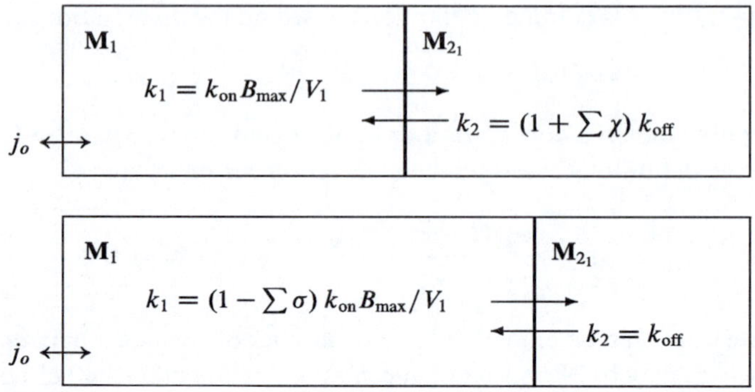


**Fig. 2.15** Partial steady-states of free and bound ligand compartments established by receptor protein occupied by ligand $m_1$. Depending on individual steady-states, either *top* or *bottom* panel defines relaxation constants. Note that receptor protein may be occupied by other ligands also. *Top panel* requires constant ligand concentrations, *bottom panel* constant receptor occupancy

The main applications of the protein–ligand association equations for receptor binding are the determination of the maximum number of receptor sites, the affinity of the sites for the ligand, or the ligand's occupancy of the receptors. These determinations all involve the binding potential. The binding potential is obtained from the steady-state receptor binding given by the product of the steady-state ligand concentration and volume of distribution (2.21),

$$M_{2_1} = V_1 \left(\frac{k_1}{k_2}\right) C_1 = \frac{k_{on} B_{max}\, C_1}{k_{off}(1 + \sum \chi)} = \frac{B_{max}\, C_1}{K_{d_1}(1 + \sum \chi)} = \frac{B_{max}\, \chi_1}{1 + \sum \chi} = \sigma_1\, B_{max} \tag{2.179}$$

which is the formulation of the Michaelis–Menten equation for the binding of a ligand to its receptor. The ratio $k_1/k_2$ is the ligand's binding potential, according to (2.165),

$$p_{B_1}(\chi_1) = \frac{k_1}{k_2} \tag{2.180}$$

and its steady-state volume of distribution is $V_1\ (1 + [k_1/k_2])$. The maximum number of available sites, $B_{max}$, is related to the binding potential by rearrangement of (2.166),

$$M_{2_1} = B_{max} - [V_1\ K_{d_1}\ (1 + \chi_3)]\ p_{B_1}(\chi_1) \tag{2.181}$$

which is the Eadie–Hofstee equation of a line with slope $-V_1\ K_{d_1}\ (1 + \chi_3)$ and ordinate intercept $B_{max}$ but only when the concentrations of all ligands other than $m_1$ are constant or effectively nil. The slope reflects the dissociation constant of the binding, the volume of distribution of the unbound ligand, and the concentrations (constant or nil) of all other ligands interacting with the receptors. The presence of other ligands affect the affinity and, therefore, the binding potential of the ligand in question.

### *2.2.6 Facilitated Diffusion*

A protein can facilitate the diffusion of the ligand across a membrane when $k_{\text{in}}$ is not negligible relative to $k_{\text{off}}$. The ligand usually remains intact during the translocation, but in principle may also undergo a chemical change. This process is important when the unassisted diffusion is slow, i.e., when the protein spans a membrane in which the diffusion coefficient of the ligand is low. Association and release obey the same equation as receptor binding but the predominant decay of the compartment occurs by release of the intact ligand from the compartment other than the one whence it originated. The translocation is unidirectional if the binding sites are accessible from only one compartment ("side"), here symbolized by $\mathbf{M}_1$. Spontaneous conformational change of the transporter protein makes the transporter accessible from alternating sides but the basic model has two compartments with a sink (Fig. 2.16, adapted from Fig. 2.3).

The flux and, hence, the relaxation constant depend on the concentration of the ligand. The maximum transport rate $T_{\max}$ is defined as $k_{\text{in}} B_{\max}$. Hence the rate constants are,

$$k_1 = \left(1 - \sum \sigma\right) \frac{k_{\text{on}} B_{\max}}{V_1} = \left(1 - \sum \sigma\right) \frac{k_{\text{on}} T_{\max}}{k_{\text{in}} V_1} = \frac{\left(1 - \sum \sigma\right) T_{\max}}{V_1 \left(K_{\text{M}_1} - K_{d_1}\right)}, \tag{2.182}$$

where the magnitude of $\sum \sigma$ depends on the simultaneous accessibility of the transporter from both sides of the membrane. The steady-state relaxation constant is a function of the partition coefficient as expressed in (2.33),

$$k_1' = k_{\text{in}} \, p_{\text{B}_1}(\chi_1) = \frac{\left(1 - \sum \sigma\right) k_{\text{in}} \, B_{\max}}{V_1 K_{\text{M}_1}} = \frac{\left(1 - \sum \sigma\right) T_{\max}}{V_1 \, K_{\text{M}_1}} \tag{2.183}$$

provided $\sum \sigma$ reflects uniformly distributed ligands in true compartments. Because the presence of the transporter establishes an effective permeability for the ligand,

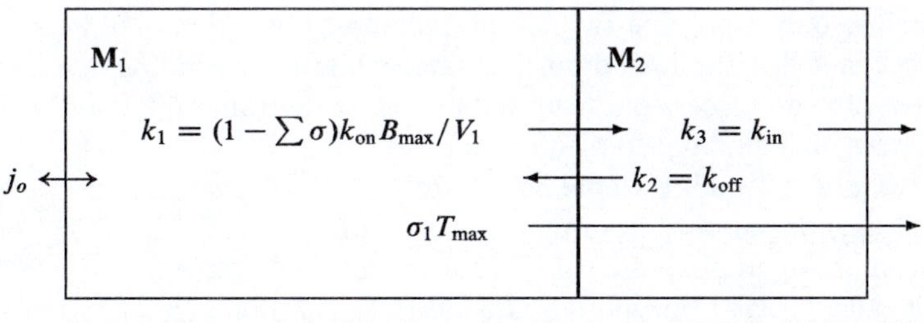


**Fig. 2.16** Steady-state model of metabolite interaction with facilitating transport protein ($\mathbf{M}_2$). When $p_{\text{B}_1}$ is binding potential, steady-state $k_1'$ equals $k_{\text{in}} \, p_{\text{B}_1}$ and reaches $k_1 = k_{\text{on}} B_{\max}/V_1$ when $k_{\text{in}} \gg k_{\text{off}}$

the steady-state relaxation constant of compartment $\mathbf{M}_1$ defines an apparent permeability in the direction of the net transport, $P_1'$,

$$P_1'S = V_1\, k_1' = V_1\, k_{\text{in}}\, p_{\text{B}_1} = \left(1 - \sum \sigma\right) \frac{T_{\max}}{K_{\text{M}_1}}, \tag{2.184}$$

where the binding potential is now given by:

$$p_{\text{B}_1} = \frac{k_1}{k_2 + k_3} \tag{2.185}$$

such that the apparent permeability-surface area product now is,

$$P_1'S = V_1 \left(\frac{k_1 k_3}{k_2 + k_3}\right) \tag{2.186}$$

and the steady-state flux is a simple function of the ligand's occupancy,

$$J_1' = P_1'SC_1 = \left(1 - \sum \sigma\right) \left[\frac{T_{\max}}{K_{\text{M}_1}}\right] C_1 = \sigma_1\, T_{\max}, \tag{2.187}$$

where the evaluation of the influence of the ligand's concentration, i.e., the magnitude of the occupancy, uses the linear rearrangement of the Eadie–Hofstee (2.166),

$$J_1' = T_{\max} - \left[V_1\, K_{\text{M}_1} \left(1 + \sum \chi - \chi_1\right)\right] k_1' = T_{\max} - \left[K_{\text{M}_1} \left(1 + \sum \chi - \chi_1\right)\right] P_1'S, \tag{2.188}$$

where $T_{\max}$ is the ordinate intercept and $-V_1\, K_{\text{M}_1}\, (1 + \sum \chi - \chi_1)$ is the slope, provided the interactions are competitive.

### Net Transport of Single Ligand

The net flux depends on the simultaneous accessibility of the transporter from both sides of the membrane. In the conventional *unidirectional* case, the transporter sites are accessible only from one side of the membrane at a time and hence enjoy a higher occupancy of the ligand on that side, while in the *bidirectional* case, the transporter sites are accessible from both sides at the same time and hence have lower degrees of saturation from either side.

In the case of the unidirectional transporter, the transporter sites can be accessed by several ligands but only from the same side of the membrane at any one time. If only a single ligand is present in the precursor compartment, as in the case of glucose and the GLUT1 transporter (Gjedde 1992), (2.184) is modified to,

$$P_1'S = \frac{(1 - \sigma_1)\, T_{\max}}{K_{\text{M}_1}} = \frac{T_{\max}}{K_{\text{M}_1} + C_1} \tag{2.189}$$

and the flux of the uniformly distributed substrate, facilitated by the unidirectional action of the transporter, equals,

$$J_1' = V_1\, k_1'\, C_1 = P_1' S\, C_1 = \sigma_1\, T_{\max} \tag{2.190}$$

and in the opposite direction, provided $k_{\text{off}} = k_{\text{in}}$, i.e., when $T_{\max}$ is the same in the two directions,

$$J_3' = V_3\, k_4'\, C_3 = P_3' S\, C_3 = \frac{T_{\max}\, C_3}{K_{M_3} + C_3} = \sigma_3\, T_{\max} \tag{2.191}$$

which implies a different apparent permeability, dictated by the magnitude of $C_3$ relative to $C_1$, and the difference in turn affects the magnitude of the distribution volume. The net transport is the difference between the fluxes in the two directions,

$$J_1' = (\sigma_1 - \sigma_3)\, T_{\max} = f_b{}'\, \sigma_1\, T_{\max}, \tag{2.192}$$

where $1 - f_b{}'$ is now the ratio between the different occupancies of the ligand on the two sides of the interface, established by the saturable nature of the binding to the transporter, rather than simply being the ratio between the concentrations as in (2.119).

Facilitated diffusion is in principle passive, i.e., non-energy requiring. In case of active, energy-requiring transport against a concentration gradient, the ligand can be translocated by association with sites activated by another ligand which itself may or may not be translocated. The occupancy of the ligand in question may be low, yet net transport proceeds against a concentration gradient because more sites are made available for the ligand at the low concentration than for the ligand at the high concentration on the opposite side of the interface. The activation is reflected in the definition of $f_b{}'$, when $\sigma_1\, \varsigma_1 > \sigma_3\, \varsigma_3$,

$$f_b{}' = 1 - \frac{\sigma_3\, \varsigma_3}{\sigma_1\, \varsigma_1}$$

in which the low occupancy of $M_1$ ($\sigma_1$) is compensated by the high occupancy ($\varsigma_1$) of the activating ligand (Fig. 2.17).

### Flow-Limited Net Transport of Single Ligand

Equation (2.189) raises the issue of the distribution of the ligand in the delivery compartment, when this compartment is the vascular bed. Not only are the apparent permeabilities different for the two directions of transport but they also vary along the length of the capillary because of the decline of the ligand concentration, as the ligand is delivered to the exchange compartment. Gjedde (1980) treated the case of the varying apparent permeability of the blood–brain barrier to glucose and showed

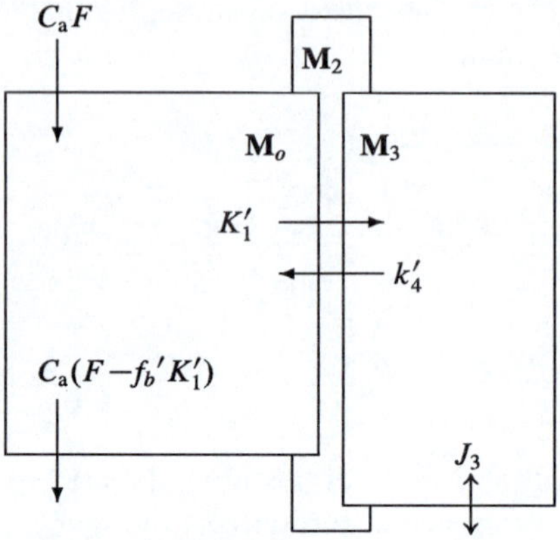


**Fig. 2.17** Compartmental model of capillary and tissue compartments in which tissue compartment clears solute such as glucose from plasma flowing through capillary compartment by means of unidirectional GLUT1 transporters. The fraction $1 - f_b{}'$ is the ratio between effective concentrations of glucose in tissue and capillary compartments, caused by higher apparent permeability in direction from tissue to capillary because of lower glucose concentration. Primes refer to net transport through membranes of $\mathbf{M}_2$

that the effect can be ignored when the extraction is low (Gjedde 1980). The analysis showed that the decrease of the ligand concentration, as blood passes along the capillaries, has little influence on the magnitude of the apparent permeability when the ratio $J_1'/[F(K_{M_1} + C_1)]$ is less than 0.1, where $F$ is blood flow, i.e., when $f_b{}'\sigma_1 P_1' S/F < 0.1$. Thus, it is approximately correct to modify (2.192) for the apparent permeability generated by the facilitated diffusion, as illustrated earlier,

$$J_1' \cong f_b{}' C_a\, F\left(1 - e^{-P_1' S/F}\right) = f_b{}'\, K_1'\, C_a\,, \tag{2.193}$$

where $C_a$ is the arterial glucose concentration, and $K_1'$ is the apparent clearance established by the presence of the permeability symbolized by $P_1' S$, the prime referring to the net transport through the two membranes of the endothelium.

The error incurred by ignoring the effect of the changing substrate concentration is revealed by the formula for the average capillary concentration of the ligand (2.151),

$$\overline{C}_1 = C_a\left[1 - f_b{}'\left(1 - \frac{1 - e^{-P_1' S/F}}{P_1' S/F}\right)\right], \tag{2.194}$$

where the relationship between $\overline{C}_1$ and $C_a$ depends on the magnitudes of $f_b{}'$ and the $P_1' S/F$ ratio. The ratio of the occupancies of the ligand on the two sides of the interface can be inferred from the ratio between the "unidirectional" and net fluxes at steady-state,

$$f_b{}' = \frac{J_3}{C_a\, K_1'} = \frac{K'}{K_1'}, \tag{2.195}$$

where $K'$ is the net clearance. This ratio is approximately 0.5 for the transport of D-glucose across the human blood–brain barrier, indicating that the occupancy on the tissue side of the interface is half of that in plasma and the concentration in the tissue water one-third of that in plasma (Gjedde 1995a).

#### Flow-Limited Net Transport of Multiple Ligands

Equation (2.193) is valid also when multiple ligands are present in the precursor compartment, as in the case of the transporter of large neutral amino acids (LeFauconnier 1992). Changes of the apparent permeabilities on the two sides are then buffered by the large number of ligands. The net flux is given by the equation,

$$J_1 \cong K_1' \, C_a \left(1 - \frac{\sigma_3}{\sigma_1}\right) = f_b{}' \, K_1' \, C_a, \tag{2.196}$$

where, to recapitulate, $1 - f_b{}'$ is the ratio between the occupancies on the two sides of the interface (the prime referring to facilitated rather than simple diffusion), $K_1'$ the rate of clearance from the virtual compartment $\mathbf{M}_o$ through compartment $\mathbf{M}_2$ to compartment $\mathbf{M}_3$ in the direction from vessel to tissue, and $C_a$ is the arterial concentration of the ligand, equal to the uniform concentration of the ligand in $\mathbf{M}_0$. The net translocation ceases when the occupancies are the same on the two sides. Multiple ligands, on the other hand, may influence the translocation indirectly by changing the occupancy of the ligand on one or both sides.

### 2.2.7 *Enzymatic Reactions*

The protein is said to be an enzyme when the product is different from the precursor and physical translocation of the product is not a main function of the protein. The action of the protein may or may not involve translocation, and the sites usually are accessible by both product and precursor, although not at the same time. The model is adapted from Fig. 2.5 (Fig. 2.18).

The maximum rate of the enzymatic reaction ($V_{max}$) is $k_{in} B_{max}$. Thus,

$$k_1 = \left(1 - \sum \sigma\right) \frac{k_{on} B_{max}}{V_1} = \left(1 - \sum \sigma\right) \frac{k_{on} V_{max_1}}{k_{in} V_1} = \frac{(1 - \sum \sigma) \, V_{max_1}}{V_1 \, (K_{M_1} - K_{d_1})}, \tag{2.197}$$

where the magnitude of $\sum \sigma$ depends on the number of substrates and inhibitors accessed by the enzyme. The steady-state relaxation constant in the direction from

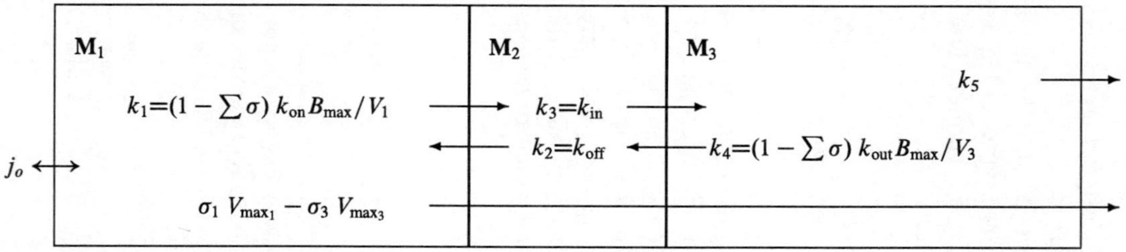


**Fig. 2.18** Model of precursor and product compartments separated by enzyme (**M**$_2$) occupied by precursor ($m_1$) and product ($m_3$). Reaction sites are not accessed by precursor and product at same time

precursor to product is a function of the affinity of the enzyme for the precursor as expressed in (2.167),

$$k_1' = \frac{(1 - \sum \sigma)\, k_{\text{in}}\, B_{\text{max}}}{V_1\, K_{\text{M}_1}} = \frac{(1 - \sum \sigma)\, V_{\text{max}\,1}}{V_1\, K_{\text{M}_1}} \tag{2.198}$$

provided the precursor is uniformly distributed in the compartment. Likewise, the steady-state relaxation constant in the direction from product to precursor is a function of the affinity of the enzyme for the product,

$$k_4' = \frac{k_{\text{off}}\, B_{\text{max}}}{V_3\, K_{\text{M}_3}} = \frac{(1 - \sum \sigma)\, V_{\text{max}\,3}}{V_3\, K_{\text{M}_3}} \tag{2.199}$$

provided the product is also uniformly distributed in the compartment. The relaxation constant for an enzymatic reaction is often referred to as the enzyme activity, i.e., the maximum rate relative to the $K_{\text{M}}$ and corrected for volume of distribution ($V_1$) and presence of competitors ($\sum \sigma$). The actual rate of the enzymatic reaction is then,

$$J_1' = \sigma_1\, V_{\text{max}_1} - \sigma_3\, V_{\text{max}_3} = f_b{}'\, \sigma_1\, V_{\text{max}_1}, \tag{2.200}$$

where $1 - f_b{}'$ is now the ratio $\sigma_3 V_{\text{max}\,3}/[\sigma_1 V_{\text{max}\,1}]$, that is, not only a function of the affinity of the enzyme and concentration of the ligand but also of the maximum rate in both directions of the transformation. This is the general form of the Michaelis–Menten equation.[2]

If the net flux is close to zero, this relationship may be difficult to establish. The reaction then is said to be near-equilibrium. In this case, the ratio of the precursor and the substrate is a simple function of the ratio between the dissociation constants at steady-state,

$$\frac{C_3}{C_1} = \frac{K_{\text{M}_3} V_{\text{max}\,1}}{K_{\text{M}_1} V_{\text{max}\,3}} = \frac{k_{\text{in}}\, k_{\text{on}}}{k_{\text{out}}\, k_{\text{off}}} = \frac{K_{d_3}}{K_{d_1}} \tag{2.201}$$

which may differ for different precursors and substrates, as both $K_{\text{M}}$ and $V_{\text{max}}$ magnitudes vary among precursors and substrates. At near-equilibrium, fluxes in the two directions are of similar magnitude as well as much greater than the net reaction rate. This contrasts with nonequilibrium steady-state, where concentrations are constant and fluxes in the two directions are very different.

---

[2] Although $f_b{}'$ varies with the occupancies of precursor and product, magnitudes of $V_{\text{max}_1}$ and an apparent $K_{\text{M}_1}$ can be related by the Eadie–Hofstee equation (Eadie 1952; Hofstee 1952),

$$J_1' = V_{\text{max}_1} - \left[ V_1\, K_{\text{M}_1} \left( 1 + \left[ \frac{C_3}{K_{d_3}} \right] \right) \right] k_1',$$

where $V_{\text{max}_1}$ is the ordinate intercept and $-V_1\, K_{\text{M}_1}(1 + [C_3/K_{d_3}])$ is the slope, provided $C_3$ is constant. The relationship implies that the bidirectional exchange influences the apparent Michaelis constant.

## 2.3 Determination of Relaxation Constants

### 2.3.1 Stimulus-Response Relations

The purpose of the kinetic analysis is to deconvolve the characteristic constants of the impulse response function from the relationship between a perturbation of the steady-state of a system of compartments and the time-course of its return to steady-state. The perturbation of the steady-state usually consists in an induced change of the contents of the delivery compartment ("stimulus"). The subsequent return to steady-state ("response") is recorded as timed measurements of the contents of the entire system under investigation. Under certain limited circumstances, the use of tracers allows investigators to estimate the magnitude of relaxation constants and derived variables (clearance and volume of distribution) by deconvolving the impulse response function from one or more solutions of the differential equations which predict the response of the compartments to the stimulus.

Generally, tracers are labeled molecules that enter compartments without affecting the magnitude of the relaxation constants which define the compartments. Tracers are designed to have the same physical properties as the native molecules but to exist in such low concentration that the total number of molecules in the compartment is constant. This characteristic is not always fulfilled, however, because the tracer molecules may differ from the native molecules in their chemical properties, particularly with respect to the magnitude of affinities, and dissociation and Michaelis constants. For this reason, it is often necessary to carefully distinguish between relaxation constants, which follow the exchanges of the native molecules, and relaxation constants which describe the exchanges of the tracers.

When tracer molecules enter a compartment, the underlying assumption is that they are chemically indistinguishable from the native members of the compartment from the point of view of the mechanism responsible for the relaxation. Here, the tracer molecules in any compartment are symbolized by $m^*$ and their concentration in the compartment by $c^*$.

By definition, tracers rarely reach steady-state. Eventually, they disappear by decay or washout, unless they are continuously supplied, or neither decay nor escape from the compartments where they end up. The transient solutions are therefore useful to the estimation of relaxation constants, provided the predicted solutions of the differential equations are sufficiently accurate. The significance of this accuracy is the object of the regression analysis.

The methods fall into three categories, the single-bolus nonlinear deconvolutions, the steady-state programmed infusions, and the multiple-time graphical analyses. The deconvolutions are accurate but potentially imprecise and offer no directly verifiable results. The infusions are designed to let the tracer reach steady-state in extended periods of time. The multiple-time graphical analyses combine the advantage of the bolus injection with the visibility of the infusion results.

### 2.3.2 *Regression Analysis*

**Regression vs. Function Analysis**

The purpose of regression analysis is to establish the relations between two sets of variables, in the current context between the timed measurements of tracer concentration or content in a delivery compartment, say $\mathbf{M}_1$, and the timed measurements of tracer concentration in a number of tissue compartments, $\mathbf{M}_2$ to $\mathbf{M}_n$, often including the delivery compartment. The relationship between the two sets of variables is established in the form of a model with a number of constants, parameters, the magnitudes of which are unknown (model-based regression), or in the form of a class of equations (basis-functions) with known coefficients applicable to a system of compartments, the number and structure of which must be deduced from the data (data-based function analysis). The regression relation is based on the variations of both the "true" values of the variables and the random errors to which the observations of these variables are subject. The functional relations, on the other hand, depend on the variation of "true" values of these variables only, stripped of the random errors associated with their observation.

**Regression Analysis** The conventional regression analysis assumes a model solution and delivers estimates of the magnitude of the parameters that are most consistent with the known model. The regression analysis indicates with a certain statistical probability the range of values of the parameters which are consistent with the chosen model. As the relationship in principle is either nonlinear or linear (i.e., uni- or multilinear), regression analyses are likewise either nonlinear or linear (i.e., uni- or multilinear). Non-linear regression analysis is still an imperfect art with many pitfalls, while linear regression analysis both conceptually and mathematically is simpler. While nonlinear regression analysis proceeds by iteration until a given threshold of accuracy is reached, linear regression analysis often can be completed in a single analytical step. The disadvantage of linear regression, on the other hand, is that it can be subject to important bias, which must be taken into account.

The importance of the parameter estimates must be established by statistical tests, which are frequently misinterpreted. The problem is that a parameter estimate may have been established within satisfactorily narrow limits, yet may express no more than the noise inherent in the measured variables. Another parameter estimate may have been established within fairly wide limits, yet may represent a very accurate estimate of the biological variability of the measured variables. For these and other reasons, it is important to distinguish between precision and accuracy, also in terms of their practical applicability. Accuracy may be useless if the variables indeed are subject to unavoidable biological variation, while precision may be useful as an indicator of the difference among several sets of data, despite the bias of the estimates.

**Basis Function Analysis** The data-driven basis-function analysis assumes a set of parameters of known magnitude and chooses the model configuration most consistent with the data. This approach is a search for the optimal structural relations among compartments which predict the form of the response function. As an intriguing consequence it is possible to fit combinations of models to the data. These model solutions are "probabilistic" in the sense that the resulting model structure can be established only with a finite probability of consistency with the data. The analysis is not an actual deconvolution but rather a segmentation of the data in probablistic clusters of model structure.

### Model-Based Deconvolution

In seeking the regression relation between two sets of tracer measurements, $c^*(t)$ and $m^*(t)$, the values of the dependent variable $m^*$ are presumed to be randomly distributed about the regression function, so that its *expected* values are known to be a specific function of the *observed* values of the independent variable $c^*$. The expected values of the dependent variable $m^*$ are the function $\overline{m}(c^*(t), t, p_1, ..., p_n)$, where $p_1, ..., p_n$ refers to the $n$ parameters of the functional relation. If the distribution of $m^*(T)$ (where $T$ is a time of observation) about $\overline{m}(c^*(T), T, p_1, ..., p_n)$ is normal for all $T$, the method of Least Squares is the most efficient means of estimating the parameters. The sum of squares to be minimized is,

$$S_{\text{sq}} = \sum_{i=a}^{b} \left[m^*(T_i) - \overline{m}(c^*(T_i), T_i, K_1, k_2, ..., k_n)\right]^2 , \tag{2.202}$$

where $a$ and $b$ define a range of measurements chosen to ensure sufficient degrees of freedom in relation to the number of parameters. The minimization is carried out by means of differential calculus but it can be performed by elementary algebra in the cases of linear and multilinear regression in a way which gives simultaneously the estimates of the parameters and the minimal sum of the squares. These methods will not be discussed here, as a large number of commercially available programs efficiently carry out the necessary computations.

In regression analyses of model parameters, it is common to refer to the "virtual" first compartment, $\mathbf{M}_o$, as the *delivery compartment* and the second compartment, $\mathbf{M}_2$, as the *precursor* or exchange compartment. The subsequent compartments, $\mathbf{M}_3$-$\mathbf{M}_n$, are then the *product compartments*. The concentration of the tracer precursor, $c_1^*(0)$, in the delivery compartment is $c_{\mathrm{a}}^*$, which denotes the arterial concentration when the delivery compartment is the capillary bed. The tracer content of the precursor compartment is then $m_{\mathrm{e}}^*$ and the content of all the subsequent compartments is $m_p^*$. The sum of all compartments, $V_o c_{\mathrm{a}}^* + m_{\mathrm{e}}^* + m_p^*$, is then $m^*$. For three compartments, the tracer content obeys (2.43). The regression analysis then yields the parameter estimates, which minimize the sum of squares, and determines the residual sum of squares.

### Data-Driven Basis-Function Analysis

In general, linear first-order differential equations define a system with an infinite number of compartments distinguished by the magnitude of their exponentials. An experimentally relevant subset of the compartment can be identified by means of least-squares fitting of the basis functions to the data (Gunn et al. 2001). The convolution of the impulse response function with the vascular or tissue forcing function leads to the actually observed tissue curve predicted by (2.16). The underlying impulse response function is assumed to be represented as a cluster of elements picked from a sufficiently large ("overcomplete") pool of pairs of preselected coefficients ($K1_h$) and relaxation constants ($k2_h$) assumed to include all of the actual compartments of the system. As standard least-squares analysis does not apply when the number of parameters exceeds the number of measurements, addition of a penalty function reduces the number of permitted elements to one which minimizes the penalty function. The penalty function depends on the variability of the data, expressed in the magnitude of the regularization term $\mu$. The variability coefficient can be determined by any one of several methods of "denoising" or "smoothing" (Gunn et al. 2002). Greater variability of the data allows fewer elements in the cluster of compartments and prevents both overfitting and underfitting of the data (Shao 1993; Hjorth 1994),

$$S_{\text{sq}} = \sum_{i,h=a,1}^{b,n-1} \left( \left[ m^*(T_i) - \overline{m}(c^*(T_i), T_i, K1_h, k2_h) \right]^2 + \mu_h \right), \tag{2.203}$$

where $\mu$ is the regularization term.

## 2.3.3 Deconvolution of Response Function by Differentiation

In the cases of more than three or four compartments, the number of parameters is so great that regression analysis fails to identify a meaningful and unique set of estimates. In these cases it is possible to use the regression analysis to identify a precursor compartment because it regulates the influx to all of the subsequent compartments.

Using the substitution $m^*$ for the total contents of the precursor and product compartment, $c_{\text{a}}^*$ for $c_1(0)$, and $V_o$ for the volume of the "virtual" capillary compartment defined in (2.150), the total flux of a labeled substance into any closed system of compartments is given by (2.73),

$$\frac{\mathrm{d}m^*}{\mathrm{d}t} = V_o \, \frac{\mathrm{d}c_{\text{a}}^*}{\mathrm{d}t} + K_1 \, c_{\text{a}}^* - k_2 \, m_{\text{e}}^*, \tag{2.204}$$

where asterisks denote the labeled substance ("tracer"). When tracers are introduced into living tissue, the tracer precursor and its products distribute in separate compartments, which cannot be probed individually. The purpose of the regression is to determine the individual time courses of the precursor and the products. Among other applications, this regression is used to determine true steady-state values of a binding potential ($p_B$), and the corresponding steady-state magnitude of bound tracer. The regression is based on the claim that the tracer flux into the tissue is driven by the concentration difference between the tracer contrations in the capillary circulation and the tissue with which the capillary is directly exchanging (precursor compartment). Equation (2.204) can be solved for the tracer content of the precursor (exchange) compartment,

$$m_e^* = V_e \left[ c_a^* - \frac{1}{K_1} \left( \frac{dm^*}{dt} - V_o \frac{dc_a^*}{dt} \right) \right], \tag{2.205}$$

where $V_e = K_1/k_2$ is the partition volume. To obtain an estimate of the quantity of exchangeable tracer, it is necessary to determine the derivatives of $m^*$ and $c_a^*$, and the magnitudes of $V_o$, $V_e$, and $K_1$.

To differentiate the function underlying the observed values $m^*$, the function must first be optimized by any suitably realistic formula, which minimizes the sum of squares but also has the property that the observed values maintain normal distribution about the optimized values of this formula. A degree of smoothing must be performed to allow differentiation with minimal variability. Equations (2.43) and (2.48) are recommended for this smoothing. With (2.43), the smoothing is accomplished by ordinary six-parameter nonlinear regression to the measured pairs of $m^*$ vs. $c_a^*$. With (2.48), the best estimate of the expected ("smoothed") value $\overline{m}$ is calculated by an appropriate method of "denoising." One approach is to apply a differentiable function directly,

$$\overline{m} = a_1 c_a^* + a_2 \int_o^T c_a^* \, dt + a_3 \int_o^T \int_o^u c_a^* \, dt \, du + a_4 \int_o^T m^* \, dt + a_5 \int_o^T \int_o^u m^* \, dt \, du, \tag{2.206}$$

where $\overline{m}$ is the optimized value, and the coefficients are those defined for (2.48).

**Analytic Differentiation** The nonlinear smoothing by means of (2.43) has the advantage that the derivatives of the function are analytically known in advance. Using the parameter estimates $\overline{k}_2$, $\overline{k}_3$, and $\overline{k}_4$, the expected quantity of tracer precursor is calculated from the relationship given in (2.39),

$$\overline{m}_e = V_e \, \overline{k}_2 \, \overline{\zeta}(t), \tag{2.207}$$

where $V_e$ is the partition volume measured in a reference region and $\overline{\zeta}(t)$ is the expected function $\overline{m}_e/K_1$ calculated from (2.39),

$$\overline{\zeta}(t) = \left( \frac{\overline{q}_2 - \overline{k}_4}{\overline{q}_2 - \overline{q}_1} \right) \int_o^T c_a^* \, e^{-\overline{q}_2 \, (T-t)} \, dt - \left( \frac{\overline{q}_1 - \overline{k}_4}{\overline{q}_2 - \overline{q}_1} \right) \int_o^T c_a^* \, e^{-\overline{q}_1 \, (T-t)} \, dt, \tag{2.208}$$

where $\overline{q}_1$ and $\overline{q}_2$ are calculated from the estimates $\overline{k}_2$, $\overline{k}_3$, and $\overline{k}_4$ by means of (2.40) and (2.41).

**Numerical Differentiation** Simple differentiation of (2.206) (Wong et al. 1998a) yields the flux into the precursor and product compartments required for determination of $\overline{m}_{\mathrm{e}}$,

$$\overline{j}_1 = \frac{\mathrm{d}m^*}{\mathrm{d}t} - a_1 \frac{\mathrm{d}c_{\mathrm{a}}^*}{\mathrm{d}t} = a_2\, c_a^* + a_3 \int_o^T c_{\mathrm{a}}^*\, \mathrm{d}t + a_4\, m^* + a_5 \int_o^T m^*\, \mathrm{d}t \quad (2.209)$$

such that (2.205) becomes,

$$\overline{m}_{\mathrm{e}} = V_{\mathrm{e}} \left[ c_a^* - \frac{\overline{j}_1}{a_1 a_4 + a_2} \right], \quad (2.210)$$

where $V_{\mathrm{e}}$ is known from separate measurement of the tracer's partition volume $(K_1/k_2)$ in a reference region, and where the relations following (2.48) have been used to derive the denominator, $a_1 a_4 + a_2$.

**Determination of Precursor–Product Ratio (Binding Potential)** The tracer product is the precursor subtracted from the total,

$$\overline{m}_p = \overline{m} - \overline{m}_{\mathrm{e}} - a_1\, c_{\mathrm{a}}^*,$$

where $\overline{m}$ is the time-activity curve in the region of interest, smoothed according to (2.206)(Wong et al. 1998a). At the peak of $\overline{m}_p$, $\mathrm{d}\overline{m}_p/\mathrm{d}t$ is zero and $\overline{M}_p$ is at steady-state (transient equilibrium). The precursor–product ratio (steady-state binding potential), $p_{\mathrm{B}}$, is the ratio between $\overline{M}_p$ and $\overline{M}_e$ at this time only (indicated by upper-case symbols),

$$p_{\mathrm{B}} = \frac{\overline{M}_p}{\overline{M}_e} = \frac{\overline{M} - V_o C_{\mathrm{a}}^*}{\overline{M}_e} - 1, \quad (2.211)$$

where $V_o$ is the estimate of $a_1$ in the case of the multilinear regression by (2.206) as shown in Wong et al. (1998a). The quantity of tracer product at the peak is determined as the ratio between $\overline{M}_p$ and the specific activity $A$ (Fig. 2.19),

$$M_p = \frac{\overline{M}_p}{A}.$$

### 2.3.4 *Deconvolution by Temporal Transformation*

#### Constant Forcing Function

The deconvolution of the response function by means of temporal transformation is based on the properties of (2.74) when $c_{\mathrm{a}}^*$ is maintained constant (i.e., symbolized by $C_{\mathrm{a}}^*$),

$$v^*(T) \equiv \sum_{i=1}^{n} \frac{m_i^*(T)}{C_{\mathrm{a}}^*} = V_o + \sum_{h=1}^{n-1} \left[ \frac{K_{1_h}}{k_{2_h}} \left( 1 - \mathrm{e}^{-k_{2_h} T} \right) \right],$$

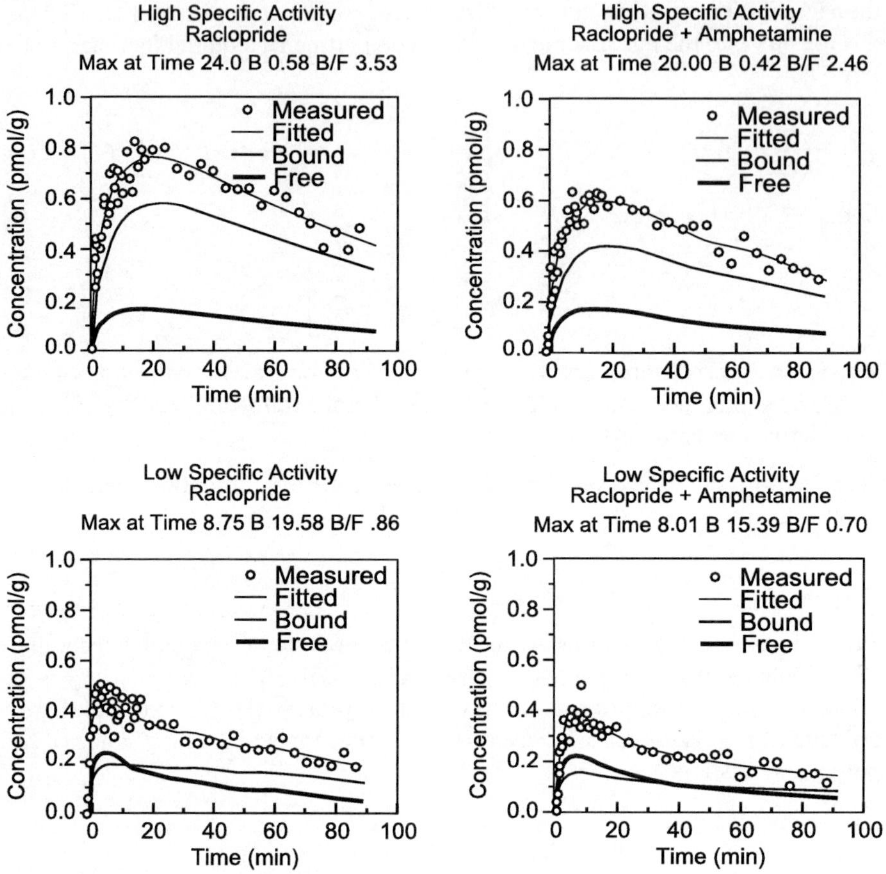


**Fig. 2.19** Precursor and product curves at high (*top panels*) and low (*bottom panels*) specific activies before (*left panels*) and after (*right panels*) competitor (dopamine released amphetamine) action. Resulting estimates for the upper left-hand panel are listed in Table 2.2. From Wong et al. (1998a)

where $v^*(T)$ is the apparent volume of distribution as a function of time ($T$). The equation simplifies to the equation,

$$v^*(T) = V - \sum_{h=1}^{n-1} \left[ \frac{e^{-k_{2_h} T}}{k_{2_h}} \right],$$

where $V$ is the total steady-state volume of distribution. The behavior of the equation depends on the relationship between $k_{2_h}$ and $T$. For values of $T \ll 1/k_{2_h}$ for all $h$,

$$v^*(T) = V_o + T \sum_{h=1}^{n-1} K_{1_h} = V_o + K_1 T$$

which has the properties of a straight line and is a special case of (2.21) for constant $c_1$. For values of $T \gg 1/k_{2_h}$ for all $h$,

$$v^*(T) = V_o + \sum_{h=1}^{n-1} \frac{K_{1_h}}{k_{2_h}} = V = V_o + \sum_{h=1}^{n-1} V_h$$

which is a steady-state plateau equivalent to the level defined by (2.22). For intermediate values of $T$, $1/k_{2_h} \ll T$ for $h < g$ and $1/k_{2_h} \gg T$ for $h \geq g$, the intermediate equation results,

$$v^*(T) = V_o + \sum_{h=1}^{g-1} V_h + T \sum_{h=g}^{n-1} K_{1_h} = V_o + \sum_{h=1}^{g-1} V_h + K\,T \qquad (2.212)$$

which is a straight line with an ordinate intercept and a slope of $K$, formally equivalent to (2.84). The equivalence indicates that normalization of the integrated forcing function compensates for nonlinearity of the forcing function. The normalization has unit of time and represents a temporal transformation.

The use of the transformed time as independent variable, and the volume $v^*$ as the dependent variable, changes the time course of the response and allows inference ("graphical analysis") to be made of the steady-state clearance of the tracer.

### Variable Forcing Function

In analogy with (2.21) and (2.84), the primary temporal transformation creates a new time variable in the case of a variable forcing function,

$$\Theta^*(T) = \frac{\int_o^T c_a^* \, dt}{c_a^*(T)}$$

as does the secondary temporal transformation,

$$\Theta'^*(T) = \frac{\int_o^T \int_o^u c_a^* \, dt \, du}{\int_o^T c_a^* \, dt}. \qquad (2.213)$$

Also judging from the equivalence with (2.21) and (2.84), the primary dependent variable is an apparent partition volume,

$$v^*(T) = V_o + \sum_{i=2}^{n} v_i^* = \sum_{i=1}^{n} \frac{m_i^*}{c_a^*}$$

and the secondary dependent variable an apparent partition volume which approaches the steady-state volume of distribution, i.e., the integral of the impulse response function, when time approaches infinity,

$$v'^*(T) = V_o + \sum_{i=2}^{n} v_i^* = V_o + \sum_{i=2}^{n} \frac{\int_o^T m_i^* \, dt}{\int_o^T c_a^* \, dt} \tag{2.214}$$

which has the special significance of being the ratio of the total areas under the curves of the response and the stimulus. The fact that this ratio approaches the steady-state volume of distribution as $T$ approaches infinity, is the *stimulus-response theorem* (Lassen and Perl 1979).

### Graphical Analysis of Primary Temporal Transformation

For $n$ compartments, the expected results of the multilinear solution to (2.204) are given by (2.74):

$$m^*(c_a^*, m_e^*) = V_o \, c_a^* + K_1 \int_o^T c_a^* \, dt - k_2 \int_o^T m_e^* \, dt,$$

where $m_e^*$ is the tracer content of the precursor compartment. The earlier equation rearranges to,

$$\frac{m^*}{c_a^*} = V_o + K_1 \int_o^T \left[\frac{c_a^*(t)}{c_a^*(T)}\right] dt \left(1 - \left[\frac{k_2 \int_o^T m_e^* \, dt}{K_1 \int_o^T c_a^* \, dt}\right]\right)$$

and, thus, to the temporally transformed equation,

$$v^*(T) = V_o + K_1 \, \Theta^* \left(1 - \left[\frac{v_e'^*(T)}{V_e}\right]\right), \tag{2.215}$$

where $v_e'^*(T)$ is the apparent partition volume at time $T$.

**Two Compartments** For two compartments, (2.215) reduces to the Gjedde–Patlak plot (2.21) for all times $T$ for which $v_e'^*(T) \ll V_e$,

$$v^*(T) = V_o + K_1 \, \Theta^*. \tag{2.216}$$

**Multiple Compartments** With continued incremental accumulation of at least one product ($m_p^*$), Gjedde (1982) showed that (2.216) reduces to (2.84) also for multiple compartments, provided steady-state exists between the delivery and precursor compartments. Thus, when it is true that the ratio $V_e'^*(T)/V_e$ maintains a constant ratio less than unity but greater than zero, it is also true that

$$m_p^*(T_2) - m_p^*(T_1) = K \; (\Theta^*(T_2) - \Theta^*(T_1)), \tag{2.217}$$

where $K$ is the net clearance of the precursor to at least one of several product compartments, relative to the concentration of the precursor in the circulation. The general Rutland–Gjedde–Patlak Equation (2.218) is then valid,

$$v^*(T) = V_o + \sum_{i=2}^{n} V_i + K\,\Theta^* = V_n + K\,\Theta^*, \tag{2.218}$$

where $V_n$ is a "virtual" precursor volume of distribution. The precursor volume is "virtual" because it does not actually contain the precursor from the onset of the uptake but from the onset of the steady-state (secular equilibrium) between the delivery and precursor compartments.

The lumped variables (2.213) define linear relationships applicable to results from in vivo or ex vivo tomography, including autoradiography, positron emission tomography, and magnetic resonance spectroscopy.

### Graphical Analysis of Secondary Temporal Transformation

The apparent partition volume $v'^*(T)$ is the ratio of the areas under the response function and the forcing (stimulus) function curves (AUC). The analysis extends the graphical analysis, which can be useful also in the absence of measurements of a proper vascular forcing function. The method extends the graphical analysis by integrating delivery and precursor–product curves prior to the normalization. As with the simple ratios, the independent variable has unit of time, and the new dependent variable is the ratio between the integrals of the total content of the precursor and product compartments and the integral of the tracer concentration of the delivery compartment.

**Vascular Forcing Function** The approach to the ratio of integrated compartments employs the temporally transformed independent variable $\Theta'$, equal to $\int_o^T \int_o^u c_a^* \mathrm{d}t\,\mathrm{d}u / \int_o^T c_a^* \mathrm{d}t$. The AUC analysis is applicable to models with any number of compartments, but the interpretation of the ratios of the areas under the curves depends on the model.

The model shown in Fig. 2.20 is an example of a model which can be analyzed by the AUC analysis. $\mathbf{M}_o$ is the virtual delivery compartment, in which the tracer concentration is $c_a^*$. It masks the real delivery compartment, which does not fulfill the fundamental compartment criteria as discussed earlier. $\mathbf{M}_2$ is the precursor compartment and $\mathbf{M}_{3_1}$ and $\mathbf{M}_{3_2}$ are the irreversible and reversible accumulation compartments, respectively. It is further assumed that the relaxation constants $k_{3_2}$ and $k_{4_2}$ are of such great magnitude that compartments $\mathbf{M}_2$ and $\mathbf{M}_{3_2}$ maintain an approximately constant ratio (secular equilibrium), equal to a binding potential, $p_{\mathrm{B}}$. The constant ratio has the effect of reducing the three compartments $\mathbf{M}_2$, $\mathbf{M}_{3_1}$, and $\mathbf{M}_{3_2}$ to two with the relaxation constants $k'_2$ and $k'_{3_1}$,

$$k'_2 = \frac{k_2}{1 + p_{\mathrm{B}}} \quad \text{and} \quad k'_{3_1} = \frac{k_{3_1}}{1 + p_{\mathrm{B}}}, \tag{2.219}$$

where $p_{\mathrm{B}}$ is the ratio $k_{3_2}/k_{4_2}$.

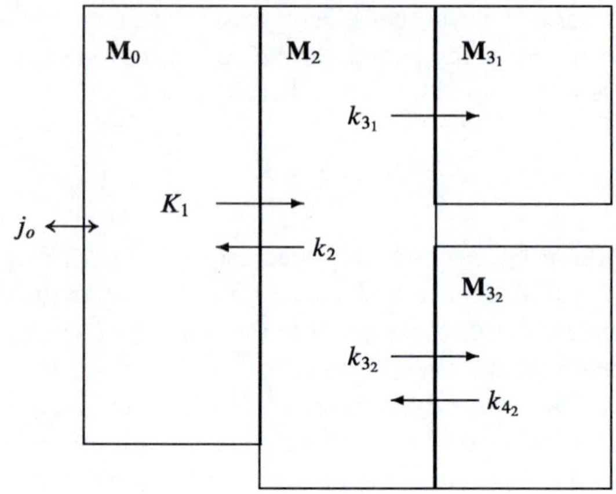


**Fig. 2.20** Model of multiple compartments with two parallel compartments and a vascular delivery compartment

In this model, the total uptake of tracer is given by the Blomqvist equation (2.49),

$$m^* = V_o c_a^* + \left[K_1 + V_o(k_2' + k_{3_1}')\right] \int_o^T c_a^* \, \mathrm{d}t$$
$$+ K_1 k_{3_1}' \int_o^T \int_o^u c_a^* \, \mathrm{d}t \, \mathrm{d}u - (k_2' + k_{3_1}') \int_o^T m^* \, \mathrm{d}t, \qquad (2.220)$$

where the symbols have their usual meaning. Equation (2.220) rearranges to the ratio of the areas under the curves as a function of time $T$,

$$v'^*(T) = \frac{\int_o^T m^* \, \mathrm{d}t}{\int_o^T c_a^* \, \mathrm{d}t} = \left[\frac{K_1}{k_2' + k_{3_1}'} + V_o\right] + \frac{K_1 k_{3_1}'}{k_2' + k_{3_1}'} \left(\frac{\int_o^T \int_o^u c_a^* \, \mathrm{d}t \, \mathrm{d}u}{\int_o^T c_a^* \, \mathrm{d}t}\right)$$
$$- \left[\frac{m^* - V_o \, c_a^*}{(k_2' + k_{3_1}') \int_o^T c_a^* \, \mathrm{d}t}\right], \qquad (2.221)$$

where the ratio of the areas under the time-activity curves is $v'^*$, $K_1 k_{3_1}'/(k_2' + k_{3_1}')$ is $K$, the net clearance of the ligand, $\int_o^T \int_o^u c_a^* \, \mathrm{d}t \, \mathrm{d}u / \int_o^T c_a \, \mathrm{d}t$ is the new time variable $\Theta'^*$, and $(K_1 \int_o^T c_a^* \, \mathrm{d}t - m)/[(k_2' + k_{3_1}') \int_o^T c_a^* \, \mathrm{d}t]$ represents the mono-exponential approach to the time-variable ordinate-intercept $V_g(1 + p_\mathrm{B})$, equal to $K_1 k_2'/(k_2' + k_{3_1}')^2$, where $V_g$ is defined as $K_1 k_2/(k_2 + k_{3_1})^2$,

$$v'^* = K \, \Theta'^* + V_o + V_g \, (1 + p_\mathrm{B}) \, (1 - \mathrm{e}^{-\alpha \, \Theta'^*}) \qquad (2.222)$$

which defines an early phase of slope $K/\alpha$ and ordinate intercept $V_o$, a late phase of slope $K$ and ordinate intercept $V_o + V_g$, and a monoexponentially changing transition between the two phases. For $k_{3_1} = 0$, (2.220) shrinks to,

$$v'^* = V_o + V_\mathrm{e} \, (1 + p_\mathrm{B}) \, (1 - \mathrm{e}^{-\alpha \, \Theta'^*}), \qquad (2.223)$$

where $V_o + V_e(1 + p_B)$ is the steady-state partition volume $V$, such that $p_B = [(V - V_o)/V_e] - 1$. For $k_{3_1} = 0$ and $p_B = 0$, (2.220) further reduces to,

$$v'^* = V_o + V_e\,(1 - e^{-\alpha\,\Theta'^*}) \tag{2.224}$$

which yields the partition volume $V = V_e + V_o$ at steady-state.

**Tissue (Indirect) Forcing Function** Although safe, the use of arterial sampling can be an inconvenience to experimental subjects and, in case of patients, can be impossible. Several methods have been proposed for the estimation of binding potentials, using tracer uptake curves in regions of little or no specific binding.

The reference tissue area-under-curves ratio analysis extends the ordinary area-under-curves ratio analysis by integrating the reference region and region of interest tracer uptake curves. The reference region takes the place of the combined delivery and precursor compartments by reversely estimating the tracer concentration in the delivery compartment which yielded the tracer content of the reference region precursor compartment. The new independent variable still has unit of time, and the new dependent variable is the ratio between the integrals of the regions of interest and the integral of the reference region tracer uptake curves (Fig. 2.21).

In the region of reference, the term $m^*_{2_2}$ denotes the tracer content of the precursor compartment of the reference region, $\mathbf{M}_{2_2}$, as function of time. The term $m^*_{[2_1+3_{1_1}+3_{1_2}]}$ denotes the tracer content of the precursor and product compartments

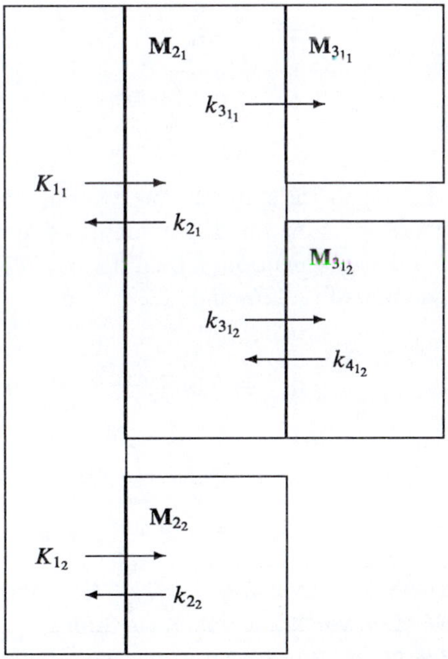


**Fig. 2.21** Model of four compartments with two parallel compartments in region of interest, and two compartments in reference region

of the region of interest as a function of time. In the reference region, let the tracer uptake as a function of time be given by (2.18), provided $V_o$ can be claimed to be negligible,

$$m^*_{2_2} = K_{1_2} \int_o^T c^*_\mathrm{a}\, \mathrm{d}t - k_{2_2} \int_o^T m^*_{2_2}\, \mathrm{d}t, \tag{2.225}$$

where the symbols have their usual meaning. Note that no initial volume of distribution was assumed in this region. The equation was rearranged to yield an expression of the integral of $c_\mathrm{a}$,

$$\int_o^T c^*_\mathrm{a}\, \mathrm{d}t = \frac{m^*_{2_2}}{K_{1_2}} + \frac{k_{2_2}}{K_{1_2}} \int_o^T m^*_{2_2}\, \mathrm{d}t \tag{2.226}$$

which is integrated once more to yield,

$$\int_o^u \int_o^T c^*_\mathrm{a}\, \mathrm{d}t\, \mathrm{d}T = \frac{\int_o^u m^*_{2_2}\, \mathrm{d}T}{K_{1_2}} + \frac{k_{2_2}}{K_{1_2}} \int_o^u \int_o^T m^*_{2_2}\, \mathrm{d}t\, \mathrm{d}T \tag{2.227}$$

for insertion into the equation later.

In the region of interest, a region of specific but reversible accumulation of tracer, the uptake as a function of time is expected to obey the solution to the three-compartment model, (2.49), provided $V_o$ can be claimed to be negligible,

$$m_{[2_1+3_{1_1}+3_{1_2}]} = K_{1_1} \int_o^T c^*_\mathrm{a}\, \mathrm{d}t + K_{1_1} k'_{3_1} \int_o^T \int_o^u c^*_\mathrm{a}\, \mathrm{d}t\, \mathrm{d}u$$
$$-(k'_{2_1} + k'_{3_1}) \int_o^T m^*_{[2_1+3_{1_1}+3_{1_2}]}\, \mathrm{d}t, \tag{2.228}$$

where again the symbols have their usual meaning (for example $k'_{2_1} = k_{2_1}/(1+p_\mathrm{B})$ where $p_\mathrm{B}$ is the binding potential of the tracer for rapidly reversible accumulation in a compartment other than the one responsible for the relaxation constant $k_{3_{1_1}}$). Insertion of (2.226) and (2.227) yields:

$$m_{[2_1+3_{1_1}+3_{1_2}]} = R_1 m^*_{2_2} + R_1 (k_{2_2} + k'_{3_{1_1}}) \int_o^T m^*_{2_2}\, \mathrm{d}t + R_1\, k_{2_2}\, k'_{3_{1_1}}$$
$$\times \int_o^T \int_o^u m^*_{2_2}\, \mathrm{d}t\, \mathrm{d}u - (k'_{2_1} + k'_{3_{1_1}}) \int_o^T m^*_{[2_1+3_{1_1}+3_{1_2}]}\, \mathrm{d}t, \tag{2.229}$$

where $R_1$ symbolizes the $K_{1_1}/K_{1_2}$ ratio. The following derivations assume a short phase in which tracer concentrations, $c^*_a$, in the vascular bed are significant. During this phase, transfer from the vascular bed to tissue is assumed to dominate, and

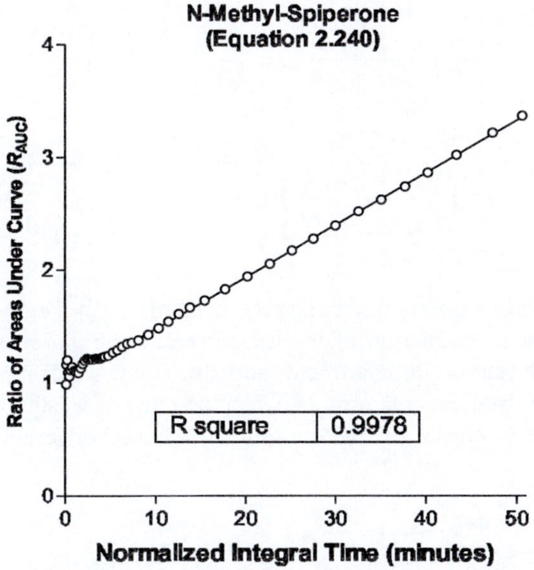


**Fig. 2.22** Reference region areas under curves ratio analysis according to (2.240) of tracer $N$-methylspiperone (NMSP) uptake into human striatum in vivo, assumed to be mediated by high-affinity binding to dopamine receptors. Slope is 0.0460 min$^{-1}$ (s.e. $\pm$0.0005), $R_1$ is 1.11 $\pm$0.01, and $\alpha$ is 0.2$\pm$0.06

other transfers are assumed negligible. Thus, during this phase, the ratio of tracer mass in the region of interest to that in the reference region is $K_{1_1}/K_{1_2}$, or $R_1$. Following this initial phase, is a second in which $c_a^*$ is assumed negligible and the processes ignored in the first phase are assumed to dominate. The second phase, called washout, begins at $t = 0$ with $m^*_{[2_1+3_{1_1}+3_{1_2}]}(0)/m_2^*(0)$ equal to $R_1$. It is further assumed that $m^*_{3_{1_1}}(0) = 0$ and that $k_{2_1}/k_{2_2} = K_{1_1}/K_{1_2}$ (Fig. 2.22).

To compute the ratio of the contents of tissue of interest relative to the reference tissue as a function of time, transformed time variables are generated from the reference tissue curve in analogy with the definitions of $\Theta^*$ and $\Theta'^*$ (2.213),

$$\theta^* = \left[\frac{\int_o^T m^*_{2_2}\,\mathrm{d}t}{m^*_{2_2}}\right] \tag{2.230}$$

and

$$\theta'^* = \left[\frac{\int_o^T \int_o^u m^*_{2_2}\,\mathrm{d}t\,\mathrm{d}u}{\int_o^T m^*_{2_2}\,\mathrm{d}t}\right] \tag{2.231}$$

and dividing by the integral of the tracer accumulation in the reference region, the following analog of (2.240) is obtained,

$$\rho^* = \frac{\int_o^T m^*_{[2_1+3_{1_1}+3_{1_2}]}\,\mathrm{d}t}{\int_o^T m^*_{2_2}\,\mathrm{d}t} = \left(\frac{k_{2_1}k_{3_{1_1}}}{k_{2_1}+k_{3_{1_1}}}\right)\theta'^* + R_1\left(\frac{k_{2_2}+k'_{3_{1_1}}}{k'_{2_1}+k'_{3_{1_1}}}\right)\left[1-\frac{\left(\frac{m^*_{[2_1+3_{1_1}+3_{1_2}]}}{R_1\,m^*_{2_2}}-1\right)}{\left(k_{2_2}+k'_{3_{1_1}}\right)\theta^*}\right], \tag{2.232}$$

where $\rho^*$ is defined as the ratio between the contents of the region of interest and the reference region as function of the transformed time variable. The following substitutions are based on compartment modeling, which will reduce (2.232) to a more manageable form. For the ratio between the rates of washout from the region of interest and the reference region, the following linked differential equations are solved,

$$\frac{\mathrm{d}m^*_{[2_1+3_{1_2}]}}{\mathrm{d}t} = -(k'_{2_1}+k'_{3_{1_1}})\,m^*_{[2_1+3_{1_2}]},$$

$$\frac{\mathrm{d}m^*_{3_{1_1}}}{\mathrm{d}t} = k'_{3_{1_1}}\,m^*_{2_1}$$

and

$$\frac{\mathrm{d}m^*_{2_2}}{\mathrm{d}t} = -k_{2_2}\,m^*_{2_2} \tag{2.233}$$

from which follows the solution for $m^*_{[2_1+3_{1_1}+3_{1_2}]} = m^*_{[2_1+3_{1_2}]} + m^*_{3_{1_1}}$,

$$\frac{m^*_{[2_1+3_{1_1}+3_{1_2}]}}{m^*_{2_2}} - R_1 = R_1\left[\left(\frac{k'_{3_{1_1}}}{k'_{2_1}+k'_{3_{1_1}}}\right)\mathrm{e}^{k_{2_2}T} + \left(\frac{k'_{2_1}}{k'_{2_1}+k'_{3_{1_1}}}\right)\mathrm{e}^{(k_{2_2}-k'_{2_1}-k'_{3_{1_1}})\,T} - 1\right]. \tag{2.234}$$

The second substitution is the result of the wash-out of $m^*_{2_2}$, such that

$$\theta^* \cong \frac{\mathrm{e}^{k_{2_2}T}-1}{k_{2_2}}, \tag{2.235}$$

where $k_{2_2}$ is the wash-out rate or fractional clearance of tracer from the reference region, which is valid when (2.233) hold. With the substitutions, the following intermediate equation arises,

$$\rho^* = \left(\frac{k_{2_1} k_{3_{1_1}}}{k_{2_1} + k_{3_{1_1}}}\right) \theta'^* + R_1 \frac{k_{2_2} + k'_{3_{1_1}}}{k'_{2_1} + k'_{3_{1_1}}} \times \left[1 - \frac{k_{2_2}\, k'_{2_1}}{(k_{2_2} + k'_{3_{1_1}})(k'_{2_1} + k'_{3_{1_1}})} \right.$$
$$\left. \times \left(\frac{1 - \mathrm{e}^{-(k_{2_2} - k'_{2_1} - k'_{3_{1_1}})\,T}}{1 - \mathrm{e}^{-k_{2_2} T)}}\right) \mathrm{e}^{-(k'_{2_1} + k'_{3_{1_1}})\,T} - \frac{k_{2_2}\, k'_{3_{1_1}}}{(k_{2_2} + k'_{3_{1_1}})(k'_{2_1} + k'_{3_{1_1}})}\right]. \quad (2.236)$$

By making the substitution, justified by the empirical observation that, for $T$ between 0 and 100, for the radiotracer raclopride, and for $k_{2_2}$, $k'_{2_1}$, and $k'_{3_{1_1}}$ within two standard deviations of the averages shown in Table 2.2, the left hand side is log linear,

$$\left(\frac{k_{2_2} - k'_{2_1} - k'_{3_{1_1}}}{k_{2_2}}\right) \mathrm{e}^{-\alpha\, \theta'^*} \cong \left(\frac{1 - \mathrm{e}^{-(k_{2_2} - k'_{2_1} - k'_{3_{1_1}})\,T}}{1 - \mathrm{e}^{-k_{2_2}\, T)}}\right) \mathrm{e}^{-(k'_{2_1} + k'_{3_{1_1}})\,T} \quad (2.237)$$

the operational equation is derived,

$$\rho^* = \left(\frac{k_{2_1} k_{3_{1_1}}}{k_{2_1} + k_{3_{1_1}}}\right) \theta'^* - \frac{R_1\, k_{2_2}\, k'_{3_{1_1}}}{(k'_{2_1} + k'_{3_{1_1}})^2}$$
$$+ \frac{R_1}{k'_{2_1} + k'_{3_{1_1}}} \left[k_{2_2} + k'_{3_{1_1}} - k'_{2_1} \left(\frac{k_{2_2}}{k'_{2_1} + k'_{3_{1_1}}} - 1\right) \mathrm{e}^{-\alpha\, \theta'^*}\right]. \quad (2.238)$$

**Irreversible Accumulation** For $p_B = 0$, (2.238) reduces to,

$$\rho^* = \left(\frac{k_{2_1} k_{3_{1_1}}}{k_{2_1} + k_{3_{1_1}}}\right) \theta'^* + R_1 \left[\frac{k_{3_{1_1}}}{k_{2_1} + k_{3_{1_1}}} + \left(\frac{k_{2_1}}{k_{2_1} + k_{3_{1_1}}}\right) \mathrm{e}^{-\alpha\, \theta'^*}\right]$$
$$+ \left(\frac{k_{2_1}}{k_{2_1} + k_{3_{1_1}}}\right)^2 \left[1 - \mathrm{e}^{-\alpha\, \theta'^*}\right] \quad (2.239)$$

which is symmetrical in $k_{2_1}$ and $k_{3_{1_1}}$ when $R_1$ is close to unity and $\theta'^*$ exceeds a certain minimum. Therefore, the relation does not accurately identify the individual estimates of the two rate constants, which must be inferred from other information.

The steady-state ordinate intercept is an indicator of the relative magnitude of the two rate constants, as shown on the left panel of the graph: A lower intercept is indicative of rate constants of similar magnitudes. The closer to unity the steady-state ordinate intercept is, the more dissimilar the magnitudes of the rate constants are. Thus, as shown in Fig. 2.22, for $k_{3_{1_1}} \ll k_{2_1}$,

$$\rho^* \simeq k_{3_{1_1}}\, \theta'^* + 1 - (1 - R_1)\, \mathrm{e}^{-\alpha\, \theta'^*}. \quad (2.240)$$

On the other hand, for $k_{2_1} \ll k_{3_{1_1}}$, as shown for the tracer N-[$^{11}C$] methyl-spiperone (NMSP) on the right panel of graph 4.4.4.3., the slope loses its sensitivity to $k_3$ but the steady-state ordinate intercept remains close to $R_1$,

$$\rho^* \simeq k_{2_1}\, \theta'^* + R_1 \left(1 + \frac{k_{2_1}}{k_{3_{1_1}}} \left[\frac{k_{2_1}}{k_{3_{1_1}}} + \left(1 - \frac{k_{2_1}}{k_{3_{1_1}}}\right) \mathrm{e}^{-\alpha\, \theta'^*}\right]\right). \tag{2.241}$$

Thus, only for significantly reduced binding of NMSP, e.g., by blockade of the binding by a pharmacologically active dose of an antagonist, is it possible to show any relation of the slope to the receptor density.

Judged from other studies (Gjedde and Wong 2001), (2.22) shows that the slope is not too far from the known value of $k_2 = K_1/V_e = 0.1/3\ \mathrm{min}^{-1}$ for NMSP and thus has lost its sensitivity to the magnitude of $k_3$. A partial but inadequate solution to the problem of estimating $k_3$ under these circumstances is to use the estimate of $\alpha$, which is close to the value of $k_{2_1} + k_3$, as shown in (2.237). Assuming this to be the case, $k_{2_1}$ is close to 0.051 min$^{-1}$ and $k_3$ close to 0.47 min$^{-1}$, almost an order of magnitude greater. In principle, the ratio could be the inverse, although this would not be consistent with known estimates of $K_1$ for NMSP. In this case, it appears to be approximately correct to further simplify (2.240) and (2.241),

$$\rho^* \simeq k\, \theta'^* + 1, \tag{2.242}$$

where it is unknown whether $k$ is closest to $k_{2_1}$ or to $k_{3_{1_1}}$ when $R_1$ is unity and $\theta'^*$ exceeds a certain minimum.

**Reversible Accumulation** When it is known a priori that the tracer is subject to completely reversible accumulation, the relaxation constant $k_{3_{1_1}}$ is zero. As shown in Fig. 2.23, the comprehensive Equation (2.238) then reduces to,

$$\rho^* = 1 + p_\mathrm{B}\left(1 - \mathrm{e}^{-\alpha\, \theta'^*}\right) + (R_1 - 1)\ \mathrm{e}^{-\alpha\, \theta'^*} \tag{2.243}$$

which for $R_1 = 1$ becomes,

$$\rho^* = 1 + p_\mathrm{B}\left(1 - \mathrm{e}^{-\alpha\, \theta'^*}\right) \tag{2.244}$$

which is shown in the graph later (Fig. 2.23 for raclopride. The results of applying this analysis to multiple subjects is listed in Table 2.4. Although the result is visually satisfying, there is reason to be cautious about the many assumptions underlying the analysis and plot, known colloquially as Estimation of Reversible Ligand Binding and Receptor Density (ERLiBiRD).

### *2.3.5 Deconvolution of Response Function by Linearization*

The descriptive graphical analyses reduce the number of apparent variables by lumping the primary variables. They provide precise and often reasonably accurate

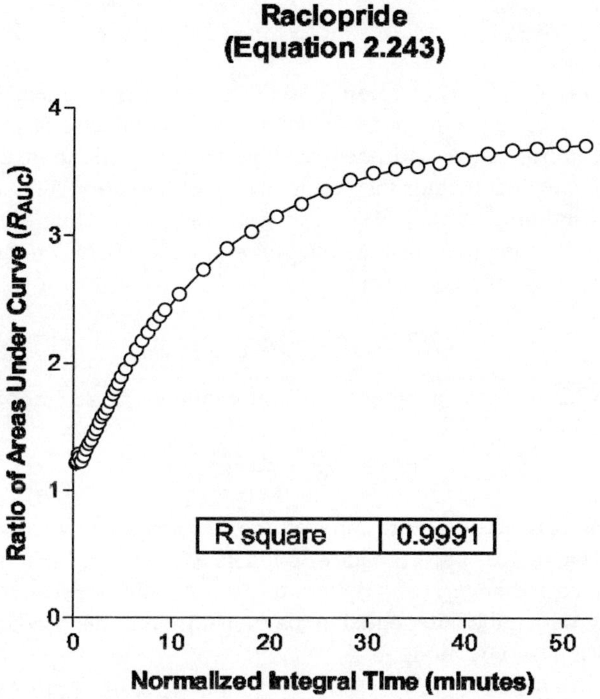


**Fig. 2.23** Reference region areas-under-curves ratio analysis of reversible tracer raclopride uptake into human striatum in vivo, assumed to be mediated by medium-affinity binding to dopamine receptors. Binding potential estimate is 2.8 (s.e., $\pm 0.01$, $R_1$ is $1.08 \pm 0.01$, and $\alpha$ is $0.072 \pm 0.001$)

illustrations of transient kinetic processes of brain uptake and metabolism. One great advantage of the methods is the ease with which they reveal key features of kinetic processing, which can be invisible in more conservative presentations of uptake data. A disadvantage is the compounding comminution of dependent and independent variables, which yields a potential bias (Logan et al. 2001).

The apparent clearance as a function of time of circulation is defined as:

$$\kappa_i^* = \frac{m_i^*(T)}{\int_o^T c_a^* \, dt},$$

and the apparent residence time in the precursor or exchange compartment has previously been defined as:

$$\theta_i^* = \frac{\int_o^T m_i^* \, dt}{m_i^*(T)}$$

which are useful when they can be determined directly from the timed measurements of $c_a^*$ and $m^*$ as functions of time, as described earlier.

### Vascular Reference

**Single-Compartment Linearization** For the single-compartment tracer model, the linear plots arising from the application of the lumped variables to (2.17) roughly divide into negative and positive slope plots. In this model, $m^* = m^*_{2_1}$. The *negative-slope* plots include the "clearance" plot (equation 37 in Gjedde 1982; equation 3 in Cumming et al. 1993), Also see footnote 8. which relates the ratio between the product and the substrate integral to the ratio between the product and substrate integrals, as in the equation,

$$\kappa_e^*(T) = K_{1_1} - k_{2_1}\, v_e'^*(T), \tag{2.245}$$

where $v_e'^*(T)$ (2.214) is the apparent partition volume. The reciprocal "ratio" plot (Gjedde et al. 2000),

$$v_e'^*(T) = V_e - \frac{1}{k_{2_1}}\, \kappa_e^*(T), \tag{2.246}$$

where $V_e$ is the precursor's partition volume. The *positive-slope* plots include the "reciprocal clearance" plot (equation 36 in Gjedde 1982), which relates the ratio between the precursor integral and the product to the ratio between the product integral and the product, the latter equal to the transit time of the product through the product pool, as in the equation,

$$\frac{1}{\kappa_e^*(T)} = \frac{\theta_e^*(T)}{V_e} + \frac{1}{K_{1_1}}, \tag{2.247}$$

where $1/\kappa^*$ is a reciprocal clearance and $k_{2_1}/K_{1_1}$ is the reciprocal partition volume. The equation was reversed by Logan et al. (1990) as the Logan plot (Logan et al. 1990),

$$\theta_e^*(T) = \frac{V_e}{\kappa_e^*(T)} - \frac{1}{k_{2_1}}, \tag{2.248}$$

where the positive slope is the partition volume. The third positive-slope equation is the "reciprocal time" plot (equation 2 in Reith et al. 1990), which relates the ratio between the precursor and the product integral to the ratio between the integrals of the precursor and product (Fig. 2.24),

$$\frac{1}{\theta_e^*} = \frac{K_{1_1}}{v_e'^*(T)} - k_{2_1}, \tag{2.249}$$

where the slope is the rate constant of the unidirectional clearance. The most commonly used of these pseudo-linear relations are the Logan plot for the estimation of steady-state volumes of distribution, and the plot of Gjedde (1982) and Cumming et al. (1993) for the estimation of the rates of enzymatic reaction. The plots were originally derived to allow information from multiple experiments to be jointly analyzed, because the ratios eliminate differences of scale among the experiments. The advent of positron emission tomography made the use of the unilinearized plots obsolete, except in the cases in which the number of regressions made the non- or multilinear regressions computationally prohibitive.

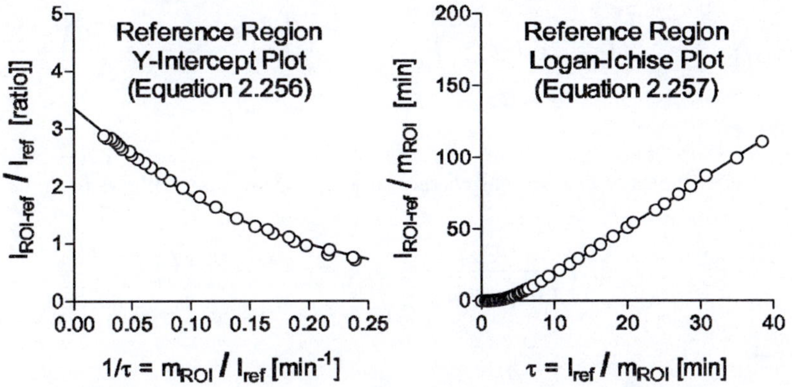


**Fig. 2.24** Pseudo-linear reference region plots of binding of tracer raclopride to dopamine receptors in striatum are linear only after steady-state between precursor and delivery compartments, averages of 13 healthy volunteers. In both cases, $\alpha$ is 0.30 min$^{-1}$ and $k'_{2_1}$ is 0.68 min$^{-1}$, with binding potentials are listed in Table 2.3

**Two-Compartment Linearization** For the two-compartment models, (2.74) and (2.216) are replaced by (2.18),

$$\frac{\int_o^T m^*\,\mathrm{d}t}{\int_o^T c_\mathrm{a}^*\,\mathrm{d}t} = \left(\frac{K_{1_1}}{k_{2_1}} + V_o\right) - \left(\frac{m^* - V_o\, c_\mathrm{a}^*}{k_{2_1}\int_o^T c_\mathrm{a}^*\,\mathrm{d}t}\right) \tag{2.250}$$

which yields the negative slope "ratio" plot (Gjedde et al. 2000) for times great enough to reduce the term $V_o/\Theta^*$ to zero,

$$v^*(T) = (V_\mathrm{e} + V_o) - \left(\frac{1}{k_{2_1}}\right)\kappa^*, \tag{2.251}$$

where the ordinate intercept is $V_\mathrm{e} + V_o$ and the slope is $-1/k_{2_1}$ when the term $V_o/v^*$ approximates zero. The alternative slope or Logan plot follows from the rearrangement of (2.251),

$$\theta^*(T) = (V_\mathrm{e} + V_o)\left(\frac{1}{\kappa^*(T)}\right) - \left(\frac{1}{k_{2_1}}\right) \tag{2.252}$$

which has the slope $V_\mathrm{e} + V_o$ and the ordinate intercept $-1/k_{2_1}$.

### Tissue Reference

The reparameterization allows the deconvolution to proceed by linearization. Thus, for $k_{3_{1_1}} = 0$, provided $K_{1_1}/k_{2_1} = K_{1_2}/k_{2_2}$, (2.238) reduces to,

$$\int_o^T m^*_{[2_1+3_{1_1}+3_{1_2}]}\, dt = (1+p_B)\int_o^T m^*_{2_2}\, dt - \left[\frac{m^*_{[2_1+3_{1_1}+3_{1_2}]} - R_1\, m^*_{2_2}}{k'_{2_1}}\right] \tag{2.253}$$

which can be linearized in several ways. For $k_{3_{1_1}} = 0$, it also follows from (2.234) that,

$$1 - R_1 \frac{m^*_{2_2}}{m^*_{[2_1+3_{1_1}+3_{1_2}]}} = 1 - e^{-(k_{2_2}-k'_{2_1})\,T}, \tag{2.254}$$

such that,

$$\int_o^T m^*_{[2_1+3_{1_1}+3_{1_2}]}\, dt = (1+p_B)\int_o^T m^*_{2_2}\, dt - m^*_{[2_1+3_{1_1}+3_{1_2}]}\left[\frac{1 - e^{-(k_{2_2}-k'_{2_1})\,T}}{k'_{2_1}}\right]. \tag{2.255}$$

With its three parameters $(p_B, R, k'_{2_1})$, (2.253) can be fitted with nonlinear regression analysis, but (2.254) and (2.255) can be further rearranged to yield approximately linear relationships, as shown in Fig. 2.24.

**Ordinate-Intercept Plot** The Ordinate Intercept Plot is obtained from (2.255) by subtracting and dividing with $\int_o^T m^*_{2_2}\, dt$,

$$\frac{\int_o^T (m^*_{[2_1+3_{1_1}+3_{1_2}]} - m^*_{2_2})\, dt}{\int_o^T m^*_{2_2}\, dt} = p_B - \frac{m^*_{[2_1+3_{1_1}+3_{1_2}]}}{\int_o^T m^*_{2_2}\, dt}\left[\frac{1 - e^{-\alpha\tau}}{k'_{2_1}}\right], \tag{2.256}$$

where $\tau$ is $\int_o^T m^*_{2_2}\, dt / m^*_{[2_1+3_{1_1}+3_{1_2}]}$, and $\alpha = 1/\tau$ which is linear with an ordinate intercept of $p_B$ and a slope of $-1/k'_{2_1}$ for times great enough to render the exponential term negligible.

**Logan-Ichise Plot** Alternatively, (2.255) can be rearranged to yield the slope plot (reminiscent of the tissue reference region version of the Logan plot) by subtracting and dividing with $m^*_{[2_1+3_{1_1}+3_{1_2}]}$,

$$\frac{\int_o^T (m^*_{[2_1+3_{1_1}+3_{1_2}]} - m^*_{2_2})\, dt}{m^*_{[2_1+3_{1_1}+3_{1_2}]}} = p_B\left[\frac{\int_o^T m^*_{2_2}\, dt}{m^*_{[2_1+3_{1_1}+3_{1_2}]}}\right] - \frac{1 - e^{-\alpha\tau}}{k'_{2_1}} \tag{2.257}$$

which is linear with a slope of $p_B$ and an ordinate intercept of $-k'_{2_1}$ for times great enough to render the exponential term negligible.

**Table 2.1** Kinetics of enzyme-substrate compound of peroxidase

| Symbol | Mean | Unit |
|---|---|---|
| $k_{on}$ | 0.6 | $min^{-1}\ nM^{-1}$ |
| $k_{off}$ | 12 | $min^{-1}$ |
| $k_{in}$ | 252 | $min^{-1}$ |
| $K_d$ | 20 | nM |
| $K_M$ | 440 | nM |

## 2.4 Application of Relaxation Constants

### *2.4.1 Peroxidation*

The original Michaelis–Menten formulation of enzymatic action required approximate steady-state between the ligands in compartments $\mathbf{M}_1$ and $\mathbf{M}_2$. An approximate steady-state exists between these two compartments when $k_{in} \ll k_{off}$. Briggs and Haldane (1925) showed that the formulation is valid also when $k_{in} \gg k_{off}$, which appears to be the case for many enzymes. Estimates of the relaxation constants tend to confirm that magnitudes of $K_M$ greatly exceed the magnitudes of $K_d$ for most enzymes. Chance (1943) studied the enzyme peroxidase and observed that the dissociation constant was only 4.5% of the Michaelis constant, as shown in Table 2.1, indicating that only this small fraction of the molecules do not change into the product. This observation means that the maximum rate is different for the precursor and the product.

### *2.4.2 Dopaminergic Neurotransmission*

#### Dopamine Receptor Binding Constants

Analysis of tracer [$^{11}$C]raclopride binding to dopamine receptors in the putamen region of the neostriatum of healthy volunteer subjects by means of standard nonlinear regression of (2.43) (assuming $V_1$ to have the value zero) to measurements of the time courses of tracer accumulation in arterial blood and a region of interest in the brain ("putamen") yielded the estimates listed in Table 2.2. The estimates of $V_e = K_1/k_2$ (2.51) and $p_B = V_e\,[1 + (k_3/k_4)]$ (2.52) were calculated from the in-

**Table 2.2** Tracer raclopride compartmental relaxation constants ($n = 13$) for binding to dopamine $D_{2,3}$ receptors in putamen of human neostriatum (Gjedde and Wong 1990)

| Constant | $K_1$ | $k_2$ | $k_3$ | $k_4$ | $V_e$ | $p_B$ |
|---|---|---|---|---|---|---|
| Unit | $ml\ g^{-1}\ min^{-1}$ | $min^{-1}$ | $min^{-1}$ | $min^{-1}$ | $ml\ g^{-1}$ | ratio |
| Mean estimate | 0.16 | 0.49 | 0.37 | 0.100 | 0.34 | 3.8 |
| Standard error | 0.01 | 0.03 | 0.02 | 0.008 | 0.02 | 0.2 |

**Table 2.3** Tracer raclopride binding potential at dopamine $D_{2,3}$ receptors in human striatum, estimated by multiple methods

| Analysis | Regression | | Differentiation | | Temporal transformation | | Linearization | |
|---|---|---|---|---|---|---|---|---|
| Equation | 2.43 | 2.253 | 2.208 | 2.210 | 2.243 | 2.243 | 2.256 | 2.257 |
| Material | Individual subjects ($n = 13$) | | | | | Population average | | |
| Mean estimate | 3.8 | 3.2 | 3.3 | 3.5 | 3.0 | 2.9 | 3.3 | 3.3 |
| Standard error | 0.2 | 0.1 | 0.1 | 0.2 | 0.1 | 0.01 | 0.02 | 0.01 |

dividual estimates of the clearance and relaxation constants. When calculated from the average estimates, the binding potential ($p_B$) had the value 3.5, indicating some noise-dependent bias also of the conventional nonlinear regression analysis. It may be appropriate to regard this value as the unbiased "gold" standard.

### Dopamine Receptor Binding Potential

Other analyses of tracer [$^{11}$C]raclopride binding to dopamine receptors in the putamen region of the neostriatum of the healthy volunteer subjects yielded slightly different values of the binding potential at the dopamine $D_2$ and $D_3$ receptors, to which raclopride binds. The results of the different approaches are shown in Table 2.3. The table confirms that estimates vary according to the regression method used to obtain the results. Compared to the previously identified "gold" standard of a binding potential of 3.5, the Differentiation and Linearization methods, the latter applied to population averages, yield the closest results.[3]

### Comprehensive Model of Dopaminergic Neurotransmission

To examine the ability of compartmental modelling to reveal the role of individual relaxation constants on the regulation of the turn-over of molecules in a specific system of brain biochemistry, we have combined the known elements of dopaminergic neurotransmission in the striatum of the brain. The model consists of the following compartments: The tyrosine compartment being fed from the blood stream by the blood–brain barrier facilitated diffusion transporter, the dopa compartment being fed by tyrosine hydroxylase and decaying both to the blood stream, to 3-*O*-methyldopa, and to the main conduit in the form of dopa decarboxylase to dopamine. The intracellular dopamine department is also fed by the dopamine transporters linking the extracellular and intracellular dopamine compartments. One fraction of the intracellular dopamine is transported into and concentrated in the vesicles and subsequently

[3] "Individual subjects" and "population average" refer to results averaged from regression to 13 individual observations ("individual subjects"), or single result from population averaged observation ("population average"). Averaging removes noise and leads to substantially but unrealistically lowered standard errors.

**Table 2.4** Relaxation constants of dopamine turnover

| Enzyme or transporter | Relaxation constant | Estimate ($min^{-1}$) |
|---|---|---|
| Tyrosine hydroxylase (*secondary flux generator*) | $k_{TH}$ | 0.005 |
| DOPA decarboxylase | $k_{DDC}$ | 1 |
| Monoamine oxidase | $k_{MAO}$ | 20 |
| Dopamine vesicular transporter | $k_{VET}$ | 50 |
| Dopamine release (*primary flux generator*) | $k_{rel}$ | 0.015 |
| Dopamine diffusion from synapse | $k_D$ | 125 |
| Dopamine plasma membrane transporter | $k_{DAT}$ | 1,000 |
| LNAA transporter (Tyrosine) | $k_{AT_{TH}}$ | 0.06 |
| LNAA transporter (DOPA) | $k_{AT_{DOPA}}$ | 0.05 |
| LNAA transporter (3-*O*-Methyl-DOPA) | $k_{AT_{3OMD}}$ | 0.2 |
| Catechol-*O*-Methyl-transferase (DOPA) | $k_{COMT_{DOPA}}$ | 0.03 |
| Catechol-*O*-Methyl-transferase (DOPAC) | $k_{COMT_{DOPAC}}$ | 0.03 |
| Choriod plexus anion transporter | $k_{CSF}$ | 0.04 |

released to the intrasynaptic cleft, whence the dopamine diffuses to the extrasynaptic space. From the extrasynaptic space it returns to the intracellular space by means of the dopamine transporters' facilitated diffusion. Another fraction of the intracellular dopamine is converted to DOPAC by the monoamine oxidase enzyme. DOPAC in turn is 3-*O*-methylated to homovanillic acid (HVA).

At steady-state, the system is fully described by two out of three variables, e.g., the relaxation constants and the flux through the system, which together define the compartment contents. To calculate the compartment contents, and thus to describe the system in full, tracer measurements of relaxation constants and net flux through the dopamine metabolic pathway were combined to yield the missing compartment contents. The relaxation constants applied to the analysis of dopamine turnover are listed in Table 2.4, and the resulting compartment contents and concentrations are listed in Table 2.5.

The values of the relaxation constants were acquired or calculated on the basis of their definitions. Note that the flux generating steps have relaxation constants of the lowest magnitude of the steps which constitute the pathway. In the definitions later, $f_b{}'$ symbolizes the bidirectionality of the flux, assuming the value of unity for unidirectional flux and the value of zero for a net flux of zero. The symbol $\sigma$ refers to the occupancy of the substance, and the symbol $\varsigma$ to the occupancy of a noncompetitive activator in the cases of active transport or an agonist in the cases of receptor binding.

**Tyrosine** The large neutral amino acid (LNAA) transporter of the blood–brain barrier transports tyrosine by facilitated diffusion from the circulation. The transporter is 95% saturated with the many LNAA in the circulation. However, this competition does not affect the dopamine turnover, as most of tyrosine continues into proteins, and as the entry into the monoamine synthetic pathways is regulated by the activity

**Table 2.5** Contents and concentrations of dopaminergic metabolites

| Metabolite | Content (nmol $g^{-1}$) | Concentration (μM) | Volume (ml $g^{-1}$) |
|---|---|---|---|
| Tyrosine | 56 | 190 | 0.300 |
| DOPA | 0.3 | 0.4 | 0.800 |
| 3OMD | 0.06 | 0.075 | 0.800 |
| Free intracellular DA | 0.015 | 2 | 0.0075 |
| Vesicular DA | 48 | 60, 000 | 0.0008 |
| Intrasynaptic DA | 0.006 | 7.5 | 0.0008 |
| Extrasynaptic DA | 0.00075 | 0.005 | 0.150 |
| Total DA | 48 | | |
| DA dissociation constant ($K_{d_{DA}}$) | | 30 | |
| DA maximum binding capacity ($B_{max}$) | 0.03 | | |
| Bound DA | 0.006 | | |
| DOPAC | 10 | | |
| HVA | 7.5 | | |

of tyrosine hydroxylase (TH), which is saturated with tyrosine (2.198). The relaxation constant hence is inversely proportional to the content of tyrosine,

$$k_{TH} \cong \frac{V_{max}^{TH}}{M_{Tyr}} \tag{2.258}$$

which in brain tissue is close to 60 nmol $g^{-1}$ in brain tissue. The rate of dopamine turnover has been calculated to be 0.3 nmol $g^{-1}$ $min^{-1}$, equal to the $V_{max}$ of TH. Thus, the magnitude of $k_{TH}$ is close to 0.005 $min^{-1}$ (Cumming et al. 1998) (Fig. 2.25).

**DOPA** DOPA (di-hydroxy-phenylalanine) is the product of tyrosine hydroxylation. DOPA is the substrate of aromatic amino acid decarboxylase, also called DOPA decarboxylase (DDC), as well as of the large neutral amino acid transporter of the blood–brain barrier and the enzyme catechol-*O*-methyltransferase (COMT). DDC is unsaturated and the DDC reaction is irreversible, with the relaxation constant,

$$k_{DDC} \cong \frac{V_{max}^{DDC}}{V_{DOPA} K_M^{DDC}}. \tag{2.259}$$

As the enzyme is unsaturated, it is difficult to determine its true activity in vivo. Instead, the relaxation constant of DDC in mammalian brain has been measured directly with labeled substrates of the enzyme in vivo and its magnitude varies from region to region, but the results are suspected of being low because of loss of tracer. In striatum, the measured value ranges from 0.05 to 0.1 $min^{-1}$ but the true value may be as high as 1 $min^{-1}$ (Cumming et al. 1995). In addition to the decarboxylation, DOPA is also subject to transport by the LNAA transporter. The relaxation constant of this process, $k_{AT}$, averaged 0.05 $min^{-1}$ in vivo (Cumming et al. 1993),

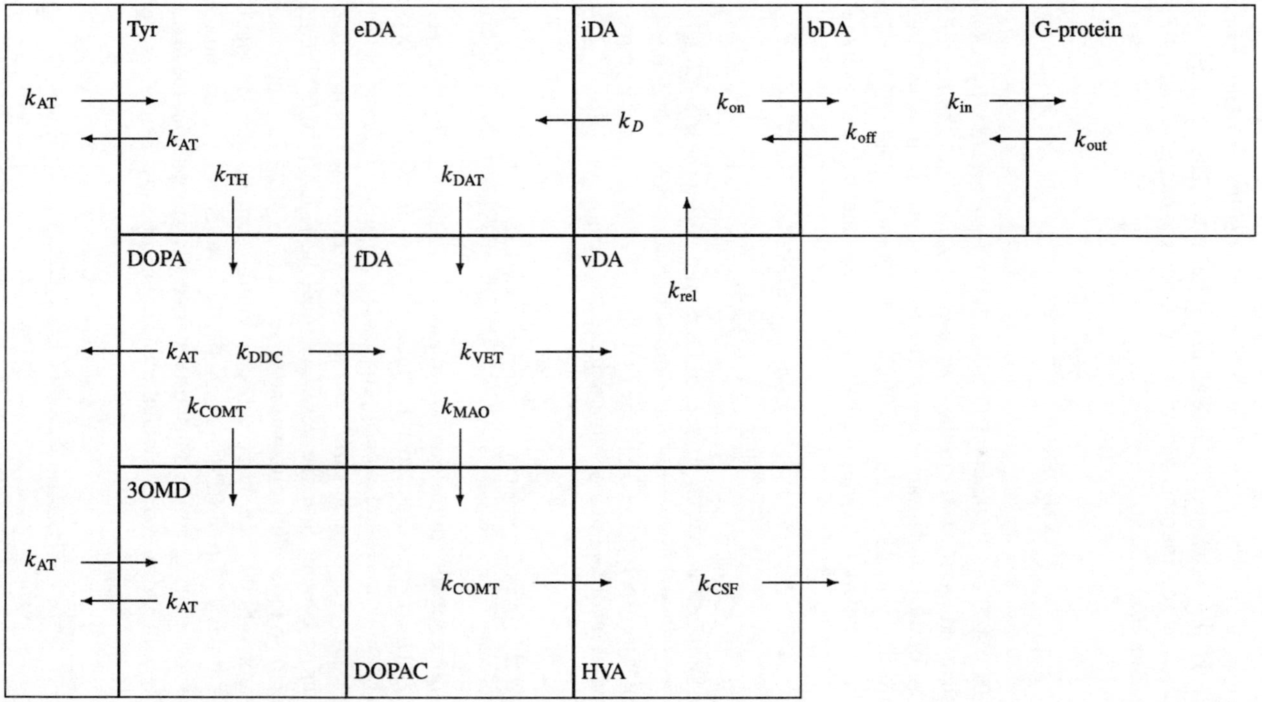


**Fig. 2.25** Model of neurotransmitter compartments separated by enzymes (TH, COMT, DDC, and MAO), transporters (AT, VET, and DAT), vesicular release (REL), and intra- to extrasynaptic permeability by passive diffusion ($D$)

$$k_{\mathrm{AT(DOPA)}} = \frac{f_b{}' \, \sigma_{\mathrm{DOPA}}^{\mathrm{LNAA}} T_{\max}^{\mathrm{LNAA}}}{M_{\mathrm{DOPA}}}, \tag{2.260}$$

where $f_b{}'$ is close to unity because of the unidirectional efflux of DOPA from brain tissue. Finally, DOPA is subject to 3-*O*-methylation in the liver and other tissues including brain. The relaxation constant of this process averages about 0.03 $\mathrm{min}^{-1}$ in vivo (Cumming et al. 1993),

$$k_{\mathrm{COMT(DOPA)}} = \frac{\sigma_{\mathrm{DOPA}}^{\mathrm{COMT}} V_{\max}^{\mathrm{COMT}}}{M_{\mathrm{DOPA}}}. \tag{2.261}$$

**3-*O*-methyl-DOPA** The 3-*O*-methyl-DOPA (3OMD) generated by COMT in tissues is exchanged with 3OMD in the circulation by means of the neutral amino acid transporter of the blood–brain-barrier. The concentrations of 3OMD in plasma and tissues are likely to be similar because of the comparatively high relaxation constant of the transporter, $k_{\mathrm{AT(3OMD)}}$, which has been measured to average about 0.2 $\mathrm{min}^{-1}$ (2.194),

$$k_{\mathrm{AT(3OMD)}} = \frac{f_b{}' \, \sigma_{\mathrm{3OMD}} \, T_{\max}^{\mathrm{LNAA}}}{M_{\mathrm{3OMD}}}. \tag{2.262}$$

**Intracellular Dopamine** As the product of DDC and dopamine transporter action, intracellular dopamine (fDA) is the junction of two metabolic paths, one involving the vesicular transporter (VET) in the membranes of vesicles, driven by proton antiport, the other involving monoamine oxidase (MAO),

$$k_{\mathrm{MAO}} = \frac{\sigma_{\mathrm{fDA}}^{\mathrm{MAO}} V_{\max}^{\mathrm{MAO}}}{M_{\mathrm{fDA}}}, \tag{2.263}$$

where the unidirectionality of the reaction renders $f_b{}'$ equal to unity. The relaxation constant of MAO has been estimated as the ratio between the $V_{\max}$ and the $K_{\mathrm{M}}$ in tissue homogenates and was found to be approximately 20 $\mathrm{min}^{-1}$ (Azzaro et al. 1985). To match the known dopamine turnover rate of 0.3 nmol $\mathrm{g}^{-1}$ $\mathrm{min}^{-1}$, the intracellular dopamine content (outside vesicles) must be equal to 0.015 nmol $\mathrm{g}^{-1}$, corresponding to an intracellular dopamine concentration of 2 μM. This concentration is consistent with the free intracellular dopamine concentration in giant squid neurons of the order of 1 μM (Chien et al. 1990). Also acting on the intracellular dopamine, the vesicular dopamine transporter actively trasfers dopamine to the vesicles in symport with hydrogen ions, which act as noncompetitive activators (Johnson and Scarpa 1979),

$$k_{\mathrm{VET}} = \frac{f_b{}' \, \sigma_{\mathrm{fDA}}^{\mathrm{VET}} \, \varsigma_{\mathrm{H+}}^{\mathrm{VET}} \, T_{\max}^{\mathrm{VET}}}{M_{\mathrm{fDA}}}. \tag{2.264}$$

The relaxation constant of the vesicular transporter (VET) was calculated from the rate of release of dopamine from the vesicles. To be consistent with a lower limit of dopamine release from vesicles (see later), $k_{\mathrm{VET}}$ must average at least 50 $\mathrm{min}^{-1}$, acting on the intracellular content of 0.015 nmol $g^{-1}$ for a flux of 0.75 nmol $g^{-1}$ $min^{-1}$. Then, the total influx of dopamine into the free intracellular compartment must be at least 1.05 nmol $g^{-1}$ $min^{-1}$, of which an amount of 0.75 nmol $g^{-1}$ $min^{-1}$ reenters the vesicles and an amount of 0.3 nmol $g^{-1}$ $min^{-1}$ undergoes monoamine oxidation.

**Vesicular Dopamine** Most of the tissue dopamine, 48 nmol $g^{-1}$, is believed to reside in vesicles in the dopaminergic terminals. Quanta of the vesicular dopamine are released to the intrasynaptic space in response to action potential arrival. The relaxation constant of the release of dopamine from vesicles in the baseline state is the ratio between the dopamine release rate of 0.75 nmol $g^{-1}$ $min^{-1}$ and the vesicular dopamine (vDA) content of 48 nmol $g^{-1}$ (see (2.111)),

$$k_{\mathrm{rel}} = \frac{J_{\mathrm{vDA}}^{\mathrm{rel}}}{M_{\mathrm{vDA}}}, \tag{2.265}$$

where the baseline state is considered the lower limit of the half-life of total dopamine in the tissues with active dopaminergic neurotransmission. This halflife is close to 45 min, (Cumming et al. 1999), corresponding to a rate constant of 0.015 $min^{-1}$.

**Intrasynaptic Dopamine** Intrasynaptic dopamine (iDA) is balanced among the processes of vesicular release, receptor binding, and passive diffusion to the extrasynaptic space. The passive diffusion is driven by the concentration gradient established by the dopamine transporters. The compartment decays by diffusion to the extrasynaptic space, with the relaxation constant,

$$k_D = \frac{A\sqrt{kD}}{V_{\mathrm{iDA}}}, \tag{2.266}$$

where $k$ is the rate constant of the concentration dependent removal of the dopamine, catalyzed by the dopamine transporter facing the extrasynaptic space (see later), $D$ the dopamine diffusion coefficient, $A$ the area of the interface between the intra- and extrasynaptic compartments, and $V_{\mathrm{iDA}}$ is the volume of the intrasynaptic compartment. The dopamine content of the compartment was assumed to be 6 pmol $g^{-1}$ in agreement with the binding potential of unity derived from one study of dopamine binding (see later), corresponding to a relaxation constant of 125 $min^{-1}$ for a baseline dopamine flux of 0.75 nmol $g^{-1}$ $min^{-1}$.

**Extrasynaptic Dopamine** Extrasynaptic dopamine (eDA) is a weighted average of diffusion profiles established by the clearance of dopamine from the intrasynaptic space and facilitated diffusion mediated by the dopamine transporters surrounding the extrasynaptic space. The dopamine transport actively transports dopamine in proportion to the sodium ion gradient. The sodium ion concentration provides the noncompetitive activation which raises the maximum transport capacity in proportion to the sodium ion concentration, as expressed operationally by the difference between the magnitudes of the sodium ion occupancy ($\varsigma_{\mathrm{Na}^+}$) on the two sides of the membrane. The relaxation constant of this compartment is (2.138) and 2.194),

$$k_{\mathrm{DAT}} = f_b{}' \frac{A\sqrt{kD}}{V_{\mathrm{eDA}}} \left[ \frac{1 - \mathrm{e}^{-V_{\mathrm{eDA}}\sqrt{k/D}/A}}{1 - f_b{}' \left(1 - \frac{1 - \mathrm{e}^{-[V_{\mathrm{eDA}}\sqrt{k/D}/A]}}{V_{\mathrm{eDA}}\sqrt{k/D}/A}\right)} \right], \tag{2.267}$$

where $1 - f_b{}'$ is the ratio between the occupancies of the transporter ligand on the two sides of the transporting membrane at the onset of diffusion, i.e., at the interface between the intra- and extrasynaptic spaces (see (2.138)), $A$ is the area of the interface between the intra- and extrasynaptic compartments, $V_{\mathrm{eDA}}$ the volume of the extrasynaptoc space, $D$ the dopamine diffusion coefficient in the extrasynaptic space, and $k$ is the relaxation constant of the facilitated transport, defined as (2.128),

$$k = \frac{\overline{P}'_{\mathrm{DAT}} S}{V_{\mathrm{eDA}}} \tag{2.268}$$

and equal to (2.183),

$$k = \frac{\overline{f_b}' \, \overline{\sigma}_{\mathrm{eDA}} \, \overline{\varsigma}_{\mathrm{Na}^+} \, T_{\max}}{M_{\mathrm{eDA}}} = \frac{J_{\mathrm{DAT}}}{M_{\mathrm{eDA}}}, \tag{2.269}$$

where $\overline{f_b}'$ is the weighted average magnitude of $f_b{}'$, $\overline{\sigma}_{\mathrm{eDA}}$ the weighted average extracellular dopamine occupancy of the transporter, $\overline{\varsigma}_{\mathrm{Na}^+}$ the weighted average occupancy of the noncompetitive activator ($\mathrm{Na}^+$), and $M_{\mathrm{eDA}}$ is the weighted average extrasynaptic dopamine content. The magnitude was calculated to be 1,000 $\mathrm{min}^{-1}$ from the relationships derived later. Because $1 - f_b{}'$ refers to the ratio between the concentrations of free dopamine in the dopaminergic neurons and dopamine in the intrasynaptic space, $f_b{}'$ must be close to unity. When $f_b{}' = 1$, and the steady-state volume ratio $V_{\mathrm{eDA}}\sqrt{k/D}/A = V_{\mathrm{eDA}}/(A\, l_{\mathrm{D}}$ is represented by the symbol $\lambda$, (2.267) reduces to,

$$k_{\mathrm{DAT}} = \lambda \left( \frac{A\sqrt{kD}}{V_{\mathrm{eDA}}} \right) = \frac{\lambda\, K_o}{V_{\mathrm{eDA}}}, \tag{2.270}$$

where $\lambda$ is a measure of how much larger the extracellular space is than the extrasynaptic volume reached by dopamine diffusion and hence an index of the degree to which the dopamine transport raises the gradient of the diffusion of dopamine in the extracellular space by a factor, which is also the reciprocal of the fraction of the extracellular space reached by dopamine diffusing from the sites of release.

**Receptor-Bound Dopamine (bDA)** The dopamine occupancy of dopamine's receptors is a function of the intrasynaptic concentration and the receptor dissociation constants, and the amount of bound dopamine is the product of the maximum binding capacity and the occupancy (2.165),

$$\rho_{B_{\mathrm{iDA}}} = \frac{\sigma'_{\mathrm{iDA}}\, B_{\max}^{\mathrm{DAR}}}{M_{\mathrm{iDA}}}, \tag{2.271}$$

where $\sigma'_{\mathrm{iDA}}$ symbolizes the increased binding due to the higher affinity caused by the binding of GTP-free G-protein. The binding of dopamine to G-protein linked receptors acts as a competitive activation of the G-protein binding, which in turn activates second-messenger systems in the cells (2.177) as in Gjedde and Wong (2001) and Cumming et al. (2002). The average normal dopamine occupancy was assumed to be 20%. With equal amounts of bound and free dopamine, 6 pmol $g^{-1}$ (Borbely et al. 1999), the binding potential is unity.

**Dihydroxy-phenyl-acetic acid** Dihydroxy-phenyl-acetic acid (DOPAC) is the product of the reaction of dopamine with monoamine oxidase. In turn, DOPAC is the substrate of an irreversible reaction catalyzed by COMT. The relaxation constant of the COMT reaction was argued earlier to be close to 0.03 $\mathrm{min}^{-1}$ in vivo (Cumming et al. 1993).

$$k_{\mathrm{COMT(DOPAC)}} = \frac{\sigma_{\mathrm{DOPAC}}\, V_{\max}^{\mathrm{COMT}}}{M_{\mathrm{DOPA}}}. \tag{2.272}$$

**Homovanillic Acid** Homovanillic acid (HVA) is the product of the reaction between DOPAC and COMT. HVA is cleared from brain tissue by anionic transport across the choroid epithelium, at a rate estimated to be 0.04 $\mathrm{min}^{-1}$ (Cumming et al. 1992),

$$k_{\mathrm{CSF}} = \frac{f_b'\, \sigma_{\mathrm{HVA}}^{\mathrm{CSF}}\, \varsigma_{\mathrm{OH^-}}^{\mathrm{CSF}}\, T_{\max}^{\mathrm{CSF}}}{M_{\mathrm{HVA}}}. \tag{2.273}$$

### Dopamine Turnover

**Normal Dopamine Turnover** To determine the magnitudes of the relaxation constants and compartments participating in the dopamine turnover, it is necessary to measure the relaxation constants or contents, or both, in vivo. However, not all constants or contents are known yet, in humans or other mammals. The relationships can be used to infer values, which subsequently undergo experimental verification. The values listed in Tables 2.4 and 2.5 are presented as internally consistent examples, to be adjusted in the future, with the aid of the following ratios. The steady-state sum of unbound dopamine in dopaminergec neurons, relative to tyrosine, is,

$$\frac{M_{\mathrm{DA}}}{M_{\mathrm{Tyr}}} = \left(\frac{k_{\mathrm{DDC}}}{k_{\mathrm{MAO}}}\right)\left[\frac{k_{\mathrm{TH}}}{k_{\mathrm{DDC}} + k_{\mathrm{AT}} + k_{\mathrm{COMT}}}\right]\left(1 + \frac{k_{\mathrm{VET}}}{k_{\mathrm{rel}}} + \frac{k_{\mathrm{VET}}}{k_D} + \frac{k_{\mathrm{VET}}}{k_{\mathrm{DAT}}}\right), \tag{2.274}$$

where the free intracellular dopamine is given by:

$$M_{\mathrm{fDA}} = \left(\frac{k_{\mathrm{DDC}}}{k_{\mathrm{MAO}}}\right)\left[\frac{k_{\mathrm{TH}}}{k_{\mathrm{DDC}} + k_{\mathrm{AT}} + k_{\mathrm{COMT}}}\right] M_{\mathrm{Tyr}} = f_{\mathrm{f}}\, M_{\mathrm{Tyr}}, \tag{2.275}$$

the vesicular dopamine by,

$$M_{\mathrm{vDA}} = \left(\frac{k_{\mathrm{DDC}}}{k_{\mathrm{MAO}}}\right)\left[\frac{k_{\mathrm{TH}}}{k_{\mathrm{DDC}} + k_{\mathrm{AT}} + k_{\mathrm{COMT}}}\right]\left(\frac{k_{\mathrm{VET}}}{k_{\mathrm{rel}}}\right) M_{\mathrm{Tyr}} = f_{\mathrm{v}}\, M_{\mathrm{Tyr}}, \tag{2.276}$$

and the intrasynaptic dopamine is given by:

$$M_{\mathrm{iDA}} = \left(\frac{k_{\mathrm{DDC}}}{k_{\mathrm{MAO}}}\right)\left[\frac{k_{\mathrm{TH}}}{k_{\mathrm{DDC}} + k_{\mathrm{AT}} + k_{\mathrm{COMT}}}\right]\left(\frac{k_{\mathrm{VET}}}{k_{D}}\right) M_{\mathrm{Tyr}} = f_{\mathrm{i}}\, M_{\mathrm{Tyr}} \tag{2.277}$$

such that the bound dopamine is,

$$M_{\mathrm{bDA}} = \frac{B_{\max}\, M_{\mathrm{Tyr}}}{M_{\mathrm{Tyr}} + \left(\frac{k_{\mathrm{MAO}}}{k_{\mathrm{DDC}}}\right)\left[\frac{k_{\mathrm{TH}}+k_{\mathrm{AT}}+k_{\mathrm{COMT}}}{k_{\mathrm{TH}}}\right]\left(\frac{k_{D}}{k_{\mathrm{VET}}}\right) K_{\mathrm{d_{DA}}}} = \frac{B_{\max}\, M_{\mathrm{Tyr}}}{M_{\mathrm{Tyr}} + (K_{\mathrm{d_{DA}}}/f_{\mathrm{i}})} \tag{2.278}$$

and the extrasynaptic dopamine is,

$$M_{\mathrm{eDA}} = \left(\frac{k_{\mathrm{DDC}}}{k_{\mathrm{MAO}}}\right)\left[\frac{k_{\mathrm{TH}}}{k_{\mathrm{DDC}} + k_{\mathrm{AT}} + k_{\mathrm{COMT}}}\right]\left(\frac{k_{\mathrm{VET}}}{k_{\mathrm{DAT}}}\right) M_{\mathrm{Tyr}} = f_{\mathrm{e}}\, M_{\mathrm{Tyr}} \tag{2.279}$$

which, as confirmation, yields the correct ratio between intra- and extrasynaptic dopamine,

$$\frac{M_{\mathrm{iDA}}}{M_{\mathrm{eDA}}} = \frac{k_{\mathrm{DAT}}}{k_{D}} = \lambda\, \frac{V_{\mathrm{iDA}}}{V_{\mathrm{eDA}}} \tag{2.280}$$

and hence the ratio between the average bconcentrations of intra- and extrasynaptic dopamine,

$$\frac{C_{\mathrm{iDA}}}{\overline{C}_{\mathrm{eDA}}} = \lambda = 1,500, \tag{2.281}$$

where $\lambda$ is,

$$\lambda = \frac{L}{l_{\mathrm{D}}} = \sqrt{\frac{\overline{\sigma}_{\mathrm{eDA}}\, \varsigma_{\mathrm{Na^+}}\, T_{\max}}{\overline{C}_{\mathrm{eDA}}\, DA/L}} = \sqrt{\frac{J_{\mathrm{DAT}}}{\overline{C}_{\mathrm{eDA}}\, DA/L}} = \sqrt{\frac{D_{\mathrm{app}}}{D}} \tag{2.282}$$

according to which the ratio rises and falls with the occupancy of the dopamine transporter and the length of the diffusion path. The value of lambda suggests that the dopamine transporter accelerates the rate of dopamine removal from the extrasynaptic space by a factor of $2.25 \times 10^6$.

**Inhibition of Dopamine Transporter** Inhibition of dopamine transport will raise both intra- and extrasynaptic dopamine. The change depends on the degree of inhibition of the transporter and hence on the lowering of the apparent dopamine diffusion from the intrasynaptic pool, according to (2.282). A lowering of the effective transport rate to 1% of the normal average will raise intrasynaptic dopamine by a factor of 10 and extrasynaptic dopamine by a factor of 100, as well as extending the extracellular space reached by dopamine by a factor of 10 as well.

# Chapter 3
# Analysis of Neuroreceptor Binding In Vivo*

## 3.1 The Receptor Concept

A number of biologically important drugs and chemicals that act in the brain do so specifically, that is, by association with membrane proteins designed to receive endogenous neurotransmitters and neuromodulators, although the endogenous ligands may not always be known. The existence of chemical receptors was suspected before the turn of the century in the form of Langley's concept of a "receptive" substance (Langley 1878) and Ehrlich's "side-chain" theory (Ehrlich 1909), but modeling did not begin until the discovery of the binding of oxygen to specialized proteins, which made direct observation of receptor–ligand interaction possible. When the binding of oxygen to myoglobin was first observed to be a function of the oxygen tension in the solution (Bohr 1885), the relationship between the quantities of bound and free oxygen corresponded exactly to the relationship between the invertase reaction velocity and substrate concentration later derived by Michaelis and Menten (1913). Bohr (1904) published an equation for the interaction between hemoglobin and oxygen that antedated the equation suggested by Hill (1910) for receptor sites that interact cooperatively.

Receptors can be studied in vitro in tissue samples or in tissue slices, using the in vitro autoradiographic method of Young and Kuhar (1980) for regional differentiation of binding at the resolution offered by light microscopy. In the in vitro study, the ligand is applied to the receptors after the death of the subject and fixation of the tissue. The autoradiographic method represents a major technical advance, and has resulted in the synthesis of ligands labeled with radioactive isotopes that cover innumerable different receptor and neurotransmitter systems in the rat brain and almost as many in the human brain. As a result of these studies, receptor changes have been implicated in several diseases.

In vitro, neuroreceptor findings may be difficult to relate to functional states that existed prior to the death of the subject. Fortunately many ligands can also be studied

*Adapted from Gjedde and Wong (1990) Compartmental analysis, in *Quantitative Imaging: Neuroreceptors and Neurotransmitters*, edited by J. James Frost and Henry N. Wagner, Jr., New York: Raven Press, Ltd. with permission from Raven Press, Ltd.

A. Gjedde et al., *Neurokinetics: The Dynamics of Neurobiology In Vivo*,
DOI 10.1007/978-1-4419-7409-9_3, 

**Table 3.1** Early neuroreceptor radioligands labeled for positron tomography (for a more complete list, see Table 4.3)

| | Ligand | | |
|---|---|---|---|
| Receptor | [$^{11}$C]labeled | [$^{18}$F]labeled | [$^{75}$Br]labeled |
| *Cholinergic* | | | |
| Muscarinic | Dexetimide | | |
| | Ouinuclidinyl benzilate | | |
| *Monoaminergic* | | | |
| $\alpha_1$-Adrenergic | Prazosin | | |
| $\beta$-Adrenergic | Practolol | | |
| | Pindolol | | |
| Dopamine $D_1$ | SCH-23 390 | | |
| Dopamine $D_2$ | Raclopride | Haloperidol | |
| | Spiperone | Spiperone | Bromospiperone |
| | Methylspiperone | Fluoromethylspiperone | |
| | | Fluoroethylspiperone | |
| | | Fluoropropylspiperone | |
| Serotonin $HT_2$ | Methyl-LSD | | Bromo-LSD |
| | Ketanserin | Ritanserin | |
| | Methylketanserin | | |
| *Amino acid* | | | |
| Benzodiazepine | Flunitrazepam | | |
| | Flumazenil (RO-15 1788) | | |
| | PK-11 195 | | |
| *Peptide* | | | |
| Opiate | Etorphine | Acetylcyclofoxy | |
| | Diprenorphine | | |
| | Carfentanil | | |

in vivo, that is, injected into the living subject and allowed to circulate for a specific period of time before determination of brain contents of tracer, either by exposure of brain sections to x-ray film or by external detection of radiation. Table 3.1 lists some of the early radioligands available for positron emission tomography, but the list has grown rapidly. For a more recent list, see Table 4.3.

Studies with labeled ligands in vivo often yield images that closely match regional distributions obtained in vitro. Unlike the in vitro images, however, the in vivo distribution is time-dependent. At early times, the distribution reflects delivery of the drug by the circulation, and at later times the label is released by the receptors and cleared from the brain. In vitro, the analysis of neuroreceptor-ligand interaction is carried out with high specific activity and excess ligand. In vivo, the unbound tracer obscures the bound tracer when the specific activity is low, or the receptors "empty" the free ligand pool when the specific activity is very high and the mass of the radioligand low, yielding an image continuously reflecting delivery by the circulation or transport across the blood–brain barrier.

Thus, attempts to apply in vitro methods of analysis to ligand binding in vivo are frustrated by important kinetic differences between the in vivo and in vitro situations. These differences have created a need for in vivo receptor kinetic modeling of positron tomographic data. The presentation below essentially is a synthesis of the kinetic models developed for this purpose by Sokoloff et al. (1977), Gjedde (1982, 1987), Lassen and Gjedde (1983), Mintun et al. (1984), Gjedde et al. (1986), Wong et al. (1986a, b [JCBFM]), and Logan et al. (1987). Kinetically, the most intensely studied receptors are the dopamine $D_2$-like sites in neostriatum, first imaged by Wagner et al. (1983) and Wong et al. (1984). They will be discussed as a model system from which the principles relevant to the analysis of other receptor systems can be derived.

The measurement of the accumulated tracer in the brain is often called "residue" detection, as opposed to the older methods of "inflow–outflow" detection, in which arterial and venous concentrations were measured. PET is one of the most advanced techniques of "residue" detection, but other techniques exist, notably single-photon emission computed tomography (SPECT), in which the single photon emission from selected isotopes is recorded tomographically by modified rotating gamma cameras. Many of the principles discussed later, therefore, also apply to SPECT.

## 3.2 The Compartment Concept

### *3.2.1 Compartmental Analysis*

Several organs have been studied by PET, including the brain, heart, liver, and lungs, but other organs may also qualify for this particular method of examination. In conventional kinetic models, an organ consists of a number of compartments corresponding to the different states of a tracer. The compartments reflect the fate of the tracer and represent a specific theory of the biochemistry of an organ. The concept of compartments can be ascribed to Sheppard (1948), who defined compartments as volumes, real or kinetic, in which the concentration of a tracer or its derivatives everywhere remains the same. In kinetic terms, Rescigno and Beck (1972a, b) interpreted this definition to mean that a compartment must obey the expression,

$$\frac{\mathrm{d}m}{\mathrm{d}t} = j(t) - km, \tag{3.1}$$

where $m$ is the quantity of tracer in the compartment, $k$ a constant, and $j(t)$ an arbitrary flux as a function of time. The concentration of the tracer is determined as the ratio between $m$ and the volume in which it is dissolved. In this context, a tracer is any compound that can be radiolabeled. All concentration gradients are placed at the interfaces between compartments. Normally, these interfaces are cell membranes or chemical reactions involving transporter, receptor, or enzyme proteins.

The purpose of the kinetic analysis of tracer uptake is to measure the size of the compartments and the velocity of exchanges between the compartments. The number and definition of compartments relevant to each tracer must be known before attempts can be made to quantify the tracer exchanges between compartments.

The individual barriers defining a series of rate-limiting steps can be distinguished only when their resistances are not too different, because resistant barriers tend to obscure less resistant barriers. Thus, by analyzing the brain uptake of suitable tracers as a function of time, only a very limited number of compartments can be analyzed.

### 3.2.2 The Basic Equation

The transfer of a tracer, applied at time zero, from one compartment to another observes a simple mathematical expression of conservation of mass. The basis of the formalism is the use of transfer coefficients ($k$) that are less rigidly defined than in classical compartmental analysis. The transfer coefficients operate on concentrations ($c$), measured in volumes ($V$). In combination, the basic variables define clearances ($K = kV$), masses ($m = Vc$), or fluxes ($j = kVc$). The variables are sometimes numbered consecutively, unlike the transfer constants of traditional compartmental analysis, which are numbered on the basis of their destination and origin. In the present description, variables belonging to general cases will be numbered and variables belonging to specific cases will be lettered.

Although $m$ symbolizes a mass, it can also conveniently be thought of as a measured quantity of tracer in a sample of brain. This usage is necessitated by the limitation that it is not possible, in vivo, to determine tracer concentrations per unit water volume.

According to (3.1), the fundamental expression of exchange of tracer between the two compartments of a closed system must be

$$\frac{\mathrm{d}m_2}{\mathrm{d}t} = k_1 V_1 c_1 - k_2 V_2 c_2, \tag{3.2}$$

where $m_2$ is the net gain of tracer in compartment 2 as a function of the time $t$. $V_1$ and $V_2$ are the actual volumes of compartments 1 and 2; and $k_1$ and $k_2$ are the descriptive transfer coefficients. The quantity of reference is the unit of brain tissue under study, determined by weighing or measuring the volume of the sample. This convention has profound consequences for the interpretation of the coefficients because they will be affected by the relative sizes of the compartments of the model. The interpretation of $k_1$ and $k_2$ depends on the particular process in question and must be based on the definition of the compartments analyzed in the study, as must the interpretation of V and c.

The symbols $c_1$ and $c_2$ in (3.2) represent concentrations of the tracer or its metabolites that retain the label in the respective compartments. The symbols,

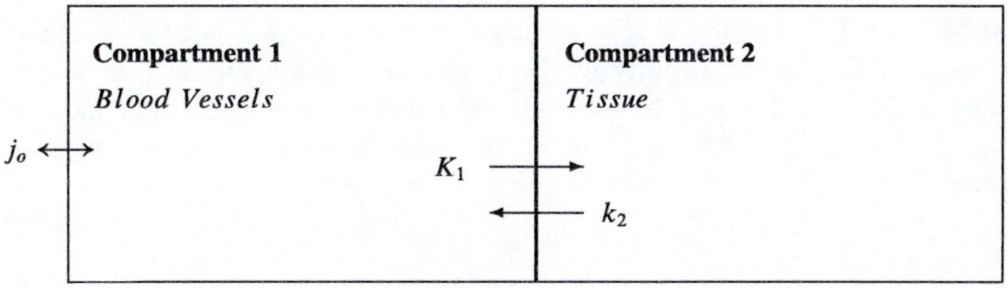


**Fig. 3.1** Model of two compartments exchanging contents consisting of a product ($M_2$) and a precursor ($M_1$) with the concentration $C_1$, replenished at the rate of $j_o$, as described by (3.3)

therefore, need not refer to the same chemical species. The product of $k_1$ and $V_1$ defines a clearance $K_1$, which reflects perfusion when there are no barriers to diffusion of the tracer between the circulation and the compartment (Fig. 3.1). The product of $V_2$ and $c_2$ defines the mass $m_2$.

### *3.2.3 The Basic Solution*

Experiments in which only a single measurement of the dependent variable is performed are referred to as "autoradiographic" because they emulate the method of quantitative autoradiography in which the radioactivity in the brain is determined only once, at the termination of the experiment. Experiments during which the dependent variable is determined several times are often referred to as "dynamic," and those are the experiments that readily lend themselves to the various forms of transient analysis discussed later.

Only one coefficient can be estimated by autoradiography because the solution of (3.2) has no degrees of freedom when only one measurement of the dependent variable is available. The general solution to (3.2) is

$$m_2 = \mathrm{e}^{-k_2 T}\left[m_2(0) + K_1 \int_0^T c_1 \mathrm{e}^{-k_2 t}\mathrm{d}t\right], \tag{3.3}$$

which is an example of a convolution integral. In principle, (3.3) is a transcendental equation that can be solved for $K_1$ only by iteration and may have zero, one, or two solutions. Special cases include $c_1 = 0$ ("wash-out") and $m_2(0) = 0$ ("fill-in").

The purpose of imaging in vivo is normally the quantitation of one or more of the coefficients in (3.2). In actual practice, one or more independent variables and one dependent variable are measured as functions of time and the desired coefficients estimated by regression analysis, using a suitable computerized optimization program. The science of regression analysis is still incomplete, however, and therefore dangerously seductive because the analysis almost always yields results but often without regard to the meaning of the estimates. "Fitting" is only meaningful if the validity of the model is independently established, and it is usually logically impossible to decide the validity of the model and obtain the best estimates of the coefficients in the same session. Several solutions presented later have been derived within the last few years to make "multidimensional" linear regression possible.

## 3.3 Two-Compartment (Permeability) Analysis

### 3.3.1 *Analysis of $K_1$ and $k_2$*

In most organs, capillaries are no barrier to the entry of small polar solutes. In the brain, however, they do constitute such a barrier, as noted earlier. In the brain, the concentration difference of many tracers between the two sides of the endothelium is initially so great that the endothelium often is the only significant barrier that these tracers meet in the brain. For these tracers, the brain is an organ of only two compartments, the vascular compartment and the tissue space, separated by the blood–brain barrier.

It is the purpose of the two-compartment analysis to estimate the unidirectional clearance ($K_1$) from compartment 1 to compartment 2 in the brain, a process that occurs inside the brain where the residues of compartments 1 and 2 cannot be detected separately. The exact interpretation of the physiological or pathophysiological meaning of $K_1$depends on the known properties of the tracer in relation to the permeability of the endothelium, blood flow to the regions of the brain, and binding of the tracer to proteins or other agents in blood plasma.

**Positron Tomographic Solution** Equation (3.2) can be integrated to yield $m$, the sum of $m_1$ and $m_2$, when $m_1(0) = 0$,

$$m = V_\mathrm{a} c_\mathrm{a} + (K_1 + k_2 V_\mathrm{a}) \int_0^T c_\mathrm{a}\,\mathrm{d}t - k_3 \int_0^T m\,\mathrm{d}t, \tag{3.4}$$

where $c_\mathrm{a}$ is the arterial concentration, measured in whole blood or a subcompartment of blood; and $V_\mathrm{a}$ is the vascular volume of distribution of the tracer in the brain, equal to the plasma water volume in the brain when the tracer concentration $c_\mathrm{a}$ is actually distributed and measured only in plasma water. The coefficients of (3.4) can be obtained by "four-dimensional" linear regression in which the dimensions include three independent variables (i.e., $c_\mathrm{a}$ and the two integrals) and one dependent variable (i.e., $m$),

$$y = p_1 x_1 + p_2 x_2 + p_3 x_3, \tag{3.5}$$

where $p_i$ denotes the coefficient of the independent variable $x_i$.

**Autoradiographic Solution** For in vivo "autoradiographic" experiments, (3.2) can be solved directly for $K_1$,

$$K_1 = \frac{m + k_2 \int_o^T m_e \, dt - V_a c_a}{\int_0^T c_a \, dt}, \tag{3.6}$$

where $m_e$ is the total mass of tracer in the extravascular water volume, equal to $V_2 c_2$. According to (3.6), the calculation of $K_l$ on the basis of a single measurement of $M$ requires both negligible backflux and a known or negligible product of $V_a$, and $c_a$. The product of $V_a$ and $c_a$ can be measured with a second tracer that remains in compartment 1, but backflux cannot easily be ruled out on the basis of single measurement, as discussed earlier.

**Normalized Solution** Alternatively, with the same number of experiments and a single tracer, both $K_1$ and $V_a$ can be estimated by terminating the experiments at different times after administration of the tracer. In experimental animals, results of a series of experiments can be compared only after normalization of the individual observations. Normalization is the division of the radioactivity in the brain by the arterial radioactivity concentrations of tracer obtained in individual experiments,

$$V = \frac{m}{c_a} = V_a + K_1 \frac{\int_0^T c_a \, dt - \frac{\int_0^T m_e \, dt}{V_e}}{c_a}, \tag{3.7}$$

where the $m/c_a$ ratio represents the measured volume of distribution and $V_e$ is the measured volume of the extravascular pool, equal to the $K_1/k_2$ ratio. The equation can be further simplified to,

$$V = V_a + K_1 \Theta - k_2 \frac{\int_0^T m_e \, dt}{c_a} \tag{3.8}$$

in which $\Theta$ indicates the $\int_o^T c_a/c_a$ ratio. The variable $\Theta$ has unit of time and represents a modified time variable. It is apparent that $V$ remains a linear function of $\Theta$ for as long as the magnitude of the $\int_o^T m_e \, dt / V_e$ ratio is negligible compared to the magnitude of $\int_0^T c_a dt$. The requirement for absent backflux is a very large value of $V_e$ (i.e., very low tissue-blood concentration ratios). Equation (3.8) underlies the so-called slope-intercept or Gjedde–Patlak plot (Gjedde 1981b; Patlak et al. 1983).

It may be worthwhile to note that $V_e$ is different from the actual volume of compartment 2 ($V_d$) unless concentrations in the arterial samples have been measured in plasma water, and only in the absence of protein or other nonspecific binding in brain tissue, or facilitated diffusion of the tracer between blood and brain. It is an empirical observation that (3.8) often can be approximated satisfactorily by:

$$V - V_a \approx V_e \left( l - e^{-K_1 \Theta / V_e} \right) \tag{3.9}$$

particularly when the object of the study is the estimation of $V_e$. In principle, (3.9) is a monotonically rising function. Hence, significant decline of $V$ as a function of $\Theta$ at later times signifies the accumulation of labeled compounds in the circulation that do not pass into the brain. The decline may thus be used to estimate the magnitude of metabolic conversion of the tracer to nonpenetrating products (see later). Thus, when $T$ and $\Theta$ approach infinity,

$$V(\infty) \rightarrow \frac{K_1}{k_2} + V_a = V_e + V_a \tag{3.10}$$

which confirms that $V_e$ is the steady-state extravascular volume of distribution of the tracer, relative to the arterial concentration.

### 3.3.2 Physiological Definitions of $K_1$ and $k_2$

Although easy to define, endothelial permeability is measurable only as an indirect index of "clearance" ($K_1$). The quantity of tracer transported from blood to brain in a given period of time is a function of $P$ (often given in units of centimeters per second or nanometers per second and thus akin to a velocity of transfer through the barrier); $S$, the area of endothelial surface available for transport (measured in square centimeters per gram or per milliliter); and the integral of the concentration difference across the endothelium or barrier.

Unfortunately, the value of $K_1$ does not reflect the magnitude of the permeability-surface area product in a simple manner, because the tracer concentration declines in the capillary as the tracer travels from the arterial to the venous end. The definitive derivation of the relationship between transendothelial clearance, perfusion ($F$), and the permeability-surface area product (PS) of the endothelium was made by Bohr (1909) during his examination of the oxygen diffusion capacity of the pulmonary endothelium,

$$\frac{\mathrm{PS}}{F} = -\alpha \ln \frac{\Delta c_v}{\Delta c_a}, \tag{3.11}$$

where $\alpha$ is the solubility of the tracer in the sample of blood relative to that in tissue and plasma water (serves as a correction for protein binding or other sequestration in blood), $\Delta c_v$is the concentration difference between tissue and blood at the venous exit from the microvascular bed holding the endothelium, and $\Delta c_a$ is the concentration difference at the arterial entry to the microvascular bed. If $\alpha$ is not unity, and the tracer clearance is determined initially when tissue concentrations can be assumed to be negligible, Crone (1963) showed that (3.11) is equivalent to,

$$\mathrm{PS} = -\alpha F \ln(1 - E_o), \tag{3.12}$$

where $E_o$, the so-called "first pass" extraction fraction, is the $K_1/F$ ratio. In theory, (3.12) is valid only for extraction fractions measured in the presence of negligible tissue concentrations, which is also the requirement for linearity of the slope-intercept plot above. When the colloquial "first pass extraction" expression is used, negligible tissue concentration is assumed (Gjedde 1983). The reason is that the capillary concentration of the tracer falls exponentially when the transport remains unidirectional. Only when extravascular concentrations remain negligible do (3.6) and (3.12) together yield the expression,

$$K_1 = F\left(1 - e^{-PS/[\alpha F]}\right) = \frac{m - m_a}{\int_0^T c_a \, dt}, \tag{3.13}$$

where $F$ is the flow rate of the solvent in which the tracer is dissolved. It follows from the definition of the steady-state distribution volume $V_e$ that the definition of $k_2$ is,

$$k_2 = \frac{K_1}{V_e} = \frac{F}{V_e}\left(1 - e^{-PS/[\alpha F]}\right) \tag{3.14}$$

as derived by Lassen and Gjedde (1983). The measured permeabilities of the blood–brain barrier to hydrophilic substances are low, of the order of 1–10 nm/s. Most substrates of brain metabolism are hydrophilic and few lipophilic substances exist in solution in plasma water (for obvious reasons). However, facilitated diffusion allows hydrophilic nutrients to cross much more readily than do inert polar nonelectrolytes.

## 3.4 Three-Compartment (Binding) Analysis

### 3.4.1 Analysis of $k_3$ and $k_4$

Certain tracers interact in some manner with the proteins of brain tissue to escape the build-up in the interstitial fluid that leads to backflux and eventually to steady-state. The kinetic equivalent of this interaction is "trapping" of the tracer.

Trapping can be described as a process imposed by a hypothetical additional compartment that reflects transport (e.g., into the intracellular space), binding (e.g., to receptor sites), or metabolism (e.g., by chemical reactions that sequester the label for a shorter or longer time). In the compartmental analysis, the process is represented by a transfer coefficient ($k_3$) that kinetically defines the path to the additional compartment, and a transfer coefficient ($k_4$) that kinetically defines the path away from this compartment. Entry into the third compartment competes with the likelihood of transfer back from the second to the first compartment. This competition is described by two differential equations,

$$\frac{dm_2}{dt} = K_1 c_1 - (k_2 + k_3)m_2 + k_4 m_3 \tag{3.15}$$

and

$$\frac{dm_3}{dt} = k_3 m_2 - k_4 m_3, \tag{3.16}$$

where $m_2$ and $m_3$ represent the exchangeable (compartment 2) and trapped (compartment 3) quantities of tracer (Fig. 3.2).

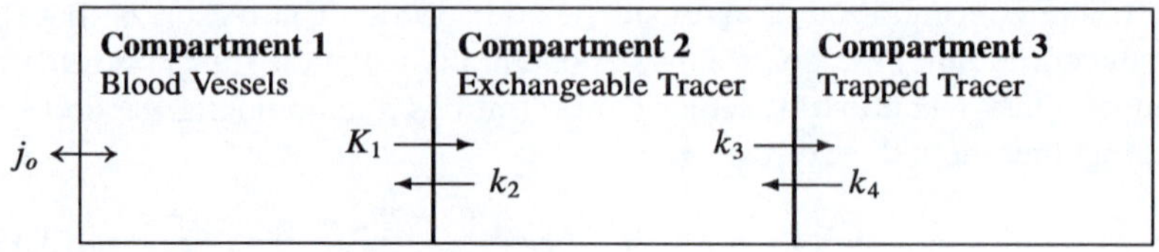


**Fig. 3.2** Closed system of three compartments

**Positron Tomographic Solution** The integrated equation for $m$, the total content of labeled material in the tissue, including the original tracer and its derived forms, has been extended to three compartments ($k_4 = 0$) by Blomqvist (1984) and further extended to include $k_4$ by Evans (1987),

$$m = V_a c_a + (K_1 + [k_2 + k_3 + k_4]\ V_a) \int_0^T c_a \, dt + (K_1[k_3 + k_4] + k_2 k_4 V_a) \int_0^T \left[ \int_0^t c_a \, du \right] dt - (k_2 + k_3 + k_4) \int_0^T m \, dt - k_2 k_4 \int_0^T \left[ \int_0^t m \, du \right] dt, \tag{3.17}$$

where $m$ is the sum of $m_1$, $m_2$, and $m_3$ in the brain, and $V_a$ and $c_a$ represent $V_1$ and $c_1$, respectively. The equation yields the coefficients by six-dimensional linear regression, in which the six dimensions include the five independent variables (i.e., $c_a$, the integrals, and the integrals of integrals) and the one dependent variable (i.e., $m$). By means of numerical integration, the calculation of the integral variables is so simple that the entire regression can be performed with a hand calculator,

$$y = p_1 x_1 + p_2 x_2 + p_3 x_3 + p_4 x_4 + p_5 x_5, \tag{3.18}$$

where $p_i$ denotes the coefficient of independent variable $x_i$.

**Autoradiographic Solution** Equations (3.15) and (3.16) can be solved for $K$, the net clearance of tracer into the third compartment, equal to the product of $k_3$ and $V_f$ where $V_f$ is the distribution volume $K_l/(k_2 + k_3)$,

$$K = \frac{m - \frac{K_l}{k_2 + k_3}\left(m_e - k_4 \int_0^T m_m \, dt\right) - V_a c_a}{\int_0^T c_a \, dt}, \tag{3.19}$$

where $m_e$ is the quantity of tracer in the extravascular, pool of exchangeable tracer, representing $m_2$; and $m_m$ is the quantity of tracer in the pool of metabolic or other trapping, representing $m_3$. The ratio $k_2/(k_2 + k_3)$ defines an "escape potential," $r_e$. From the definitions of $V_e$, $V_f$, and $r_e$, it follows that $V_f = r_e V_e$.

The following variations of the solution are of special interest: First, the magnitude of $k_3$ may be negligible compared to the magnitude of $k_2$. This is the two-compartment situation discussed earlier. Second, $k_3$ may be of the same order of magnitude as $k_2$. Backflux and binding are then equally probable. On the average, half of the tracer molecules enter the third compartment and the remaining half return to compartment 1 (e.g., the vascular compartment). The rate of net transfer to the third compartment is only 50% of the rate of initial uptake. The fraction ("entrapment") that net uptake represents of the rate of initial uptake is $1 - r_e$, where $r_e$ is the "escape potential" referred to earlier. Third, it is possible that the magnitude of the coefficient $k_3$ is much greater than the magnitude of the coefficient $k_2$. The last case permits only two-compartment analysis because the value of $k_3$ is too high to retain any influence over the accumulation of tracer in the brain, the regulation of which is relegated to the interface between compartments 1 and 2 (e.g., the blood–brain barrier).

Solution of (3.15) and (3.16) for negligible $k_4$ yields:

$$K = \frac{m - r_e m_e - m_a}{\int_0^T c_a \,dt} \tag{3.20}$$

where $m_a$ is the radioactivity in the vascular space in brain, equal to the product of $c_a$ and $V_a$. This is the solution first introduced by Sokoloff et al. (1977) for deoxyglucose and discussed in detail by Gjedde (1982, 1987).

**Normalized Solution** The complete solution of (3.15) and (3.16) for the volume of distribution of the tracer in the brain is,

$$V = \frac{m}{c_a} = K\left(\frac{\int_0^T c_a \,dt - \frac{\int_0^T m_m \,dt}{V_m}}{c_a}\right) + r_e\,\frac{m_e}{c_a} + V_a, \tag{3.21}$$

where $V_m$ is a "metabolic" or "trapping" pool equal to the ratio $(K_1k_3)/(k_2k_4)$.

The higher the value of the $K_1k_3$ product is relative to the $k_2k_4$ product (i.e., the larger $V_m$ is), the lower the probability is of escape of the tracer from the brain. When $V_m$ is sufficiently large (i.e., when $k_4$ is sufficiently small), the equation reduces to

$$V = K\Theta + r_e V_f' + V_a \tag{3.22}$$

where $V_f'$ is the measured (steady or nonsteady state) $m_e/C_a$ratio and $\Theta$ the normalized integral introduced in (3.8). When the ratio between the compound free in the brain and the compound free in the blood no longer changes rapidly, the ratio $m_e/c_a$ eventually approaches $V_f$. The graphical representation of this case is a straight line. The symbol $K$ represents the slope of the line and $r_e V_f' + V_a$ is the ordinate intercept (see "slope-intercept" or "Gjedde–Patlak" plot above). The case of irreversible trapping of the tracer N-isopropyl-p-[$^{123}$I] iodoamphetamine (Kuhl et al. 1982) in brain is shown in Fig. 3.3.

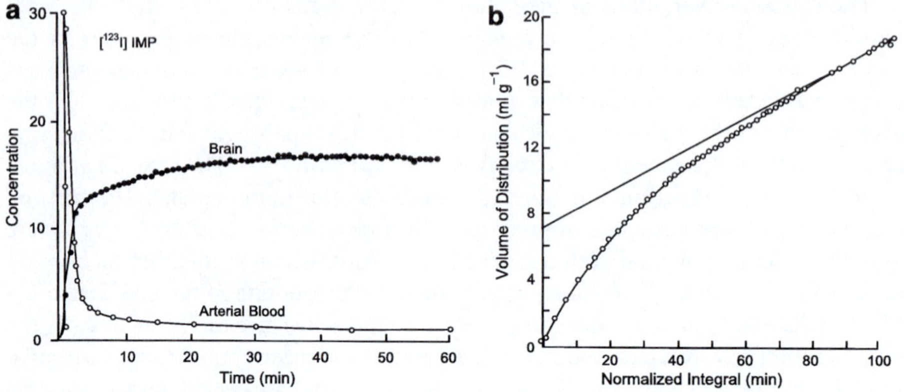


**Fig. 3.3** Trapping of N-iso-propyl-p-[$^{125}$I]iodoamphetamine (IMP) in human brain, determined by SPECT. (**a**) Time-activity curves in arterial blood (*open circles*) and brain (*filled circles*). (**b**) Same data, volume of distribution plotted vs. normalized time-concentration integral of tracer in arterial blood (*open circles*). Note irreversible trapping, either by metabolism or association with receptors (Re-drawn from Kuhl et al. 1982)

For values of $\Theta$ for which $k_4$ is still effectively nil, it is an empirical observation that (3.22) often can be simulated by:

$$V - V_\mathrm{a} \approx K\Theta + r_\mathrm{e} V_f \left(1 - \mathrm{e}^{-K_1\Theta/V_f}\right) \tag{3.23}$$

or,

$$V - V_\mathrm{a} \approx r_\mathrm{e} V_\mathrm{e} \left(k_3\Theta + r_\mathrm{e}\left[1 - \mathrm{e}^{-K_1\Theta/(r_\mathrm{e}V_\mathrm{e})}\right]\right) \tag{3.24}$$

such that, if divided by the volume of distribution in a region of no binding (3.9), the ratio eventually reaches the linear relationship,

$$\frac{V^{(1)} - V_\mathrm{a}}{V^{(2)} - V_\mathrm{a}} \approx r_\mathrm{e} k_3 \Theta + r_\mathrm{e}^2 \tag{3.25}$$

which is the basis for the so-called ratio method, which derives the value of $k_3$ from the slope and the intercept of the relationship between the ratio of the measured volumes of distribution in a region of binding ($V^{(1)}$) and a region of no binding ($V^{(1)}$).

When $k_4$ is not negligible, "trapping" becomes "binding" in which the tracer and its derivatives eventually must approach an equilibrium defined by the four coefficients of (3.15) and (3.16). At steady-state, the "equilibrium volume" of distribution is

$$V(\infty) \to \frac{K_1}{k_2}\left(1 + \frac{k_3}{k_4}\right) + V_\mathrm{a} = V_\mathrm{e} + V_\mathrm{m} + V_\mathrm{a}, \tag{3.26}$$

where the $K_1/k_2$ ratio is the volume $V_e$ and the $k_3/k_4$ ratio is the $V_m/V_e$ ratio. The $k_3/k_4$ ratio is a convenient measure of the tissue's availability of binding sites or "binding potential" ($p_B$), imposed by the presence of the binding sites (Mintun et al. 1984).

### 3.4.2 *Molecular Definitions of $k_3$ and $k_4$*

**The Michaelis–Menten Equation** Empirically, it has been found that the kinetics of neuroreceptor–radioligand interaction often is remarkably simple and identical to the Michaelis–Menten kinetic analysis of enzymatic reactions. The chief criteria of pure receptor–ligand interaction are the same as for enzyme–substrate interaction and facilitated diffusion and include saturability, stereospecificity, and competitive inhibition.

The Michaelis–Menten equation is the equilibrium solution of the differential equation describing the change per unit time of the quantity of ligand bound to the receptor sites. The differential equation is identical to (3.2) modified for binding of a ligand to a receptor. The simplest model of the binding process is in principle a two- or three-compartment model in which the ligand solution represents compartment 1, the receptor–ligand complex compartment 2, and the internalized ligand or otherwise inactivated receptor–ligand complex compartment 3. The change of quantity of bound ligand equals the difference between ligand associating with the receptor and ligand dissociating from the receptor. The association rate is the probability of each ligand molecule joining the receptor ($k_{on}$), multiplied by the number of molecules per unit volume and the number of receptor sites. The dissociation rate is the combined likelihood of dissociation by return to the ligand solution ($k_{off}$) or internalization through the membrane ($k_{in}$), multiplied by the number of receptor–ligand complexes,

$$\frac{d\mathrm{B}}{\mathrm{d}t} = k_{on}c\,(B_{max} - B) - (k_{off} + k_{in})B, \tag{3.27}$$

where $B_{max}$ is the quantity of sites available for binding, $B$ the quantity of receptor ligand complexes ("bound ligand"), and $c$ the concentration of ligand molecules in the solution. (In deference to usage, capital $B$ is used for a time-varying quantity here.) At equilibrium, the number of receptor–ligand complexes is constant in time (i.e., $d\mathrm{B}/\mathrm{d}t = 0$):

$$B = \frac{B_{max}\,C}{C + \left[\frac{k_{off}+k_{in}}{k_{on}}\right]}, \tag{3.28}$$

where the internalization or inactivation rate is $k_{in}B$. When $K_M$ replaces $[k_{off} + k_{in}]/k_{on}$, we obtain the Michaelis–Menten equation,

$$B = \frac{B_{max}C}{K_M + C}, \tag{3.29}$$

where $K_M$ is the Michaelis half-saturation concentration. Note that $C$ must have units of concentration, whereas $B$ and $B_{max}$ can be expressed in any unit of quantity of bound ligand. Also note that $K_M$ is a strict dissociation constant ($K_D$) only when $k_{in}$ is negligible compared to $k_{off}$. Hence, studies of receptor sites (for which $k_{in}$ is low) often indicate a much higher affinity between the hypothetical receptor and the ligand than described for facilitated diffusion (for which $k_{in}$ by necessity is substantial).

**The Linearized Equations** Since the Michaelis–Menten equation is a hyperbola, a nonlinear approach is needed to estimate $K_D$ and $B_{max}$ and their bias and standard errors. However, without the benefit of computerized nonlinear regression in past practice, only linear equations with two parameters were soluble with ease.

Woolf[1] derived most of the linearized forms of the Michaelis–Menten equation, which are now referred to by the names of people who later rediscovered the linearizations (related in Woolf 1951 and Haldane 1957). In receptor kinetics, the most famous of Woolf's linearizations are the so-called Scatchard and Eadie–Hofstee plots, which were reinvented no less than four times (Eadie 1942; Scatchard 1949; Hofstee 1954–1956; Rosenthal 1967). Woolf showed that (3.29) can be linearized to a rearrangement that later specifically came to be referred to as the Eadie–Hofstee plot (Eadie 1942; Hofstee 1954–1956),

$$B = B_{max} - K_D \left( \frac{B}{C} \right) \tag{3.30}$$

in which $B_{max}$ is the ordinate intercept and $-K_D$ is the slope. The ratio $B/c$ is familiar from the "bound-to-free" ratio determined in experiments on binding in vitro. Equation (3.30) shows that it also equals the "binding potential" introduced by Mintun et al. (1984). Woolf is credited with his "own" plot, the Woolf, Woolf-Hanes, or Hanes-Woolf plot,

$$\frac{c}{B} = \frac{C}{B_{max}} + \frac{K_D}{B_{max}} \tag{3.31}$$

in which the slope equals $1/B_{max}$ and the abscissa-intercept is $-K_D$ when $c/B$ is plotted vs. $c$. Nonlinear regression is still incompletely understood by some, to whom it may therefore be useful to understand the merits of each of the linearized equations (see Cressie and Keightley 1981).

**The Binding Equation In Vivo** The definitions of $k_3$ and $k_4$ follow from (3.27), which governs the association and dissociation of the ligand-receptor complex in the presence of ligand and inhibitor(s). However, the true ligand concentration is unknown in vivo. Instead, only the content of ligand in the exchangeable pool, $m_e$, can be estimated. Equation (3.27) must be modified accordingly:

[1] Barnet Woolf, MA, PhD (cantab), FRSE (1902–1983), biochemist, playwright, songwriter, poet, labor activist and founding member in 1920 of the Communist Party of Great Britain, resigned from the party in 1949; from 1926 at the Biochemical Laboratory of Cambridge University.

$$\frac{d\mathrm{B}}{\mathrm{d}t} = k_{\mathrm{on}} \left( \frac{B_{\mathrm{max}} - (B + B_i)}{V_{\mathrm{d}}} \right) m_{\mathrm{e}} - k_{\mathrm{off}}\, B, \tag{3.32}$$

where $k_{\mathrm{on}}$ is the bimolecular association rate constant, $k_{\mathrm{off}}$ the dissociation rate, $B_{\mathrm{max}}$ the maximal ligand binding capacity, and $V_{\mathrm{d}}$ is the volume of solvent in the tissue in which the ligand is dissolved. Provisionally, this volume is defined as the ratio between the content of free ligand and the concentration of free ligand in tissue water. By definition, at all times $M_e(t) = V_{\mathrm{d}}\, c(t)$, but at equilibrium $m_{\mathrm{e}}(\infty) = V_{\mathrm{d}}\, c(\infty) = V_{\mathrm{e}}\, c_{\mathrm{a}}(\infty)$.

The volume represented by $V_{\mathrm{d}}$ is one of the most difficult entities to define correctly, and the significance of its uncertainty will be discussed further later. The term $B$ represents the quantity of bound ligand, and $B_i$ is the total quantity of bound inhibitors, both assumed constant in the present analysis. This constancy is not strictly true, however, if the ligand concentration $c$ changes significantly.

The difference between $B_{\mathrm{max}}$ and as $B_i$ has been designated $B'_{\mathrm{max}}$ to indicate the number of receptor sites available for binding of the ligand in question. By simple comparison of (3.32) with (3.15) and (3.16), the transfer coefficients $k_3$ and $k_4$ can be identified as follows:

$$k_3 = k_{\mathrm{on}}(B'_{\mathrm{max}} - B)/V_{\mathrm{d}} \tag{3.33}$$

and

$$k_4 = k_{\mathrm{off}} \tag{3.34}$$

but these equations suffer from the disadvantage that $k_3$ is variable when the magnitude of $B$ is significant compared to the magnitude of $B'_{\mathrm{max}}$, i.e., when saturation is significant, in which case the equations cannot be solved. To circumvent this apparent dilemma, $k_3$ and $k_4$ are redefined such that $k_3$ combines all positive terms in the binding equation, and $k_4$ all the negative terms, (i.e., the assumption of $B \lll B'_{\mathrm{max}}$ is isolated). This typically occurs when the tracer mass occupies a significant fraction of the available receptors (i.e., in PET/SPECT studies of tracer with an insufficiently high specific activity), in which case the equations cannot be solved. To circumvent this apparent dilemma, $k_3$ and $k_4$ are redefined such that $k_3$ combines all positive terms in the binding equation, and $k_4$ all the negative terms,

$$k_3 = k_{\mathrm{on}}\, B'_{\mathrm{max}}/V_{\mathrm{d}} \tag{3.35}$$

and

$$k_4 = k_{\mathrm{off}} + k_{\mathrm{on}} \frac{M_e}{V_{\mathrm{d}}} = k_{\mathrm{off}} \left(1 + \frac{c}{K_D}\right), \tag{3.36}$$

where $m_{\mathrm{e}}$ is the quantity (mass) of unbound tracer available for binding. Because $k_3$ and $k_4$ are arbitrary assignations of the different components of the differential equations, it appears that $k_3$ is independent of the injected dose of tracer, whereas $k_4$ is a function of the free tracer concentration and therefore not a constant when $m_{\mathrm{e}}$ is not constant or negligible. This definition of $k_4$ is not new to receptor kineticists; it is known as the "observed $k_{\mathrm{obs}}$" used to calculate $k_{\mathrm{on}}$ when $k_{\mathrm{off}}$ has been determined.

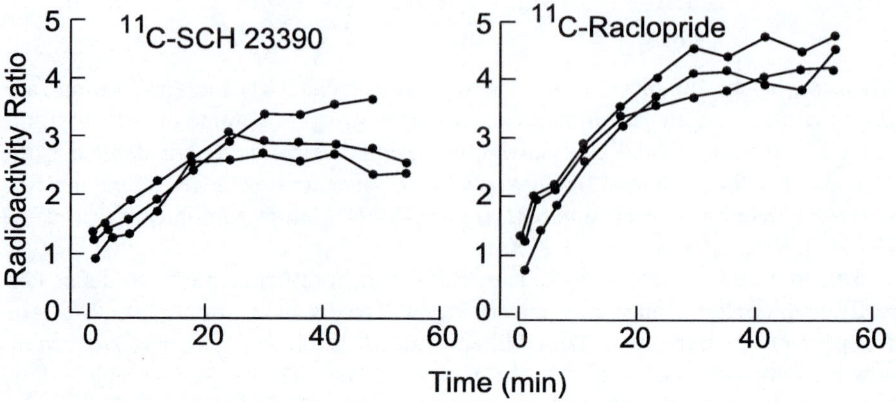


**Fig. 3.4** Approach to steady-state volume of distribution of two neuroreceptor ligands. Distribution in neostriatum was recorded relative to distribution in cerebellum and thus represents cases of the ratio method. *Left*: Dopamine $D_1$ receptor ligand SCH-23390. *Right*: Dopamine $D_2$ receptor ligand raclopride. Redrawn from Farde et al. (1987)

According to (3.35) and (3.36), the $k_3/k_4$ ratio represents the ratio between $B'_{max}/V_d$ and $K_D+c$ at steady-state (e.g., when $c$ attains a constant level). This ratio is the unitless "binding potential" $p_B$ discussed earlier (Mintun et al. 1984). Note that the magnitude of $p_B$ depends on the volume of distribution and concentration of the ligand. The higher the concentration of the ligand, the lower the magnitude of $p_B$. Thus, during the approach to equilibrium, the measured distribution space $V$, equal to the ratio $[m_a + m_e + m_b]/c$, rises to the steady-state value $V_h = V(\infty)$,

$$V_h = V(\infty) \rightarrow V_e\left(1 + \frac{B'_{max}}{V_d\,(K_D + c(\infty))}\right) + V_a, \tag{3.37}$$

where $V_a$ is the relevant vascular volume. Figure 3.4 gives two examples of radioligands that eventually reach steady-state. In PET it is often possible to do several studies in the same subject, and these studies may therefore be repeated with different total doses of radioligand. The highest ligand concentrations reduce the binding potential to zero and the equilibrium volume to its minimum value, equal to the partition coefficient $V_e + V_a$ of the ligand. By nonlinear regression, the equation allows estimates to be made of $V_e + V_a$, $B'_{max}$, and the $k_{off}/k_{on}$ ratio, or $K_D$.

### *3.4.3 Inhibition*

The quantity $B'_{max}$ represents the number of receptors available for occupation by the labeled ligand. If the receptor is blocked by a competitor, the total quantity of receptor sites in the tissue $B'_{max}$ is reduced by a factor that depends on the receptor's affinity for the competitor,

$$B'_{\max} = B_{\max} \frac{K'_I}{K'_I + C_{\mathrm{I}}}, \tag{3.38}$$

where $B_{\max}$ is the maximal quantity of bound ligand, $K'_I$ is the competitor's apparent half-saturation inhibition constant ("$IC_{50}$"), and $C_{\mathrm{I}}$ is its concentration in brain water. The apparent inhibition constant is the affinity of the competitor in the presence of the ligand (Dixon 1953),

$$K'_I = K_I \left(1 + \frac{C}{K_D}\right), \tag{3.39}$$

where $c$ is the free ligand concentration and $K_I$ is the dissociation constant of the ligand. Both of these equations follow from simple inhibition kinetics when an inhibitor competes with the ligand for binding. Note that $B_{\max}$ is defined only when $C$ and $C_{\mathrm{I}}$ are constant or negligible. Thus, the quantity of bound inhibitor is also a function of the ligand concentration,

$$B_I = \frac{B_{\max}\, C_{\mathrm{I}}}{C_{\mathrm{I}} + K_I \left(1 + \frac{C}{K_D}\right)} \tag{3.40}$$

and $K'_I$ can be a constant only when $c$ is negligible or invariant. This condition also applies to $B'_{\max}$ and $k_3$.

Blockade is the prior occupancy of the receptor by the blocking agent, unlike saturation, which is the significant occupation of the receptor by the labeled ligand during the course of the study, or displacement, which is the clearance of the labeled ligand from the receptor by an unlabeled competitor or by the unlabeled ligand itself. The effect of haloperidol blockade of N-[$^{11}$C-methyl]spiperone trapping in striatum in vivo is shown in Fig. 3.5.

The measure of blockade is saturation, defined as the ratio between bound inhibitor and the maximal binding capacity at equilibrium. In a receptor-controlled brain function, occupancy of the receptor by an agonist is the factor responsible for the magnitude of the response. Occupancy is a function of the affinity and concentrations of the ligands that compete for occupation. When a single competitor contributes to the occupancy, the degree of saturation by the inhibitor can be calculated from the ratio:

$$s = \frac{C_{\mathrm{I}}}{K'_I + C_{\mathrm{I}}} = 1 - \frac{k_3^{(I)}}{k_3}, \tag{3.41}$$

where $k_3^{(I)}$ is the value of $k_3$ measured in the presence of inhibition. The degree of saturation is also an index of the degree of occupancy established by any pharmacological agent bound to the receptor.

The apparent volume of distribution of the ligand ($V_f$) increases when binding is blocked because $k_3$ declines as less tracer is removed from the free ligand pool.

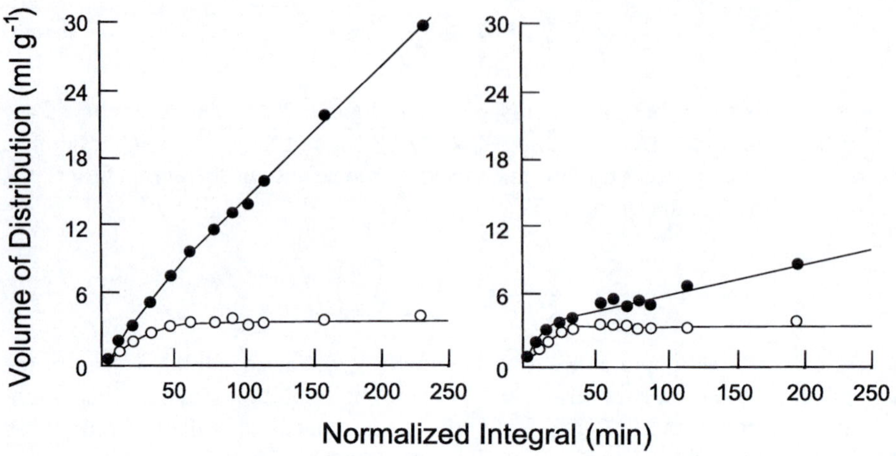


**Fig. 3.5** Detail of Fig. 1.3: N-methyl-spiperone (NMSP) binding to neostriatum (*closed circles*) and cerebellum (*open circles*) plotted as volumes-of-distribution vs. normalized integral in normal human volunteer. *Left*: Before haloperidol blockade. *Right*: After 90% blockade of NMSP binding sites in neostriatum by haloperidol at 2.5 ng/ml in plasma. Redraw from Wong et al. (1986b)

### *3.4.4 The Problem of Solubility and Nonspecific Binding*

Nonspecific binding refers to the sequestration of some fraction of the tracer by attachment to "sites" that cannot be saturated in the range of concentrations available for study. Depending on the mechanism involved, nonspecific binding can also be manifested as a change of solubility. In this sense, nonspecific binding is all binding to sites of sufficiently low affinity and high capacity to escape saturation at the chosen ligand concentrations. Some nonspecific sites may also be saturable, but nonspecific binding will be defined here as binding to sites that cannot be saturated and are of such low affinity that equilibrium is "instantaneous" for all practical purposes.

The nonspecific binding simulates an expansion of distribution space for free ligand such as to render the concentration of free ligand in tissue water lower than suspected from the sample content. The degree of nonspecific binding is practically impossible to determine under nonsteady-state conditions because the ratio between free and nonspecifically bound tracer changes as a function of time.

At equilibrium, nonspecific binding can be simulated by an expansion of the equilibrium distribution volume of the type expressed in (3.26),

$$V_{\mathrm{d}} = V_{\mathrm{w}}(1 + p_{\mathrm{NSB}}), \tag{3.42}$$

where $p_{\mathrm{NSB}}$ is the binding potential of nonspecific or low-affinity binding (see Table 3.5 later) that corresponds to the $k_3/k_4$ ratio for specific binding, and $V_{\mathrm{w}}$ is the aqueous volume of distribution measured in the absence of nonspecific binding.

At steady-state, therefore, $C$ must be calculated from $M_e$ by division with $V_d$ rather than $V_w$, and $V_w$ must be replaced by $V_d$ in all formulas in which $M_e$ is the basis for calculation of $C$.

Although the legitimacy of estimating the degree of nonspecific binding at nonsteady-state is weak, as noted, Mintun et al. (1984) did use (3.15) and (3.16) to estimate the degree of nonspecific binding of radiolabeled ligand in the brain on the basis of measurements of free fraction in the plasma in vitro. The reciprocal of the volume of distribution in the brain they denoted by the symbol for a fraction, $f_2$,

$$f_2 = \frac{1}{V_d} \tag{3.43}$$

and the fraction of free ligand in arterial samples was defined by Mintun et al. (1984) as:

$$f_1 = \frac{1}{\alpha}, \tag{3.44}$$

where $\alpha$ is the solubility of the ligand in the sample relative to water (3.11). Thus,

$$f_2 = \frac{f_1}{V_e} \tag{3.45}$$

according to which $f_2$ can be calculated when $f_1$ and $V_e$ are known. The definitions introduce a formal flaw into the solutions of (3.15) and (3.16) because $k_2$ and $k_3$ must vary if $\alpha$ and $V_d$ vary as functions of time. If the ligand distributes only in a fraction of the total tissue water (e.g., only in extracellular fluid), $V_w$ refers only to the volume of water in the extracellular space. Note that $f_1$ is a proper fraction, whereas $f_2$ strictly speaking is the reciprocal of a volume, with units of g/ml or $cm^3$/ml.

Values of $f_1$ and $f_2$ of [$^{18}$F]spiperone were measured in monkey and man by Perlmutter et al. (1986, 1987). The values were 4–6% for $f_1$ and $l-2$% for $f_2$ in monkey, and 5% for $f_1$ and 0.7% for $f_2$ in man. Values for $f_1$ for other radioligands have been reported also to be close to 5%. The effect of plasma protein binding on the measured PS-products of selected radioligands is listed in Table 3.2.

**Table 3.2** Influence of plasma protein binding on apparent permeability of radioligands

| Tracer | $PS$-product (ml/g/min) | Correction for protein binding | References |
|---|---|---|---|
| Spiperone | 2.80 | Yes | Perlmutter et al. (1987)[a] |
| Methylspiperone | 0.22 | No | Wong et al. (1986a)[a] |
| Fluorospiperone | 0.20 | No | Logan et al. (1987)[b] |
| Fluoroethylspiperone | 0.09 | No | Huang et al. (1987)[b] |

[a]Studies in humans
[b]Studies in baboons

### 3.4.5 The Problem of Labeled Metabolites

Labeled metabolites of the tracer may be generated in the circulation and then either (a) pass or (b) not pass into the brain. Labeled metabolites may also be generated in brain tissue, (c) pass back or (d) not pass back into the circulation, and/or (e) bind to the specific sites in question. If the labeled ligand is present at concentrations above tracer concentrations, (f) metabolites may further block the specific sites to a significant extent.

The six forms of interference by metabolites cannot all be completely corrected for. However, a partial correction can be made by consideration of the volume of distribution ($V_e$) recorded in a brain region of no known specific binding or a region appointed to provide baseline binding. Four findings are possible in such a region:

1. The steady-state volume of distribution, $V_e$, declines significantly during the study. Kinetically, the decline is possible only when labeled material continuously appears in the circulation without achieving steady-state distribution in the brain. As first discussed by Wong et al. (1986a), the contribution from metabolites of this nature can be accounted for by mathematically reducing the tracer concentration in the circulation to levels that yield a nondeclining steady-state volume of distribution in a nonbinding region. The presence of metabolites commensurate with this correction must be confirmed by high-performance liquid chromatographic analysis of arterial samples.
2. The steady-state volume of distribution is constant in time. In this case, any labeled metabolites in the circulation and in the brain have achieved an apparent steady-state and thus merely contribute to the magnitude of nonspecific binding estimated by the methods discussed earlier.
3. The steady-state volume of distribution increases with time. This case can be interpreted as the appearance of labeled metabolites in the brain that accumulate in a manner that mimics specific binding. Mathematical reduction of counts in the brain can again be performed to yield a nonincreasing steady-state volume of distribution.
4. Labeled metabolites only bind to specific sites in binding regions. No correction is possible for this case, but the error will chiefly affect the calculation of $K_D$ if the labeled metabolites have a different affinity, or be lumped as "ligand-like" entities if the affinity is the same, in which case no real error is committed.

## 3.5 In Vivo Analysis of Binding

### 3.5.1 Irreversible Binding: Determination of $k_3$

The main distinction observed later is between "reversible" and "irreversible" binding. It refers to the presence or absence of noticeable loss of tracer from the brain during the study. It must be determined in advance whether bound ligand has time to dissociate from binding and escape from the brain, or whether the ligand remains trapped. As explained earlier, irreversibility is the consequence of high values of the product of $V_e$ and the $k_3/k_4$ ratio (the binding potential). It can be predicted from the expected volume of distribution at equilibrium: The higher the equilibrium $M/C_a$ ratio, the lower the likelihood of reversibility in a given period of study.

Steady-state $M/C_a$ ratios of the order of 10 ml/g or more signify irreversible binding during the first hour of uptake. As a rule of thumb, the $V(\infty)$ value in ml/g of any potential radioligand indicates the order of magnitude of the half-time (in minutes) of the approach toward equilibrium. Irreversible trapping is, therefore, a characteristic of the high-affinity binding of certain radioligands (e.g., the substituted butyrophenones spiroperidol and methylspiperone) to dopamine receptors in the basal ganglia.

For this reason, the analysis of irreversible binding does not go beyond the estimation of the magnitude of the binding constant $k_3$. Complete regression analysis of the irreversible binding yields estimates of $K_1, k_2, k_3, K$, and $V_f$. As the irreversible binding, by definition, fails to reach equilibrium in the period of study, it is not logically possible to determine the equilibrium constants $B'_{\text{max}}$ and $K_D$ directly by this procedure. When the association constant $k_{\text{on}}$ is known, the value of $k_3$ can be regarded as an observed index of the receptor density that varies with $B'_{\text{max}}$ and $V_d$. The value of $B'_{\text{max}}$ may vary because the actual receptor number varies, or because some receptors are blocked by endogenous or exogenous competitors. The value of $V_d$ may vary with concentrations of protein or other low-affinity binding sites in the tissue.

Equation (3.41) predicts an inverse proportionality between the concentration of a competitive inhibitor and the value of $k_3$. Wong et al. (1986b) determined the values of $K_1, k_2$, and $k_3$ for the radioligand $N$-methylspiperone (NMSP) in neostriatum and cerebellum (where $k_3 = 0$) in human brain with the use of (3.9) and (3.24). The NMSP binding is blocked by a therapeutic dose of the neuroleptic haloperidol. Unblocked, the net clearance of NMSP ($K$) due to binding averaged was 0.094 ml/g/min, and blocked, the clearance dropped to 0.023 ml/g/min. As listed in Table 3.3, because of displacement of the radioligand from protein binding in plasma by haloperidol, the volume of distribution of NMSP ($V_f$) increased from 1.1 to 2.4 ml/g and thus approached the value of $V_e$ expected in a region of

**Table 3.3** Binding constants for N-[$^{11}$C-methyl]spiperone binding in human caudate in absence and presence of unlabeled haloperidol in vivo

| Variable | Before haloperidol | After haloperidol |
|---|---|---|
| $K_1$ (ml/g/min) | 0.15 | 0.21 |
| $k_2$ (min$^{-1}$) | 0.051 | 0.077 |
| $k_3$ (min$^{-1}$) | 0.087 | 0.010[a] |
| $V_e$ (cerebellum; ml/g) | 3.0 | 2.7 |
| $V_f$ (ml/g) | 1.1 | 2.4[a] |
| $p_{\text{NSB}}$ (relative to cerebellum; ratio) | 4.1 | 0.0[a] |

Adapted from Wong et al. (1986b).
[a]Significantly different (paired $t$-test) at $P < 0.05$

no binding (e.g., cerebellum). The corresponding value of $k_3$ declined from 0.09 to 0.01 per minute after blockade by haloperidol. Thus, at plasma concentrations of 2–3 nM, haloperidol blocked 90% of the NMSP binding sites in neostriatum. A representative example of such an experiment is shown in Fig. 3.5.

Equation (3.41) can be rearranged to yield the formula for the modified "Woolf" plot for two or more competitors of the radioligand, one an exogenously administered inhibitor, and the rest the number of endogenous agonists and antagonists that may interact with the receptors (Gjedde and Wong 2001),

$$\frac{1}{k_3^*} = \frac{V_{\mathrm{I}}}{k_{\mathrm{on}}^* B_{\mathrm{max}}} \left(1 + \sum \frac{c_j}{K_j}\right) + \frac{V_{\mathrm{I}}}{k_{\mathrm{on}}^* \, B_{\mathrm{max}}} \left(\frac{C_{\mathrm{I}}}{K_I}\right), \tag{3.46}$$

where $V_{\mathrm{I}}$ is the steady-state partition of the exogenous inhibitor between brain and plasma and $C_{\mathrm{I}}$ is the steady-state arterial concentration of the inhibitor. Regarding $c_j$ as negligible and assuming a constant value for the $V_{\mathrm{I}}/(k_{\mathrm{on}}^* K_I)$ term ("$D_W$"), Wong et al. (1986b) calculated the values of the $B_{\mathrm{max}}$ and $K_I'$ for the $D_2$ dopamine receptors and haloperidol, respectively, in human striatum from the slope and abscissa intercept of this relationship. The weakness of this indirect approach is the need to assume values for constants that cannot be verified in each experiment.

### 3.5.2 Reversible Binding: Determination of Binding Potential ($p_{\mathrm{B}}$)

Completely irreversible trapping or binding is rare. In most cases, sooner or later, an equilibrium (i.e., $\mathrm{d}c/\mathrm{d}t = 0$) is approached between the quantities of tracer in the circulation, free tracer in the brain, and tracer bound to receptors or converted to metabolites.

The purpose of the kinetic analysis of reversible binding is the estimation of the maximal value of the "binding potential" ($p_{\mathrm{B}}$), which equals the $k_3/k_4$ ratio only in the presence of truly negligible levels of the tracer, that is, when $c(\infty)$ is negligible compared to $K_D$ in (3.37) and (3.47). The values of $k_3$ and $k_4$ can be estimated by linear regression to (3.17). The analysis must therefore be extended to include a significant portion of the approach to equilibrium, although the equilibrium need not be reached completely to estimate the binding potential. Table 3.4 lists values of

**Table 3.4** Transfer coefficients for [$^{18}$F]spiperone

| Coefficient | Baboon (Logan et al. 1987) | Baboon (Perlmutter et al. 1986) | Human (Perlmutter et al. 1987) |
|---|---|---|---|
| $K_1$ (ml/g/min) | 0.15 | 0.11 | 0.12 |
| $k_2$ (min$^{-1}$) | 0.014 | 0.038 | 0.016 |
| $k_3$ (min$^{-1}$) | 0.012 | 0.024 | 0.055 |
| $k_4$ (min$^{-1}$) | 0.002 | 0.012 | 0.020 |

**Table 3.5** Binding potentials at tracer doses of selected radioligands (human and nonhuman primates)

| Ligand | Receptor | $k_3/k_4$ | Reference | Region |
|---|---|---|---|---|
| Methylspiperone | Dopamine $D_{2-4}$ | >10 | Arnett et al. (1986) | Caudate-putamen |
| Bromospiperone | Dopamine $D_{2-4}$ | 6.7 | Crawley et al. (1986) | Caudate-putamen |
| Spiperone | Dopamine $D_{2-4}$ | 6.4 | Logan et al. (1987) | Caudate-putamen |
| Carfentanil | Mu opiate | 3.4 | Frost et al. (1989) | Thalamus |
| Raclopride | Dopamine $D_{2-3}$ | 3.0 | Farde et al. (1987) | Putamen |
| | | 2.9 | Pohjalainen et al. (1998) | Striatum (age 20–40) |
| | | 2.5 | Pohjalainen et al. (1998) | Striatum (age 40–60) |
| | | 2.2 | Pohjalainen et al. (1998) | Striatum (age 60–80) |
| SCH-23390 | Dopamine $D_1$ | 2.0 | Farde et al. (1987) | Putamen |
| Carfentanil | Mu opiate | 1.8 | Frost et al. (1989) | Frontal cortex |
| Methylspiperone | Serotonin $HT_{2A}$ | 1.5 | Wong et al. (1987) | Frontal cortex |
| Haloperidol | Dopamine $D_2$ | 1.1 | Logan et al. (1987) | Caudate-putamen |

the transfer coefficients determined for the radioligand [$^{18}$F]spiperone in baboons and humans. Either from measurements of the transfer coefficients or from measurements of volumes of distribution at steady-state (see later), it is possible to calculate binding potentials for a variety of radioligands. Examples of experimental results of such calculations are given in Table 3.5.

The transfer coefficients cannot be determined by (3.15) and (3.16) when $k_4$ is not a constant. The transfer coefficient ceases to be a constant when the ligand concentration $c$ changes significantly relative to $K_D$ (3.37). Therefore, under ordinary circumstances it is not possible to determine the transfer coefficients and hence the $k_3/k_4$ ratio under nonsteady-state conditions when the ligand concentration is significant, that is, when the radioligand is not really a tracer. However, if the unlabeled ligand can be given so far in advance that its concentration in the circulation and brain tissue is approximately constant, as required by (3.36), and if the plasma concentration of the ligand can be determined, it becomes possible to determine the $k_3/k_4$ ratio at different degrees of saturation of the receptor sites and, hence, to calculate the binding constants. This type of experiment is essentially a blockade with the ligand as its own inhibitor.

There is one additional possibility. Huang et al. (1987) proposed to solve (3.15) and (3.33) directly by a two-step numerical procedure (e.g., Runge-Kutta) in which the six coefficients $V_a$, $K_1, k_2, B_{max}, k_{off}$, and $k_{on}/V_d$ are estimated simultaneously. The first step, during which ligand concentrations are negligible, yields estimates of $V_a$, $K_1, k_2, k_{on}B_{max}/V_d$, and $k_{off}$. This step is basically performed as a standard determination of $K_1$ through $k_4$ according to (3.21). The second step, during which the previously estimated coefficients are kept constant, yields $k_{on}/V_d$ and $B_{max}$ separately. The procedure does not require complete equilibrium; in theory a single low-specific-activity injection suffices, but the number of parameters may be too great to yield accurate estimates.

### 3.5.3 Equilibrium Analysis: Determination of $B_{max}$ and $K_D$

Equation (3.26) defines an equilibrium volume of distribution. According to (3.38), the equilibrium volume declines when the receptors are saturated. Thus, when the ligand concentration rises in the distribution space, the equilibrium volume falls, as shown by (3.37). The actual amounts of ligand held in the compartment can be determined by multiplying the volume of distribution with the arterial concentration of the ligand, $C_a(\infty)$. Multiplication with the arterial concentration yields the total quantity of tracer in the region at equilibrium.

$$M(\infty) = M_a + M_e + \frac{B_{\max} C}{K'_D + C}, \tag{3.47}$$

where $K'_D$ is the half-inhibition concentration of the radioligand in the presence of "hidden" endogenous or exogenous competitors, equal to $K_D(1 + \sum[C_j/K_j])$. The equation can be rearranged to yield the equation underlying the linearized plot, respectively, attributed to Rosenthal, Scatchard, Eadie, and Hofstee, as discussed earlier,

$$B = -[V_d K'_D]\, p_B + B_{\max}, \tag{3.48}$$

where $p_B$ is the binding potential equal to the $B/M_e$ ratio. The ordinate intercept is $B$ and the slope is $-V_d K'_D$. By the definition expressed as (3.32), $M_e$ and $B$ can be calculated from the equilibrium volumes of distribution $V_e$ and $V_b$,

$$M_e(\infty) = V_d C_e(\infty) = V_e C_a(\infty) \tag{3.49}$$

and

$$B = V_b C_a = p_B V_e C_a, \tag{3.50}$$

where $V_e = K_1/k_2$ and $V_b = p_B V_e$. An example of the use of (3.48) is given in Fig. 3.6. Since $k_3$ and $k_4$ cannot be estimated under nonsteady-state circumstances when the ligand concentration is significantly above tracer level, it is usually not possible to estimate $B$ and $M_e$ from the results of regression. Fortunately, $M_e$ and $B$ can be determined in several other ways. First, if a particular region of the brain is known not to contain any receptor sites for the ligand under investigation then, at steady-state, $M_e + M_a$ is by definition the tracer content in that region. Second, if no such region exists then (3.37) must be used to determine the volume of distribution of the free ligand ($V_a$) as the volume of distribution of the ligand at very high values of $C(\infty)$. The volume can be determined by nonlinear regression of (3.37) to simultaneously determined values of $V(\infty)$ and $C_a(\infty)$. Third, $M_e$ can be determined from the volume of distribution of an enantiomer of the tracer that does not bind to the receptor sites under study but does bind to all other sites in exactly the same manner and with the same physical properties as the tracer itself.

The bound quantity $B$ can similarly be determined as the difference between the quantity of tracer in a binding region and the quantity of tracer in a nonbind-

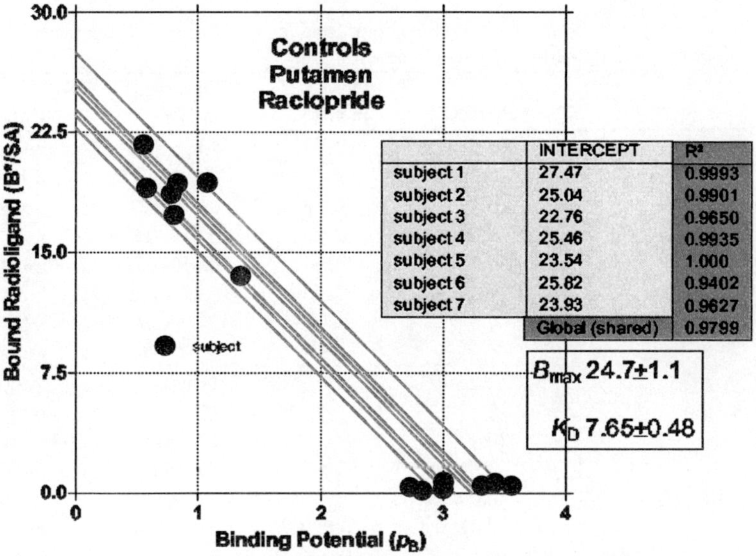


**Fig. 3.6** Eadie–Hofstee plot according to (3.48). Abscissa: Binding potential $p_B$ (bound/free ratio.) Ordinate: Quantity of bound ligand in puamen of seven normal human male volunteers injected with [$^{11}$C]raclopride at two different specific activities. See image of bound high-specific activity raclopride in Fig. 4.4

**Table 3.6** Maximum binding capacity and affinity of dopamine $D_2$ receptors in neostriatum of human and nonhuman primates

| Authors | Subjects | $B_{max}$ (pmol/g) | Method of analysis |
|---|---|---|---|
| Wong et al. (1986b) | Humans | 9 | Irreversible binding |
| Gjedde and Wong (1987) | Humans | 17 | Irreversible binding |
| Farde et al. (1986) | Humans | 25 | Equilibrium binding |
| Farde et al. (1995) | Humans | 28 | Equilibrium binding (women) |
| | | 30 | Equilibrium binding (men) |
| Phojalainen et al. (1998) | Humans | 28 | Equilibrium binding (ages 20–40) |
| | | 25 | Equilibrium binding (ages 40–60) |
| | | 24 | Equilibrium binding (ages 60–80) |
| Logan et al. (1987) | Baboons | 30 | Equilibrium binding |
| Huang et al. (1987) | Baboons | 44 | Reversible binding |
| Holden et al. (2002) | Rhesus | 23 | Reversible binding |

ing region. If a nonbinding region does not exist then the bound quantity must be determined from the volumes $V_e + V_a$ recorded at complete saturation,

$$B = M(\infty) - (M_a + M_e) = M(\infty) - (V_e + V_a)\, C_a \tag{3.51}$$

**Table 3.7** Summary of in vitro and in vivo receptor kinetic observations for interaction between generic "spiperone" and dopamine $D_2$ sites in neostriatum

| Variable | Predicted | Observed in vitro | Observed in vivo | Unit |
|---|---|---|---|---|
| $PS = -\alpha F \ln(1 - [K_1/F])$ | 3 | | 2.8 | ml/g/min |
| $F$ | 0.5 | | | ml/g/min |
| $K_1$ | 0.15 | | 0.11–0.17 | ml/g/min |
| $k_2$ | 0.06 | | 0.014–0.077 | $\text{min}^{-1}$ |
| $k_3$ | 0.05 | | 0.012–0.09 | $\text{min}^{-1}$ |
| $k_4$ | 0.01 | | 0.01–0.02 | $\text{min}^{-1}$ |
| $V_e = f_1/f_2 = K_l/k_2$ | 2.5 | | 2–15 | ml/g |
| $p_B = k_3/k_4$ | 5 | | 0.5–45 | ratio |
| $V_b = V_e\, p_B$ | 12.5 | | 1–500 | ml/g |
| $V_h = V(\infty)$ | 15 | | 12–?? | ml/g |
| $\alpha = 1/f_1$ | 20 | | 17–25 | ratio |
| $V_d = 1/f_2$ | 50 | | 50–140 | ml/g |
| $k'_{on} = k_{on}/V_d$ | 0.002 | | 0.01 | g/min/pmol |
| $k_{off}$ | 0.01 | 0.022 | 0.01–0.02 | $\text{min}^{-1}$ |
| $k_{on}$ | 0.1 | 0.086 | 0.5–1.4 | $\text{min}^{-1}$/nM |
| $K_D$ | 0.1 | 0.1 | 0.04–0.007 | nM |
| $B_{max} = k_3/k'_{on}$ | 25 | 11 | 9–44 | pmol/g |

according to which

$$V_b = \frac{M}{C_a} - (V_e + V_a), \tag{3.52}$$

where $V_b$ is the volume of distribution of the bound ligand. The corresponding Woolf plot is the reciprocal of (3.47) (Gjedde and Wong 2001),

$$\frac{1}{p_B} = \frac{M_e}{B_{max}} + \frac{V_d K'_D}{B_{max}} \tag{3.53}$$

which as the definition of the binding potential basically is identical to (3.46).

Calculation of the binding potential $p_B$ at (at least) two different ligand concentrations allowed Farde et al. (1986, 1987, 1990) to determine the maximum binding capacity ($B_{max}$) and affinity ($K'_D$) of the dopamine $D_{2,3}$ receptor ligand raclopride (see Tables 3.1 and 3.5) in parts of neostriatum. The concentration of free ligand was estimated as the quantity of tracer in the cerebellum at each dose of tracer raclopride injected. The same procedure allowed Logan et al. (1987) to calculate the neostriatal $B_{max}$ in baboons. The published estimates of these values in normal subjects are summarized in Table 3.6. It is evident that the average rose over the years, possibly because the estimates of the absolutely bound quantity of the radioligand depended on uncertain calculations of specific activity.

The discussion has focused on the practical estimation of a number of fundamental binding kinetic constants believed also to reflect the binding in vivo. The point is a conjecture that may be very difficult to prove in actual practice. For each new case or receptor system, it is necessary to determine as many of the values listed in Table 3.7 as possible before a receptor system governing neuroreceptor–radioligand interaction can be said to have been explored fully. The sample values actually listed in the table are those measured or calculated for labeled "spiperone" interacting with dopamine $D_{2,3}$ sites in neostriatum of primates or humans.

Certain differences between the in vitro and the in vivo findings remain unresolved. These differences include the remarkably irreversible binding of NMSP, which is not fully accounted for by the in vivo binding constants of spiperone, and the nature of the nonspecific binding sites in the brain that remove 99% of the ligand in the brain from the pool of ligand available for specific binding.

# Chapter 4
# Neuroreceptor Mapping In Vivo: Monoamines*

## 4.1 Introduction

With the advent of neuroreceptor and neurotransmission imaging by means of positron emission tomography (PET) and single photon emission computed tomography (SPECT) in vivo, it is now possible to directly record the normal and pathophysiological processes served by monoamines in the brain, as well as to examine the possibilities and potential impact of treatment of disorders of monoaminergic neurotransmission. This review describes the methods of noninvasive brain imaging and the benefits that this imaging can bring to the understanding of disturbed monoaminergic neurotransmission, the mechanisms underlying the disturbances, and the impact of possible therapies.

## 4.2 Monoaminergic Neurotransmission

Monoamines are neuromodulators that arise from the decarboxylation of large neutral and aromatic amino acids in brain tissue. Depending on the specific precursor, they fall into two categories, the *catecholamines* that ultimately are derived from phenylalanine but in brain only come from tyrosine, and the *indoleamines* that are derived from tryptophan. The most important catecholamines are dopamine and noradrenaline, and the most important indoleamine is serotonin.

Neuromodulation is a special kind of neurotransmission. Neurotransmission, in turn, is the term applied to the exchange among brain cells of specific information encoded in the chemicals known as transmitters. The conventional description of neurotransmission includes, at the very least, the steps of synthesis, vesicular storage, release, reuptake, and removal or breakdown. This is a simple view of

* Adapted from Wong and Gjedde (2009) Monoamines: Human Brain Imaging. In: Squire LR (ed.) Encyclopedia of Neuroscience, volume 5, pp. 939–952. Oxford: Academic Press, with permission from Oxford: Academic Press.

A. Gjedde et al., *Neurokinetics: The Dynamics of Neurobiology In Vivo*,
DOI 10.1007/978-1-4419-7409-9_4, 

the molecular interactions that neurotransmitters engage in, but it is the basis on which the molecular imaging of neurotransmission rests. The individual steps are implemented in extra- or intracellular proteins that act as enzymes, receptors, or transporters. Molecules that interact with the proteins are known as ligands because the first step of the interaction is binding, regardless of the additional functions of the protein. By necessity, all transmitter molecules are ligands, but not all ligands work as transmitters.

The binding is the first step in the processing of the ligands, the ultimate fate of which differs, according to the proteins they interact with. Receptors return the ligand to the original pool of unbound molecules after a delay that depends on the affinity of the receptor toward the ligand. Transporters relocate the ligand to a different pool of unbound ligands. Enzymes release the ligand to the same pool but in a new chemical form. Exogenous or artificial ligands may differ so much from the native ligand that transporters or enzymes fail to carry out their primary function and instead act as receptors by simply returning the ligand unchanged to its original pool. Intentionally or unintentionally, this interaction with a protein may interfere with the access of any endogenous ligand. Depending on the degree of interference, exogenous or artificial ligands may thus be classified as antagonists or partial agonists of the action of the native transmitter.

As neuromodulators, monoamines are not classical neurotransmitters. Classical neurotransmitters confine their action to receptors located in synapses or "kissing points" between cells that are also the sites of release. This type of neurotransmission is "wired"; in wired transmission, release, binding, and reuptake are linked by very short distances that prevent wide dissemination of the transmitter. In contrast, "volume" transmission is a paracrine process, in which the transmitter molecules diffuse widely and their concentrations depend less on the moment-to-moment fluctuations of release. Instead, it is the density and location of transporters in relation to sites of release that determine the specificity of action. Thus, in volume transmission, the transporters are located far from the site of release and their location determines the nature of the transmission. Hence, transmitters engaged in volume transmission carry no specific information about the sites of release. In keeping with the mechanism of neuromodulatory volume transmission, the monoamine transporters primarily are extrasynaptic (Nirenberg et al. 1997), although the monoamines may serve both wired and volume transmissions, such as in the case of dopamine.

As neuromodulators, monoamines tend to reduce the excitability of target neurons, although this action does depend on the specific receptor subtypes that are expressed locally. Because the specificity of the proteins toward exogenous or artificial ligands varies, it is now possible to image presynaptic and intracellular as well as postsynaptic sites of transmitter interaction by selecting or designing exogenous or artificial ligands that match the specificities of the proteins. The main extra- and intracellular sites of binding, reuptake, and metabolism are listed in Table 4.1.

The expression of receptors and enzymes is highly variable and depends on the intra- and extracellular environments, making all results of PET and SPECT contingent upon the specific circumstances at the time of study. This is also true of the

**Table 4.1** Variables of monoaminergic neurotransmission

| Location | Variable | Example |
|---|---|---|
| Presynaptic | Transporter availability (binding potential) | Dopamine transporters [DAT], vesicular monoamine transporter 2 [VMAT2] |
| | Enzyme activity | Dopa decarboxylase [DDC] |
| | Precursor concentration | DOPA |
| Postsynaptic | Receptor availability, density, and affinity | Dopamine $D_2$-like receptors |
| | Receptor subtype selectivity | Dopamine $D_1$- or $D_2$-like receptors |
| Intrasynaptic | Endogenous neurotransmitter occupancy | Dopamine |
| | Endogenous neurotransmitter release | Dopamine |
| | Enzyme activity | Monoamine oxidase [MAO] A or B |

transporters but when the first tropane-based PET probes for the dopamine transporters were used in baboons and humans (Wong et al. 1993), it was generally accepted that the number of transporters remained constant. However, it is now known that the number of dopamine transporters is subject to regulation by intrinsic mechanisms as well as pharmacological manipulation (Pristupa et al. 1998; Daniels and Amara 1999; Melikian and Buckley 1999; Zahniser and Sorkin 2004).

## 4.3 Methods of Neuroreceptor Mapping

Molecular imaging of neurotransmission in vivo is an extension of the principles of in vitro and ex vivo autoradiography to the living brain. Historically, much of the imaging has focused on the dopaminergic innervations of the nigrostriatal fibres to the basal ganglia, but the serotonergic, noradrenergic, and dopaminergic innervations of the limbic and cerebral cortices have since become important targets. The first reports on the use of positron emission tomography (PET) and single photon emission computed tomography (SPECT) to image aspects of neurotransmission in the basal ganglia with radiolabeled drugs appeared as early as in 1983 (Garnett et al. 1983; Wagner et al. 1983). By labeling specific drug molecules of known action, the investigators confirmed the fate of these molecules in time and space in the brain and thus recorded events that could be ascribed to the known effect of these drugs on dopaminergic neurotransmission.

Although both PET and SPECT are used in these efforts, it is important to note their different characteristics, as listed in Table 4.2. PET tracers emit positrons, and PET detects the gamma radiation that arises from the annihilation of positrons when they collide with electrons. The most common radiolabels used with PET are the positron-emitting isotope of carbon ($^{11}C$) that replaces a stable carbon atom ($^{12}C$), and the positron-emitting isotope of fluorine ($^{18}F$) that either is added to some

**Table 4.2** Properties of radionuclides used in PET or SPECT imaging

| Isotope | Modality | Decay mode (%) | Half-life (min) | Production type |
|---|---|---|---|---|
| $^{11}C$ | PET | $\beta^+$ 100 | 20.1 | C |
| $^{13}N$ | PET | $\beta^+$ 100 | 9.996 | C |
| $^{15}O$ | PET | $\beta^+$ 100 | 2.04 | C |
| $^{18}F$ | PET | $\beta^+$ 97 EC 3 | 109.8 | C/R |
| $^{62}Cu$ | PET | $\beta^+$ 97 EC 3 | 9.7 | G[$^{62}$Zn] |
| $^{68}Ga$ | PET | $\beta^+$ 89 EC11 | 68 | G[$^{68}$Ge] |
| $^{82}Rb$ | PET | $\beta^+$ 95 EC 5 | 1.3 | G[$^{82}$Sn] |
| $^{99m}Tc$ | SPECT | IT100 | 360 | G[$^{99}$Mo] |
| $^{123}I$ | SPECT | EC100 | 780 | C |
| $^{131}I$ | SPECT | $\beta^-$ 100 | 8 (days) | C/R |

Decay modes:
$\beta+$ = positron emission
$\beta-$ = electron emission
$EC$ = electron capture
$IT$ = isomeric transition

Production types:
$C$ = cyclotron
$R$ = reactor
$G$ = generator

molecules or replaces stable fluorine ($^{19}$F) in other molecules, in most cases without significantly affecting the original kinetic properties of the molecules. PET has the advantage that the half-lives of these isotopes are very short (20 min for $^{11}$C and 110 min for $^{18}$F) but the disadvantage is that the nuclides must be produced in a local cyclotron.

SPECT tracers emit gamma radiation directly, arising from photon-emitting nuclides such as $^{99m}$technetium ($^{99m}$Tc) and $^{123}$iodine ($^{123}$I). These nuclides have much longer half-lives, as listed in Table 4.2. Some SPECT tracers, therefore, have the advantage of being available in most nuclear medicine departments, and the relatively long half-life of $^{123}$I (13 h) permits tomography to take place as late as 24 h after administration of the tracer, but it is a disadvantage that this nuclide currently is available in North America only from a cyclotron facility in Vancouver (Tri-University Meson Facility or TRIUMF).

Multiple brain images in the form of dynamic series of tomographic frames are acquired in PET studies of monoaminergic neurotransmission. Typically, 30–50 frames are obtained as two- or three-dimensional images in 60–120 min. Because of the short half-lives, this duration is the limit of most studies with carbon-11 or fluorine-18. In contrast, SPECT studies with $^{123}$I may be continued for 6 or more hours, as evident from Table 4.2. In two-dimensional imaging, the tomographic record consists of 15–64 or more sections of the brain, depending on the tomograph.

In many studies, both of animals and human subjects, investigators insert a catheter in a suitable artery to determine the tracer concentration in arterial blood. The blood samples are assayed for radioactivity and six-to-eight discrete samples of arterial plasma undergo additional high performance liquid chromatography (HPLC), or thin-layer chromatography (TLC), to determine the chemical nature of all labeled products in plasma, to account for metabolism in the course of circulation, and to reconstruct the concentration of the parent compound as a function of time. The arterial samples and brain images establish the concentrations of the tracer in blood and brain as functions of time. The records then form the basis for the subsequent kinetic analysis that yields the specifics of the monoaminergic neurotransmission, as listed in Table 4.1.

The availability of the right tracers is the key to successful imaging of monoaminergic neurotransmission in animals or humans. With PET, the imaging is limited by the half-life of the tracers, which means in the case of carbon-11-labeled tracers that all information must be extracted during the first hour of recording. The usefulness of this information in turn is dictated by the pharmacology and binding properties of the tracer in relation to the relevant proteins and not solely by the properties of the ligand itself.

An important challenge to PET imaging is the necessary anesthesia of animals used in research. It has been possible in very few places to train nonhuman primates to remain awake while seated comfortably a horizontally suspended tomograph (Tsukada et al. 2004). In other places, it is necessary to use a within-subject design of drug studies to reduce the variability of the PET measures. Other factors, such as the size of brain, genotype, age, latent disease states, personality, and adaptation to stress and drugs, have not routinely been isorted out in most preclinical or clinical studies. A few studies addressed elements of psychology and personality, including introversion and social reclusiveness (Farde et al. 1997), and the effects of social anxiety (Schneier et al. 2000). Nonetheless, it is clear that interpretation of these effects is difficult, as a single variable may reflect opposite changes of monoaminergic neurotransmission.

In one study of social isolation in monkeys, the availability of dopamine $D_2$ receptors in basal ganglia, measured by PET, was shown to rise in dominant monkeys when they moved from individual to social housing (Morgan et al. 2002). Receptor "availability" is the controversial term applied to the estimate of the ratio between bound and unbound ligand quantities in a tissue. This ratio is also known as the "binding potential." It declines when the receptors approach saturation and hence can be said to be an index of the relative number of unoccupied receptor sites. The number rises either because the total number of receptors rises or because the concentrations of relevant ligands fall, or both. At the time, the result was interpreted, somewhat vaguely, as a "rise" of dopaminergic neurotransmission when dominant monkeys meet other monkeys they can dominate (Kuhar 2002) but the effect could just as well arise from a decline of dopamine, when dominant monkeys escape from the social isolation.

An important problem is sample size. Typical numbers in PET studies of humans range from 10 to 20 subjects, and a population of 30 subjects is rare. Adequate numbers are necessary to counteract the confounding influence of the tomographs' finite power of resolution of the recorded radioactivity, known as the "partial volume" effect. The effect is most prominent in small structures that typically fuse below a diameter of 2–3 mm. The effect persists for diameters as large as 2–3 times the full-width-at-half-maximum (FWHM) property of a point source for any imaging agent. Thus, the human cerebellum usually loses about 5% or 10% of its total activity when the entire structure is imaged. The resolution of PET originally was no less than 8 mm FWHM. Current tomographs enjoy resolutions as low as 2–3 mm FWHM for the High Resolution Research Tomograph (HRRT, which allows visual separation of two points with a distances of as little as low as 0.5 cm. The partial volume effect is an even greater problem with SPECT, because the attenuation correction is much greater with this method (Rousset et al. 1993, 2000).

### *4.3.1 Tracers of Monoaminergic Neurotransmission*

PET and SPECT both use marker molecules ("tracers") that bind to the proteins of monoamine synthesis, binding, and metabolism, with sufficiently high affinity and selectivity, but ordinarily with such low concentration that pharmacodynamic effects are absent. The marker molecules are tracers because the radiation that they emit can be detected externally by the tomographs.

The typical investigation of monoaminergic neurotransmission begins with the identification of an appropriate precursor of a tracer that targets a specific step in the pathway from the aromatic amino acid parent to final metabolite. Both the choice of precursor and the synthesis of the tracer from the precursor are based on the evidence obtained in vitro or ex vivo in animals or humans. When a potential tracer is labeled, its usefulness must be examined carefully, first in animals and then in humans, and at any time in this process, unexpected species differences may interfere with the imaging, for example by blocking the transfer of the tracer across the blood–brain barrier, or by altering the metabolism of the tracer in the brain tissue. The necessary safety profile in the form of toxicology and radiation dosimetry must also be established (Brady 2005). Validation in animals may involve initial studies in vitro to confirm the saturability of the binding, or the competition with chemically or pharmacologically related drugs. Once the binding and metabolism characteristics are known, the tracer may finally be deemed appropriate for use in vivo in nonprimate mammals or nonhuman or human primates.

Table 4.3 lists the large number of monoaminergic tracers that have been tested with variable success since 1983. Unfortunately, of the hundreds of thousands of compounds developed by pharmaceutical industry, university, or government institutions, only a tiny fraction avail themselves to successful labeling. Radioligands are useful only when they meet specific requirements of selectivity, affinity, and kinetics of binding, transport or metabolism. Only when these requirements are met can reliable estimates of receptor density or neurotransmitter synthesis, release, or metabolism, be obtained noninvasively.

Three tracers illustrate the historical advance of tracer development from 1983 till today, and the problems associated with the use of the tracers. In 1983, the experimental neuroleptic spiroperidol (spiperone, 8-[4-(4-fluorophenyl)-4-oxo-butyl]-1-phenyl-1,3,8-triazaspiro[4.5]decan-4-one) was the precursor by methylation of the radioligand 3-N-[$^{11}$C]methylspiperone (NMSP), for which the dopamine $D_2$-like ($D_2$, $D_3$, and $D_4$) receptors have high affinity. Spiperone is a piperazine, originally so named because of the chemical relation to piperidine, a constituent of piperine

**Table 4.3** Monoamine neuroreceptor radioligands for PET and SPECT

| Protein | Dopamine | Serotonin | Noradrenaline |
|---|---|---|---|
| Receptors | $D_{2/3}$ | $5HT_{1A}$ | $\alpha_2$ |
| | N-[$^{11}$C]methylspiperone | [$^{11}$C]WAY 100635 | [$^{11}$C]yohimbine |
| | [$^{11}$C]raclopride | [$^{11}$C]carbonyl-WAY 100635 | [$^{11}$C]mirtazapine |
| | [$^{11}$C]FLB 457 | cis-4-[$^{11}$C]methoxy-WAY | |
| | [$^{11}$C-methyl]-(-)-OSU6162 | trans-4-[$^{11}$C]methoxy-WAY | |
| | [$^{18}$F]clebopride | [$^{11}$C]LY-274601 | |
| | [$^{18}$F]fallypride | [$^{11}$C]NAD-299 | |
| | [$^{18}$F]desmethoxyfallypride | [$^{11}$C]CPC-222 | |
| | [$^{18}$F]ioxipride | [$^{11}$C]DWAY | |
| | [$^{18}$F]methylspiperone | [$^{18}$F]FBWAY | |
| | [$^{18}$F]FCP | [$^{18}$F]FCWAY | |
| | [$^{18}$F]spiperone | [$^{18}$F]Fluoroethyl-WAY | |
| | [$^{18}$F]fluoroethylspiperone | [$^{18}$F]Fluoropropyl-WAY | |
| | [$^{18}$F]fluoropropylepidepride | [$^{18}$F]MeFBWAY | |
| | N-allyl-5-[$^{18}$F]fluoropropyle-pidepride | cis-[$^{18}$F]FCWAY | |
| | [$^{18}$F]NCQ-15 | trans-[$^{18}$F]FCWAY | |
| | [$^{123}$I]IBZM | p-[$^{18}$F]MPPF | |
| | [$^{123}$I]IBF | p-[$^{123}$I]MPPI | |
| | $D_1$ | $5HT_{2A}$ | |
| | [$^{11}$C]SKF 82957 | N-[$^{11}$C]methyls piperone | |
| | [$^{11}$C]SCH 23390 | [$^{11}$C]MDL 100,907 | |
| | [$^{11}$C]NNC 756 | [$^{18}$F]altanserin | |
| | [$^{11}$C]NNC112 | [$^{18}$F]setoperone | |
| | [$^{11}$C]SCH 20098 | $5HT_{2A}$ | |
| | R [$^{11}$C]SKF 82957 | [$^{11}$C]SB207145 | |
| | [$^{18}$F]fluoropropyl SCH 38548 | | |
| Transporters | DAT | 5HTT or SERT | NET |
| | [$^{11}$C]methylphenidate | [O-methyl-$^{11}$C]venlafaxine | [O-methyl-$^{11}$C]venlafaxine |
| | [$^{11}$C]cocaine | [$^{11}$C]McN 5652 | [$^{11}$C]-SS-MeNER |
| | [$^{11}$C]CFT (WIN35, 428) | [$^{11}$C]NS2456 | [$^{11}$C]Hydroxyephedrine |
| | [$^{11}$C]SS-CIT-FE | [$^{11}$C]DASB | |
| | [$^{11}$C]PE2I | [$^{11}$C]DAPA | |
| | [$^{11}$C]altropane | [$^{11}$C]AFM | |
| | [$^{11}$C]naphthyltropane analogue | [$^{11}$C]ADAM | |
| | [$^{11}$C]phenyltropane analogues | [$^{11}$C]MADAM | |
| | [$^{11}$C]RTI 32 | [$^{123}$I]ADAM | |
| | [$^{18}$F]fluorococaine | | |
| | [$^{18}$F]CFT | | |
| | [$^{18}$F]FP-CIT | | |
| | [$^{18}$F]FE-CNT | | |
| | [$\beta$ -[$^{123}$I]CIT | | |
| | [$^{99m}$Tc]TRODAT-1 | | |

(continued)

**Table 4.3** (continued)

| Protein | Dopamine | Serotonin | Noradrenaline |
|---|---|---|---|
| | VMAT2 | VMAT2 | VMAT2 |
| | D-[$^{11}$C]tetrabenazine | D-[$^{11}$C]tetrabenazine | D-[$^{11}$C]tetrabenazine |
| Enzymes | Dopa Decarboxylase | Dopa Decarboxylase | Tyrosine hydroxylase |
| | L-[$^{11}$C]DOPA | L-[$^{11}$C]DOPA | 2-[$^{18}$F]tyrosine |
| | 6-[$^{18}$F]fluoro-L-DOPA | 5-hydroxy-L- | 4-[$^{18}$F]tyrosine |
| | [$^{18}$F]fluorodopamine | [$\beta-^{11}$C]tryptophan] | [$^{123}$I]IMT |
| | O-methyl-[$^{18}$F]fluoro-DOPA | [$^{18}$F]FMT(fluoro-m-tyrosine) | [$^{18}$F]FMT(fluoro-m-tyrosine) |

in the black pepper plant (Piper nigrum), but the radioligand is known also as a butyrophenone. The radioligand yielded the first $D_2$-like receptor images in humans with PET (Wagner et al. 1983). However, the selectivity was not exclusive for this receptor subtype, as the serotonin $5HT_{2A}$ receptors also have significant affinity. The kinetic analysis of the binding was complicated by this high affinity; the relative influences of transport across the blood–brain barrier and binding to the receptors cannot be fully resolved when a secular equilibrium of bound and unbound ligand is not approached during the period of uptake. Therefore, ligands that bind irreversibly often have delivery- rather than affinity-dependent uptake.

In contrast, useful ligands frequently have binding characteristics that include high selectivity but only moderate affinity, commensurate with binding potentials between 1 and 6. In the case of NMSP, the binding potential exceeds 6 in human subjects, although partial volume effects may reduce this value in the smaller brains of experimental animals. Interestingly, NMSP has been of use in cerebral cortex where the lower affinity of the serotonin receptors renders the binding reversible in the period available for PET, without serious interference from the lower density dopamine receptors in cortex.

The currently most common monoamine receptor ligand used with PET is [$^{11}$C]raclopride (3,5-dichloro-*N*-[(1-ethylpyrrolidin-2-yl)methyl]-2-hydroxy-6-methoxy-benzamide), which derives from another class of chemical, the benzamides (Farde et al. 1985). Benzamides are derivatives of benzoic acid, and substituted benzamides often serve as neuroleptics or antipsychotics in psychiatric practice. Two active substances from the group of benzamides are in clinical use, sulpiride and amisulpride. Raclopride binds reversibly and selectively to the dopamine $D_2$ and $D_3$ receptors, with a binding potential of 3 (Gjedde 2003), and as such is an ideal tracer of these dopamine binding sites in high density locations such as the striatum. In cerebral cortex, the lower density renders the binding potential too low to be useful. Several alternatives with higher affinity are available for this purpose. The tracer [$^{18}$F]fallypride ([5-(3-fluoropropyl)-2,3-dimedichloro-*N*-[(2S)-1-(2-propenyl)-2-pyrrolidinyl]-methyl]-benzamide) is a radioligand with binding potentials of 0.5–1.5 in cerebral cortex. The higher affinity elevates the retention in striatum where the binding potential is so high that binding is delivery-dependent (Mukherjee et al.

2002, 2004, 2005). Benzamide tracers are also available for SPECT. They include [$^{123}$I]iodobenzamide (IBZM) and [$^{123}$I]- and [$^{125}$I]epidepride.

In the case of metabolic substrates, it is necessary that the product be retained in brain as an index of the activity of the enzyme, provided the transport across the blood–brain barrier is not rate-limiting for the uptake. The first study with PET of the metabolism of the dopamine precursor [$^{18}$F]fluorodopa (FDOPA) was also published in 1983 (Garnett et al. 1983). The tracer is one of several substrates of the aromatic amino acid decarboxylase, which produces both dopamine and serotonin from appropriate amino acid precursors in respective terminals. The uptake appears to be irreversible but the appearance may be deceiving, as the retention may reflect only a fraction of the tracer.

For FDOPA, a host of different kinetic analyses have been attempted to solve the issue of the specificity of the signal. Increasingly, it is clear that the retention primarily reflects the transport of the labeled product fluorodopamine (FDA) into vesicles, and the retention depends on the irreversibility of this transport (Gjedde et al. 1991a, b). Two processes interfere with the retention: First, FDA is subject to oxidation by monoamine oxidases, and second, release of FDA from the vesicles exposes the released FDA to renewed metabolism after reuptake into the terminals. Therefore, low retention can reflect both reduced vesicular incorporation due to loss of terminals or loss of dopa decarboxylase activity, or loss or inhibition of vesicular transporters, or increased release of FDA in response to increased dopaminergic neurotransmission (Kumakura et al. 2006). As indicators of neuronal function, these interpretations have completely opposite implications for the health of the patient, yet the PET signals may look exactly the same. It should be noted that FDOPA is not specific for dopaminergic terminals as both serotonergic and noradrenergic terminals use this enzyme in the synthesis of their transmitters. However, only in dopaminergic terminals is the product FDA retained.

As markers of dopaminergic activity in presynaptic terminals, the only alternatives to FDOPA are the tracers of the monoamine reuptake sites, which are based on the tropane structure known as nortropane (8-azabicyclo[3.2.1]octane). Tropane is a nitrogenous bicyclic organic compound, mainly famous known for the large group of alkaloids that are derived from it (called tropane alkaloids). The derivatives include, among others, atropine and cocaine. These two alkaloids both contain tropinone from which tropane is derived. Tropane alkaloids occur in plants of the family solanaceae (mandrake, henbane, deadly nightshade, potato, and tomato) and in the coca plant.

Several potentially irreversible ligands for the cocaine receptor site on the dopamine transporter are derived from cocaine or 3$\beta$-phenyltropane. The cocaine analog CFT (2$\beta$-carboxymethoxy-3$\beta$-(4-fluorophenyl) tropane or WIN-35,428) was first recommended as a radiolabeled probe for PET and SPECT by Madras and co-workers (Canfield et al. 1990; Madras et al. 1989) and later developed into the PET tracer [$^{11}$C]WIN-35,428 by Kuhar and co-workers (Boja et al. 1992; Carroll et al. 1992; Scheffel et al. 1991). The tracer binds irreversibly to the cocaine site on the dopamine transporter (DAT) but also to the serotonin (5HTT or SERT) and norepinephrine (NET) transporters. In attempts to overcome the low selectivity, the basic structure has been modified repeatedly with substitutions of fluorine or small carbon chains that may reduce affinity and raise selectivity. Versions suitable for SPECT include $\beta$-[$^{123}$I]CIT ($\beta$-carbomethoxy-3-beta-(4-iodophenyl)-tropane) for the dopamine and serotonin transporters (Pirker et al. 2000), and [$^{99m}$Tc]TRODAT-1 selective for the dopamine transporter (technetium(V)-oxo-3$\beta$-(4-chlorophenyl)-8-methyl-8-azabicyclo[3.2.1]oct-2-yl[*N*-(2-mercaptoethyl), *N*-(*N*'-(2-mercaptoethyl)-2-aminoethyl)]-amino-methyl) (Kung et al. 1996).

The evidence for regulation of monoamine transporters has raised the search for additional probes of events in monoaminergic terminals. The vesicular monoamine transporter (VMAT2) is located in the membranes of terminal vesicles and can be imaged with the radioligand D-[$^{11}$C]tetrabenazine (DTBZ, Fantegrossi et al. 2004). However, not even the imaging of the VMAT2 transporter may settle the "transporter regulation issue," as VMAT2 also appears to be regulated (Zahniser and Sorkin 2004; Zucker et al. 2005).

## 4.3.2 Pharmacokinetics of Monoaminergic Neurotransmission

Generally, the characteristics of binding of radiotracers to the proteins of monoaminergic neurotransmission in vivo are consistent with insights from studies in vitro, particularly with respect to binding. In Fig. 4.1, the binding of a number of

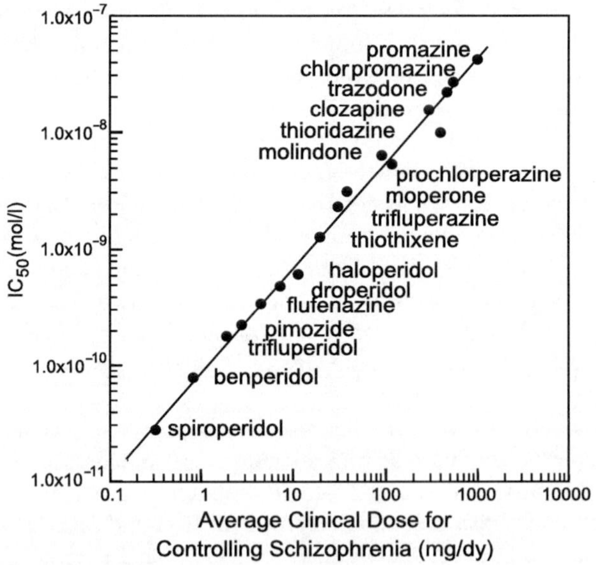


**Fig. 4.1** Relationship between $IC_{50}$ values at dopamine receptors and therapeutic dose of neuroleptics in treatment of schizophrenia. The $IC_{50}$ values are the concentrations of the antipsychotic drugs that reduce the stereospecific component of [$^{3}$H]haloperidol binding by 50%. The abscissa indicates the average values of doses used in schizophrenic patients. Redrawn from Seeman et al. (1976)

compounds is expressed as the 50% inhibitory concentration $IC_{50}$ of the tracer, which takes into account both the receptor's affinity toward the tracer and the concentrations of competitors.

The most commonly reported variable obtained with imaging of monoaminergic neurotransmission is the ratio between the density of receptors and the affinity constant, known as the binding potential. Both terms are part of Michaelis' & Menten's steady-state solution to the binding equations (Gjedde et al. 1986; Michaelis and Menten 1913),

$$B = \frac{B_{\max} C}{K_{\mathrm{D}} + C}, \tag{4.1}$$

where $B$ is the bound quantity of ligand, $B_{\max}$, the maximum number of binding sites, $C$ is the unbound ligand's concentration, and $K_{\mathrm{D}}$ is the Michaelis–Menten dissociation constant or half-saturating bindng concentration of the ligand (see Chaps. 2 and 3). The binding potential is the ratio between the bound and unbound ("free") quantities of a ligand, including an endogenous ligand, relative to the receptor's affinity for this ligand, i.e., multiplied by the Michaelis–Menten half-saturation constant of the receptor toward the ligand in question (Gjedde et al. 2005b). The two definitions of the binding potential express the potential in terms of available receptors (i.e., receptors not occupied by ligands) or in terms of the concentrations of ligands,

$$p_{\mathrm{B}} = \frac{B^*}{C^*} = \frac{B_{\max}}{C^* + K_{\mathrm{D}}^*\left(1 + \frac{C_i}{K_{\mathrm{I}}}\right)} = \frac{K_{\mathrm{I}} B_{\max}}{K_{\mathrm{D}}^*(K_{\mathrm{I}} + C_i)} = \left(\frac{K_{\mathrm{I}}}{K_{\mathrm{D}}^*}\right)\left(\frac{B_i}{C_i}\right), \tag{4.2}$$

where $p_{\mathrm{B}}$ is the binding potential and the remaining symbols represent specifically bound ligand ($B^*$), unbound tracer ($C^*$), maximum number of binding sites ($B_{\max}$), affinity toward the ligand ($K_{\mathrm{D}}^*$), and the affinity toward any competing endogenous ligand ($K_{\mathrm{I}}$) with a presence in the vicinity of the receptors ($C_i$) sufficient to occupy a significant number of receptor sites ($B_i$).

In the absence of changes of receptor density or affinities, a change of the binding potential can therefore be described either as a change of the quantities of bound ligands or as a change of concentration of one or more competitors for binding, although the two are of course interlocked aspects of the same perturbation.

Sokoloff et al. (1977) first presented the now standard three-compartment model of tracer kinetics in vivo (see Chap. 2) that typically is used to obtain the estimates of binding potential or other variables of tracer binding or metabolism (Gjedde 1982). When kinetic modeling was first proposed for the studies of neurotransmission begun in 1983 (Gjedde et al. 1986; Mintun et al. 1984), its use in pharmacology was novel because the Michaelis–Menten solution requires equilibrium of binding, which in principle is not achieved with positron emission tomography. The Michaelis–Menten equation is the solution to the differential equations of the exchange of tracer between the circulation compartment, the pool of unchanged and unbound racer in brain tissue, and the pool of bound or metabolized tracer in the

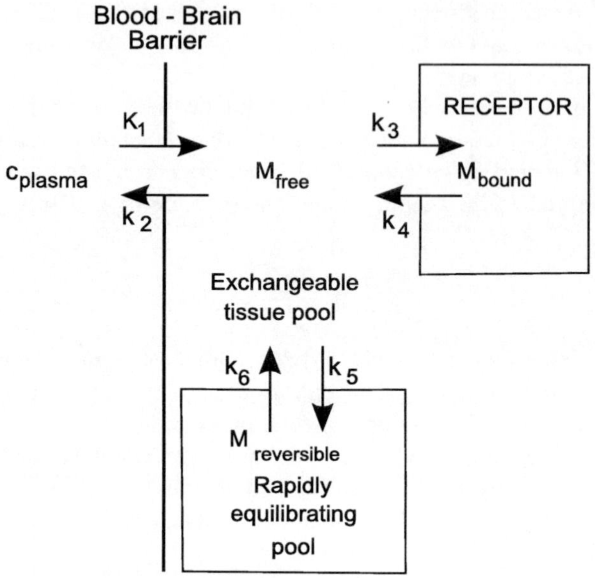


**Fig. 4.2** Three-compartment model of receptor binding, with additional compartment for non-specific binding of ligand. The transfer coefficients are clearance coefficients or rate constants, respectively. The former are given with a upper-case symbols, the latter with lower-case symbols. The coefficient $K_1$ is the blood–brain clearance, and the rate constant $k_2$ is the fractional clearance from the brain. For definition of remaining rate constants, see text. From Wong et al. (1986b)

tissue. The model assumes that the tracer or its metabolite is retained in this latter pool to such an extent that the tracer exchange with the circulation is delayed significantly.

The model shown in Fig. 4.2 includes the coefficient of clearance ($K_1$) of the tracer from the circulation to the brain, and the rate constant of fractional clearance ($k_2$) of the tracer from the exchangeable pool of unbound tracer in the brain back to the circulation. The exchange distribution volume is simply the ratio of these two terms. The rate constants $k_3$ and $k_4$ refer to the interaction with the specific proteins indicative of the retention associated with binding, transport, or metabolism. The specific definitions depend on the process and the conditions but generally the magnitude of $k_3$ is related to the quantity of protein and the concentration of the tracer while $k_4$ is the decay constant of the pool of retained substance. Depending on the steady-state, the definitions are either

$$k_3 = k_{on}\,(B_{max} - B) \tag{4.3}$$

and

$$k_4 = k_{off} \tag{4.4}$$

or

$$k_3 = k_{on}\,B_{max} \tag{4.5}$$

and

$$k_4 = k_{\text{off}} \left(1 + \frac{C}{K_{\text{D}}}\right), \tag{4.6}$$

where $k_3$ and $k_4$ define the respective association and dissociation rate constants. The ratio $k_3/k_4$ is a function of the amount of protein (such as density of receptors) and the affinity for the tracer, and the ratio is therefore an index of the tissue's potential for retention, a real "retention potential." In the case of binding, the definitions show that this ratio is also the binding potential defined earlier.

The exact solutions are given in Chap. 2 (also see Gjedde 2003). The receptor density is calculated from multiple estimates of the binding potential when the quantity of bound ligand is known for each estimate (Holden et al. 2002). The bound quantities in principle include both endogenous and exogenous ligands, such as a drug given to reduce the binding potential of the tracer by competition for binding (Farde et al. 1986; Gjedde and Wong 2001).

It is preferable to obtain the records at steady state, but the exact requirements are dictated by the pharmacology of the ligand. Ideally, the uptake would approach steady-state or equilibrium very closely. If the tracer is given as a bolus administration, true steady-state of course is never reached. If the tracer is continuously infused, it is possible to get very close to true steady-state, depending on the tracer's half-life. Equilibrium conditions are ideal, both because comparative binding in vitro in principle is in equilibrium, and because the binding transients decline to zero. Unfortunately, some of the most selective ligands have the highest affinity and do not approach secular equilibrium during the 90–120 min available for a positron emission tomographic session.

An important question is the number of frames necessary to optimize the kinetic analysis. The higher the number of frames, the lower the radioactivity recorded in each frame and the higher the random variability. The variability obeys a Poisson distribution and thus increases with the number of frames. If multiple short frames are recorded, there is a tradeoff between the number of frames and the accuracy of the kinetic analysis, expressed in terms of parameter "identifiability." The information content can be shown to decline as a function of the number of parameters as the variability of the radioactivity measurements increases. There is another trade-off between too many and too few parameters. Many researchers believe that the maximum number of parameters that can safely be derived from PET and SPECT is four, possibly as many as five, consistent with the existence of three compartments.

Estimates of the receptor density are not routine because they require significant occupation of the receptors with a ligand which in principle must then have a pharmacodynamic effect. The procedure consists of giving a pharmacologically active dose of a ligand together with the tracer (e.g., haloperidol with [$^{11}$C]NMSP or raclopride with [$^{11}$C]raclopride). As many as five different estimates of the binding potential have been made to complete Scatchard or Eadie–Hofstee plots of the binding, as shown in Fig. 4.3.

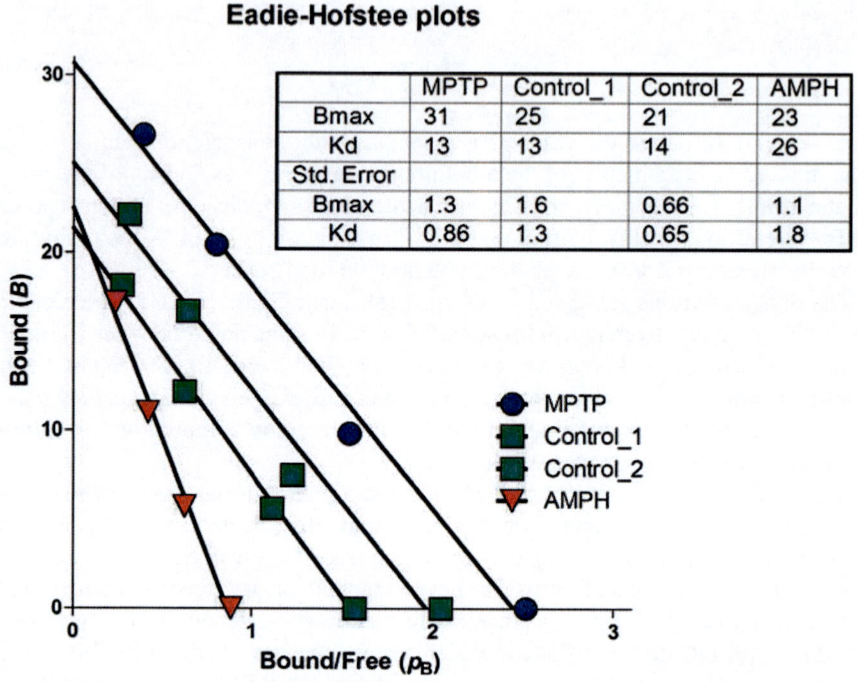


**Fig. 4.3** Eadie–Hofstee plot of dopamine receptor density and radioligand affinity, determined with [$^{11}$C]raclopride in monkeys rendered parkinsonian by means of MPTP administration, in comparison with control monkeys. Eadie–Hofstee and Scatchard plots are linear renditions of the equation that defines the binding potential. This is also the equation of the Scatchard plot, in which the abscissa is the total quantity of bound ligand () , usually determined from the specific activity and binding potential of the radioligand. The ordinate is the binding potential ($p_B$), determined as the ratio of bound to unbound ligand from binding regions-of-interest and a reference region of no binding, or from regression analysis estimates of the rate constants $k_3$ and $k_4$ shown in Fig. 4.2, or directly as their ratio $k_3/k_4$, which equals the binding potential at steady-state. In the Eadie–Hofstee plot, the axes are reversed, and the formula for a radioligand is rearranged to where is now the ordinate and is the abscissa. The Eadie–Hofstee plot is drawn from data reported by Holden et al. (2002)

Initially and rather unrealistically, it was believed that the binding of radioligands to neurotransmitter receptors would not be affected by the endogenous ligand but this appears not to be the case. Although the interference could be considered a hindrance to the proper quantification of the receptor density, it has instead permitted indirect and noninvasive investigations of the fluctuations of transmitter release, particularly with respect to the dopaminergic neurotransmission. The promise and excitement of this approach notwithstanding, the reality of the interpretations is the subject of considerable current controversy, as not only transmitter concentration but also receptor number and affinity affect the estimates of binding potentials that may change as a simple consequence of an affinity change (Gjedde et al. 2005b).

## 4.4 Altered Monoaminergic Neurotransmission

Binding potentials have been determined for a number of monoaminergic tracers in many parts of the brain in both health and disease. Figure 4.4 shows the coincident maps of average dopa decarboxylase activity, dopamine transporter binding potentials, and dopamine receptor binding potentials in normal human beings. The maps are mutually confirmatory but the individual numbers are not particularly informative in the absence of related evidence of biology or pathology. The noninvasive probing of monoaminergic neurotransmission in living human brain also promises to clarify the pharmacokinetic and pharmacodynamic mechanisms of expectation, reward, and drug effects in healthy human beings. It is now possible to map the increase of dopamine occupancy that follows when humans become aware of and assign meaning to meaningful stimuli in the environment as shown in Fig. 4.5 (Kjaer et al. 2002).

It is well-known that the pathological changes occurring in any of the three monoaminergic systems precede the presentation of overt clinical symptoms, and changes of dopaminergic, serotonergic, and noradrenergic neurotransmission are therefore actively monitored in major neurological and neuropsychiatric disorders such as Parkinson's disease, schizophrenia, stroke, depression, and dementia. Not only such studies are of great economic value but also they help avoid over- or underdosing of human subjects with the relevant medications.

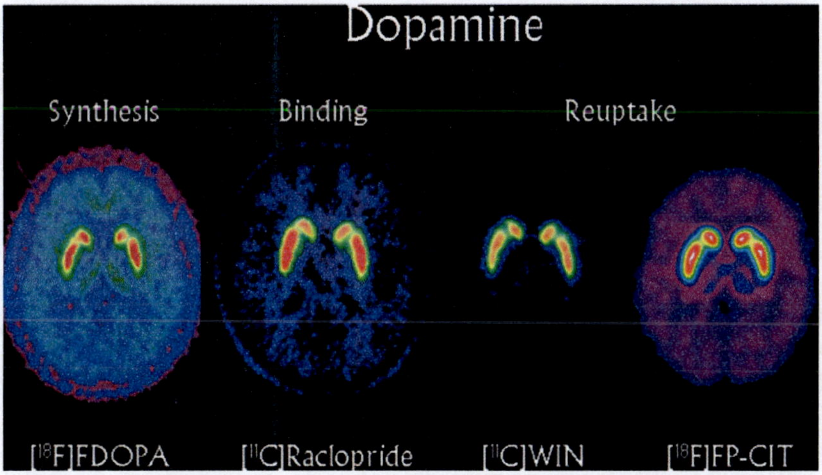


**Fig. 4.4** Images of dopaminergic neurotransmission in human striatum. The images were obtained with [$^{18}$F]FDOPA (dopamine synthesis capacity), ([$^{11}$C]raclopride (dopamine receptor availability), and ([$^{11}$C]CFT/WIN35,428 and ([$^{18}$F]FP-CIT (dopamine transporter availability). The color coding indicates values of net clearance in the case of FDOPA, and binding potentials in the cases of raclopride, CFT and CIT. Note that the images and the underlying values may be deceptive, as dopamine synthesis capacity can be translated into actual dopamine synthesis only when the relevant dopa concentration is known. Thus the capacity may be high and the actual synthesis low if the dopa concentration is very low. Likewise, binding potentials are not indicative of binding directly as they depend on the receptor density and ligand concentration. Thus, the binding potential may be high but the binding very low when the ligand concentration is very low in relation to the receptor density

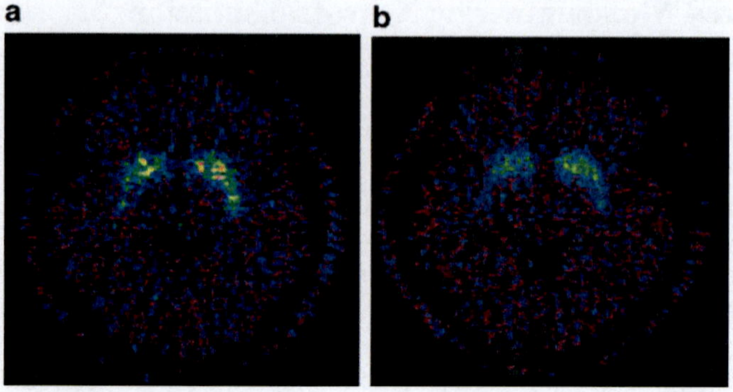


**Fig. 4.5** Reduction of dopamine receptor availability ([$^{11}$C]raclopride binding potential) in cognitive awareness. The two panels (**a**) and (**b**) depict the images of the binding potentials of raclopride as they decline in the striatum during a form of Yoga meditation. The interpretation is based on the assumption that receptor density does not change significantly during the short period from the baseline tomography to the tomography made during meditation. The decline of the binding potential is then an indication of the release of dopamine and increase of dopamine concentration in the vicinity of the receptors. It can be shown that the percentage decrease of the binding potential is a measure of the increase of dopamine's occupancy of dopamine's receptors caused by the increase of dopamine concentration. From Kjaer et al. (2002)

## *4.4.1 Dopamine*

The myriad number of sites that can now be imaged are relevant to multiple disorders, as exemplified for the dopamine systems in Table 4.4. The imaging of dopamine's receptors has shown that they vary with personality such as the propensity for sensation-seeking (Gjedde et al. 2010a) and that they are lost with normal aging (Wong et al. 1984; Pohjalainen et al. 1998). Striking clinical examples of changes include abnormalities at multiple steps in the neurotransmission of dopamine in disorders as diverse as Parkinson's disease (Gjedde et al. 1993; Hoshi et al. 1993; Varrone et al. 2001), schizophrenia (Farde et al. 1988; Wong and Gjedde 2009; Wong et al. 1986a–c), and addiction, including cocaine and methamphetamine abuse (Volkow et al. 2003).

Parkinson's disease and schizophrenia in many ways represent opposite ends of a spectrum of perturbed dopaminergic neurotransmission in striatum and midbrain. In schizophrenia, PET and SPECT reveal increased capacity of dopamine synthesis from DOPA, increased tonic and phasic dopamine release, elevated density of dopamine $D_2$-like receptors, and increased extracellular dopamine. These findings explain how antagonists of dopaminergic neurotransmission can mitigate the worst effects of the disease, at least for some subtypes with positive symptoms, but complete reversal has not been achieved, probably because the primary disorder is cortical. For antipsychotics as well as a number of antidepressants, it has been shown that occupancy of 60–80% of the dopamine receptors yields the best therapeutic effect (Farde et al. 1988). The receptor occupancy plotted against the

**Table 4.4** Changes of dopaminergic neurotransmission in vivo

| | Variable | | | | |
|---|---|---|---|---|---|
| Condition | $D_2$-like receptor availability or density | DDC enzyme activity | DA "release" or occupancy | Baseline DA concentration | DAT availability or density |
| Schizophrenia | Two to threefold increase (some studies [Tune et al. 1993; Wong et al. 1986a–c]); no change [Farde et al. 1986, others]) | Increase (multiple studies, e.g., [Ernst et al. 1999; Rousset et al. 1993]) | Increase [Breier et al. 1997; Laruelle et al. 1996] | Increase [Abi-Dargham et al. 2000; Gjedde and Wong 2001] | |
| Parkinson's disease | Increase in untreated patients [Rinne et al. 1993] | Decrease (multiple studies [Hoshi et al. 1993; Kumakura et al. 2006]) | | | Decrease [McCann et al. 1998; Varrone et al. 2001] |
| Cocaine abuse | Decrease [Volkow et al. 2003] | | Decrease (attenuation of normal increase) [Wong et al. 2006] | | Decrease [Volkow et al. 1997] |
| Attention-deficit-hyperactivity disorder | Increase [Lou et al. 2004] | | Increase [Rosa-Neto et al. 2005] | | Decrease [Volkow et al. 2007] |
| Normal aging | Decrease 7–10% per decade [Wong et al. 1984] | Decrease [Kumakura et al. 2005] | | | |
| Gilles de la Tourette's syndrome | Increase [Wong et al. 1997a–c] | | Increase [Singer et al. 2002] | | |
| Alcoholism | Decrease? | | Decrease? | | |
| Bipolar disease/Psychosis | Twofold increase [Pearlson et al. 1995] | | | Increase [Gjedde and Wong 2001] | |
| Sensation-seeking | Inverted U-shape [Gjedde et al. 2010a, b] | | | | |
| Temporal Lobe Epilepsy w/ Psychosis | | Increase [Reith et al. 1994] | | | |
| Cognitive awareness | | | Increase [Kjaer et al. 2002] | | |

plasma level of the drug essentially is a Michaelis–Menten graph of the therapeutic "window of opportunity" for a neuroleptic. This appears also to be true for partial agonists as well as antagonists of the dopamine receptors (Grunder et al. 2003).

Remarkable changes of dopamine synthetic capacity and dopamine transporter availability have been shown in Parkinson's disease (Gjedde et al. 1993; Hoshi et al. 1993; Kumakura et al. 2006; Varrone et al. 2001). The cause of the disease is not known but two different forms could exist in theory, one in which the initial insult is a decline of vesicular incorporation of dopamine, leading to neurotoxic levels of dopamine in the dopaminergic terminals, and one in which the initial insult is a loss of dopamine synthesis capacity. Dopamine will be low in both cases but the processes of synthesis and metabolism of dopamine will express opposite changes.

Attention-deficit-hyperactivity disorder (ADHD) is suspected of being a dopamine deficiency disorder related to a class of disturbances known as Reward Deficiency Syndrome. PET reveals increased dopamine receptor availability (higher than normal binding potential) in ADHD (Lou et al. 2004), consistent with lower than normal extracellular dopamine level, which can be rectified by blocking the dopamine reuptake sites with methylphenidate, as shown in Fig. 4.6 (Rosa-Neto et al. 2005). The reason for the deficiency may be abnormally high activity of the

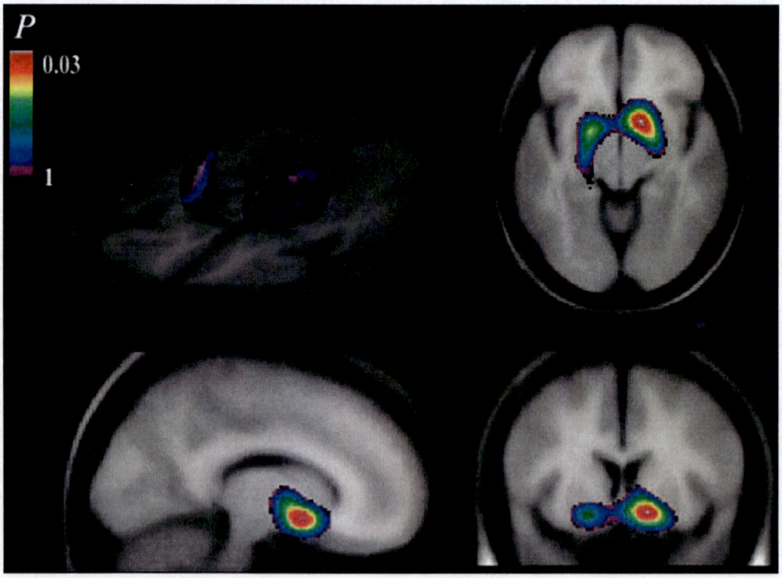


**Fig. 4.6** Reduction of dopamine receptor availability ([$^{11}$C]raclopride binding potential) by treatment with methylphenidate in patients with ADHD. The panels are color coded images of the percentage decline of raclopride's binding potential at dopamine receptors in the striatum, caused by administration of the pharmaceutical methylphenidate (Ritalin$^R$). The calculated percentage declines were superimposed on T1-weighted MR images of the anatomy of the subjects' brains. The co-registered images, therefore, show the sites of significant increase of dopamine's occupancy of dopamine's receptors. From Rosa-Neto et al. (2005)

**Table 4.5** Dopamine transporter binding potentials in striatum (From [$^{11}$C] WIN-35,428 data reported by McCann et al. 1998)

| | | Binding potentials ($k_3/k_4$) | |
|---|---|---|---|
| Group | N | Caudate nucleus | Putamen |
| Healthy volunteers | 10 | 7.5±1.7 | 7.3±1.4 |
| Methamphetamine abuse | 6 | 5.8±1.0 | 5.5±0.9 |
| Parkinson's disease | 3 | 4.0±0.9 | 2.3±0.4 |

dopamine transporters. Methylphenidates effects in ADHD may be due in part to enhancement of small DA increases from salient stimuli (Volkow et al. 2005).

Dopaminergic neuroimaging is particularly important to the investigation of substance abuse, where the challenge is to understand the interactions of receptors and enzymes with the substances abused. PET has revealed that the main effect of blockade of dopamine transporters by cocaine, amphetamine, or methamphetamine, is an increase dopamine's occupancy of its receptors because of increased extracellular dopamine, as evidenced by decline of the binding potential of a dopamine receptor ligand (Schlaepfer et al. 1997; Volkow et al. 2007). Other monoamine transporters are also affected because of the comparatively low selectivity of the transporters toward the blockers.

Many studies reveal decreased extracellular dopamine concentrations and reduced dopamine $D_{2/3}$ receptor availability and density in abstinent abusers, which would tend to exacerbate the dopamine deficiency. Indeed, the monoamine receptors tend to be the direct or indirect, intended or accidental, targets of a wide range of illicit drugs enjoyed recreationally by humans. The drugs that are abused most frequently (opiates, ethanol, nicotine, amphetamine, and cocaine) acutely increase extracellular dopamine concentrations in striatum and nucleus accumbens. In contrast, drugs that are not abused by humans do not appear to affect the dopamine occupancy, while drugs with aversive properties that are actively avoided tend to lower dopamine occupancy, indirectly proving that activation of dopaminergic neurotransmission is a fundamental property of drugs of abuse (DiChiara and Imperato 1988).

Continued abuse of cocaine or other blockers of monoamine transporters attenuates the increase of occupancy elicited by amphetamine and lowers the density of the transporters in the striatum, measured shortly after the onset of abstinence (Villemagne et al. 1998; Volkow et al. 2007; Wong et al. 1998a, b). The transporter availability rises again after longer periods of abstinence, but cue-induced craving for cocaine in abstinent abusers can raise dopamine's occupancy of its receptors without the aid of cocaine (Wong et al. 2006), probably because of cue-induced release of dopamine. Table 4.5 summarizes findings of dopamine transporter availability in three groups of subjects, as measured with the dopamine transporter ligand [$^{11}$C]WIN-35,428 (McCann et al. 1998).

### *4.4.2 Serotonin*

Serotonin receptors are most abundant in cerebral cortex, particularly in the hippocampal region and the ventromedial prefrontal cortex, where PET also shows

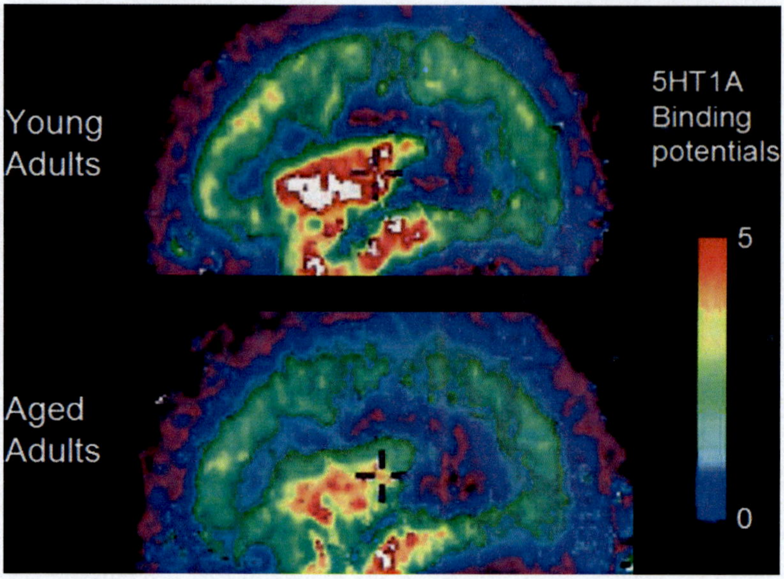


**Fig. 4.7** Reduction of serotonin $5HT_{1A}$ receptor availability ([$^{11}$C]WAY-100635 binding potential) in aging. The images show the average calculated values of the binding potentials in sagittal sections through the hippocampal region in young adults (n=7) and aged adults (n=12). The binding potentials are higher in cortical than in subcortical regions, and highest in the ventromedial prefrontal cortex, insula, and hippocampal region. As the images are the result of tomography at rest, it is the interpretation that the decline of the binding potentials reflects a loss of receptors, i.e., a decline of the receptor density. From Møller et al. (2007b)

that binding potentials at $5HT_{1A}$ receptors decline with age (Møller et al. 2007a, b) and in depression (Sargent et al. 2000), with return to normal after antidepressant treatment. The changes are somewhat equivocal because it is uncertain whether the changes are due to change of receptor density or change of serotonin level. Abnormalities of occupancy at these receptors have been difficult to show with the available radioligands because serotonin reuptake inhibitors have little effect on the binding potential. The decline of $5HT_{1A}$ binding potentials in aging is shown in Fig. 4.7. Changes attributable to loss of serotonin have been observed after stroke in patients who develop pathological crying, which is relieved instantly by administration of selective serotonin reuptake inhibitors (Møller et al. 2007a, b).

The drug MDMA (3,4-methylenedioxy-*N*-methylamphetamine), commonly known by the street name ecstasy, is a semisynthetic entactogen of the phenethylamine family, which affects the serotonin system by transport into serotonergic and dopaminergic terminals. The acute effect of MDMA appears to result from the drug's interference with dopamine and serotonin reuptake (Armstrong and Noguchi 2004). PET results obtained with the radioligands McN5652 and DASB in humans and shown in Fig. 4.8 suggest that MDMA abusers suffer a measurable reduction of serotonin transporter availability in multiple brain regions, which is relieved after abstinence, although the evidence is still equivocal (Buchert et al. 2004; McCann et al. 2005; Sabol et al. 1996).

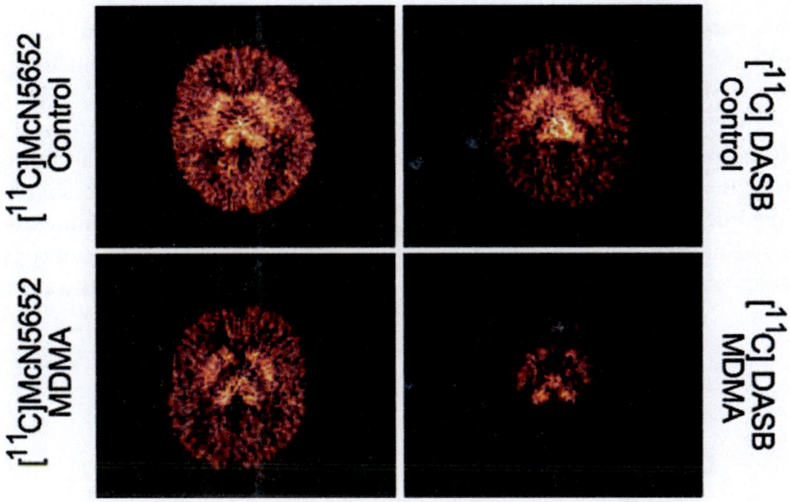


**Fig. 4.8** Effect of MDMA on serotonin transporter availability ([$^{11}$C]McN5652 and [$^{11}$C]DASB binding potentials). The images show the average calculated values of the binding potentials in transaxial sections through the basal ganglia and thalamus of healthy volunteers before and during administration of MDMA. The binding potentials at the serotonin transporters are highest in the subcortical regions. As the images show the result of an acute perturbation of serotonergic neurotransmission, the assumption is that the decline of the binding potentials in the basal ganglia and thalamus reflects the increase of serotonin concentration caused by the drug. From McCann et al. (2005)

### *4.4.3 Design of Monoaminergic Drugs*

Neuroreceptor mapping with PET and SPECT also plays an important role in the development of drugs that specifically target the abnormalities that PET and SPECT have identified in disease. The applications typically involve imaging of the biodistribution of candidate drugs labeled with [$^{11}$C] or [$^{18}$F] for PET, or [$^{123}$I] or [$^{99}$Tc] for SPECT. The areas of application include (1) therapeutic rationale, (2) justification for the study of specific biological systems, (3) mechanisms of drug action, and (4) rational drug dosing (Wong et al. 2000, Wong et al. 2009, Epub 2008). Much of this work can be carried out in neuroreceptor imaging laboratories of universities that negotiate master agreements with pharmaceutical companies for evaluations of ligands that will be used in the early stages of drug development (Wong and Pomper 2003).

## 4.5 Conclusions

Imaging approaches to the study of monoaminergic neurotransmission have relevance to both neurological and psychiatric diseases. Historically, neuroreceptor imaging focused on the pathophysiology of major neuropsychiatric disorders such as schizophrenia, Parkinson's disease, and depression, but it is now also being used to study and diagnose dementia, stroke, and drug abuse, and to assist in the development of new drugs. The method of imaged neurochemistry is emerging as an effective means of clarifying the mechanisms and neuroadaptive properties of brain disease related to monoaminergic neurotransmission. Currently, a wealth of radiotracers of the monoaminergic systems can be brought to bear on the unraveling of these disorders. The efforts help provide rational drug dosing and examine mechanisms of action of drugs that are relevant to cognitive research. This will improve objectivity in research and clarify drug mechanisms and treatment pathways in the future.

# Chapter 5
# Blood–Brain Transfer and Metabolism of Oxygen*

## 5.1 Introduction

Oxygen is essential to brain function. Disruption of oxygen delivery to brain results in complete cessation of functional activity in the human brain in 6 seconds (Hansen 1984, 1985). The reason for this dramatic effect is not known with certainty. Increased potassium conductance in neuronal membranes plays a role, but the effect also suggests that oxygen reserves in brain tissue are very low in relation to the steady-state turnover, and it implies that oxygen availability is intimately related to mechanisms that subserve consciousness, because the effect is not explained by depletion of intermediary metabolites.

The delivery of oxygen to brain tissue differs in major respects from the delivery of oxygen to other tissues. In the classic description of Krogh (1919), the standard capillary bed is a system of parallel tubes serving uniform cylinders of tissue (known as Krogh cylinders). This simple arrangement yields a quantitative expression of oxygen delivery to the tissue, suggestive of specific mechanisms of regulation. However, in brain tissue, the vascular bed differs from the vascular beds of other tissues in two important ways: Both the absent recruitment of capillaries in states of activation of neurons and the general principle of topographic arrangement of the vessels are unlike those of other vascular beds, but the influence of these differences has eluded students of the vascular physiology of the brain.

The absence of recruitment suggests that the work of the brain, measured as energy turnover, varies little over time, and the random arrangement of capillaries suggests that the geometry of the vascular network is fixed competitively by the work of the cells, rather than by the need to make frequent, orderly and substantive changes to the diffusion capacity, as in skeletal and cardiac muscles. When the geometry is always the same, there is no need for any particular order.

Thus, in brain, the vascular bed is so poorly organized that no single geometric description has general validity and hence no single prediction of oxygen

* Adapted from Gjedde (2005) Blood-brain transfer and metabolism of oxygen. In: *Blood–Brain Barriers: From Ontogeny to Artificial Interfaces* Dermietzel RS, Spray DC, Nedergaard M, eds. Hoboken: John Wiley and Sons, Ltd., volume 2, Chap. 22, pp. 523–549, with permission from John Wiley and Sons, Ltd.

A. Gjedde et al., *Neurokinetics: The Dynamics of Neurobiology In Vivo*, 
DOI 10.1007/978-1-4419-7409-9_5, 

distribution in the tissue is possible (Wang and Bassingthwaighte 2001). To predict the general properties of oxygen delivery and consumption, the valid approach is the averaging of properties across large volumes of tissue, on the strength of observations that have wide applicability, always with the provision that local and regional properties may differ significantly from the global average.

Only two claims of the capillary bed in brain appear to be indisputable, i.e., that the capillaries have multiple arterial sources but a common venous terminus, and that the density of capillaries is proportional to an average regional rate of metabolism at steady-state (also known as the "default" state, Raichle et al. 2001). From these claims arises a revised distribution of oxygen in brain tissue that is different from the classic distribution in Krogh cylinders. The revision is made possible by the simple realization, first introduced by Erwin R. Weibel in *The Pathway for Oxygen* (1984), that every segment of the capillary bed exists to supply an equivalent volume or mass of brain tissue with oxygen, i.e., that every fraction of the tissue is served by commensurate fractions of a capillary bed, having the appropriate oxygen diffusibility and accounting for equivalent fractions of the total oxygen consumption.

This approach considers the random delivery of oxygen molecules from the source in arbitrarily distributed capillaries to the sink in equally arbitrarily distributed mitochondria. The only discernible regulation is the need to make enough oxygen molecules available to the cells served by the capillaries. The approach assumes that suitable mechanisms originally served to place the capillary bed and the cells in relation to each other in such a way that this requirement would be fulfilled. It is an intriguing possibility that the astrocytic projections, in addition to other services, could provide a structural restraint on the anatomic relationship of capillaries and neurons, established to achieve the optimal match of randomly placed capillaries and cells. The current approach is not limited to capillaries per se. It is suspected that oxygen exchange with the tissue may begin in arteriolar microvessels (Duling and Berne 1970). For the purposes of this analysis, arteriolar microvessels are therefore part of the effective capillary bed.

The variable organization can be reduced to compartments of predictable behavior by making the simple assumption that the structure of the vascular bed evolved to satisfy the existing needs for oxygen, in contrast to the more traditional view that the needs for oxygen can be deduced from a specific structure of the vascular bed. It turns out that the resulting temporal and anatomical averages of the random processes of oxygen delivery achieve an appearance of orderliness that is illusory.

## 5.2 Blood–Brain Transfer of Oxygen

### *5.2.1 Capillary Model of Oxygen Transfer*

Capillary models of blood–brain transfer focus on the distribution of molecules inside microvessels. In the case of molecular oxygen, the extraction from the capillary to the tissue establishes an oxygen profile in the capillary bed that ranges from the tension associated with the more or less fully saturated hemoglobin at the arterial inlet to the tension of the more or less fully desaturated hemoglobin at the venous outlet. Between these extremes, the tension depends on the extraction of oxygen from the network of capillaries serving the tissue. The discussion earlier led to the conclusion that it is impossible to make a specific prediction of the profile in individual capillaries but it is possible to look at the ensemble of capillaries as a unit that provides the tissue with the oxygen it needs. This simple assumption leads to a solution that is close, but not identical, to the general claim that the oxygen tension declines monoexponentially from the arterial inlet to the venous outlet.

The solution is obtained by letting the oxygen extraction from the blood flowing in the general direction from the arterial inlet to the venous outlet be proportional to the fraction of the capillary bed cumulatively served by the blood flow. The fractional segment of the capillary bed is denoted by a variable $z$ that ranges from 0 to 1. In every segment, an equivalent amount of oxygen is extracted by the cells served by their respective segments of the capillary bed,

$$E_{O_2}(z) = \frac{J_{O_2}(z)}{F\, C_{O_2}^{art}} = \frac{z\, \overline{J}_{O_2}}{F\, C_{O_2}^{art}}, \tag{5.1}$$

where the oxygen extraction fraction is $E_{O_2}(z)$, $J_{O_2}(z)$ the segmental rate of cerebral metabolism of oxygen ($CMR_{O_2}$) at the segment $z$ of the capillary bed in question, $\overline{J}_{O_2}$ is the measured $CMR_{O_2}$ value per unit mass or volume of the brain region served by this particular capillary bed, $F$ the measured blood flow to the region, and $C_{O_2}^{art}$ is the measured arterial oxygen concentration at the arterial inlet. Standard properties of arterial blood are given in Table 5.1.

The extraction of oxygen from brain capillaries causes the capillary oxygen tension to decline from the arterial oxygen tension at the arterial end of the capillary bed to the venous oxygen tension at the venous end. The decline of the oxygen tension as a function of $z$ is determined by the oxygen remaining after the extraction. The loss of oxygen establishes a declining oxygen saturation of hemoglobin according to the formula:

$$S_{O_2}^{cap}(z) = S_{O_2}^{art}\,[1 - E_{O_2}(z)]\,, \tag{5.2}$$

**Table 5.1** Average properties of oxygen compartments in human brain

| Variable | Unit | Arterial | Capillary | Venous | Tissue | Mitochondria |
|---|---|---|---|---|---|---|
| $C_{O_2}$ | mM | 8.5 | 7.0 | 5.3 | | |
| $P_{O_2}$ | mmHg | 114 | 53 | 40 | 20–30 | 8.5 |
| $P_{50}$ | mmHg | 26 | 35 | 35 | | 0.5 |
| $h$ | | 2.84 | 3.5 | 3.5 | | |
| $S_{O_2}$ | ratio | 0.985 | 0.81 | 0.61 | | 0.94 |

Modified from Gjedde et al. (2005a, b)

where $S_{O_2}^{art}$ is the saturation of hemoglobin in arterial blood and $S_{O_2}^{cap}$ is the saturation in capillary blood at the segmental position $z$. The amount of oxygen remaining in the capillary bed also satisfies the Hill equation of the relationship between oxygen tension and hemoglobin saturation,

$$S_{O_2}^{cap}(z) = \frac{1}{1 + \left(\frac{P_{50}^{cap}(z)}{P_{O_2}^{cap}(z)}\right)^{h(z)}}, \tag{5.3}$$

where $P_{O_2}^{cap}(z)$ is the oxygen tension as a function of position in the capillary, $P_{50}^{cap}(z)$ the half-saturation tension of oxygen at the position $z$, and $h(z)$ is the value of the Hill coefficient at that position. The Hill equation then is solved for the capillary oxygen tension associated with the particular extraction fraction,

$$P_{O_2}^{cap}(z) = P_{50}^{cap}(z) \sqrt[h(z)]{\frac{S_{O_2}^{art}(1 - E_{O_2}(z))}{1 - S_{O_2}^{art}(1 - E_{O_2}(z))}} \tag{5.4}$$

which, on the assumption that the saturation of hemoglobin in arterial blood is negligibly different from unity, reduces to the expression,

$$P_{O_2}^{cap}(z) = P_{50}^{cap}(z) \sqrt[h(z)]{\frac{1}{E_{O_2}(z)} - 1}, \tag{5.5}$$

where $P_{O_2}^{cap}(z)$ is the pressure head of oxygen diffusion when arterial oxygen saturation is 100%.[1]

The third determinant of oxygen delivery is the oxygen diffusibility, which depends on the diffusion coefficient, the solubility, and the position of the oxygen sink or sinks. The oxygen diffusibility is given by the ratio between the oxygen flux and the difference between the pressures of oxygen in the $z$ segment of the

[1] As an alternative to the assumption of unity for $S_{O_2}^{art}$, it is possible to lump the effect of arterial desaturation into an apparent change of the half-saturation constant of hemoglobin, $P_{50}^{app}(z)$. The apparent half-saturation constant is then given by:

$$P_{50}^{app}(z) = \frac{P_{50}^{cap}(z)}{\sqrt[h(z)]{\frac{1-S_{O_2}^{ven}}{S_{O_2}^{art}-S_{O_2}^{ven}}}}$$

in which $P_{50}^{app}$ is the half-saturation tension corrected for arterial desaturation of hemoglobin and $S_{O_2}^{ven}$ is the venous oxygen saturation. The denominator assumes a value within a few percentages of unity for ordinary extraction fractions at arterial oxygen tensions above 100 mmHg. At lower oxygen tensions, the value of the denominator depends on the position in the capillary bed. As the physically dissolved oxygen represents no more than 1–2% in normoxia (although less in hypoxia), the physically dissolved oxygen "spares" some of the bound oxygen during extraction. There is indication that the half-saturation tension of arterial blood is adjusted in hypoxia to counter the effects of the low arterial hemoglobin saturation (Gjedde 2002).

capillary bed and in the mitochondria of the tissue segment served by the segment of capillary bed,

$$L(z) = \frac{\overline{J}_{O_2}}{P_{O_2}^{cap}(z) - P_{O_2}^{mit}(z)}, \tag{5.6}$$

where the oxygen diffusibility $L(z)$ is a function of the value of the segmental position $z$, and $P_{O_2}^{mit}(z)$ is the operative mitochondrial oxygen tension at the site, which a priori is not known with certainty in brain tissue, although attempts have been made to calculate the value (Gjedde et al. 2005a).

The diffusibility depends on the surface area of the capillary segment. The diffusion gradient also varies with the distribution of mitochondria in the tissue, which is not known with certainty for any specific location and, hence, must be inferred from consideration of averages across larger regions.

If the sinks are placed at the end of the diffusion path, $L(z)$ is an estimate of the diffusion coefficient, corrected for solubility, that reflects the average distance from the $z$ segment to the mitochondria served by the segment. The distinction between even and uneven distribution of metabolic sinks for oxygen will be discussed in greater detail in the subsequent section on tissue oxygen tensions.

Equations (5.1), (5.5), and (5.6) can be solved in sequence, beginning with the extraction fraction, as functions of the capillary segment $z$. The results are shown in Fig. 5.1. The extraction rises linearly as shown in (5.1), and the capillary oxygen tension declines accordingly as shown in (5.5). The decline is approximately monoexponential as predicted by conventional capillary models (Crone 1963), but the reason is not per se the simple pressure gradient between capillary and tissue but the more complex satisfaction of a constant oxygen delivery, organized as a combination of the linearly increasing extraction fraction and the sigmoidal relationship between oxygen tension and hemoglobin saturation.

The diffusibility also rises in response to the needs of the tissue segment served, to maintain the constant flux of oxygen, and the increase is not linear. A nonlinearly rising diffusibility can be achieved in several ways, including a declining distance between adjacent microvessels as well as by branching of vessels.

### *5.2.2 Compartment Model of Oxygen Transfer*

A compartment is an idealization of kinetic relationships in which all concentration gradients are placed at the interfaces among the compartments. Inside the compartments, in contravention of the true nature of gradients, concentrations are considered to be uniform. The linearity of the extraction fraction means that the capillary bed can be reduced to a single microvascular compartment with a uniform oxygen tension, $\overline{P}_{O_2}^{cap}$, derived from the linearly averaged extraction fraction. Because the extraction fraction is proportional to $z$, its average is equal to its value at $z = 0.5$. Provided the arterial saturation of hemoglobin is close to unity, the weighted average capillary oxygen tension associated with the oxygen present in all segments of

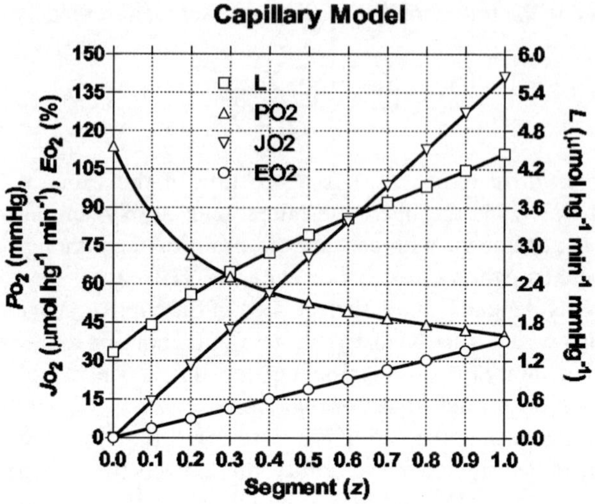


**Fig. 5.1** Oxygen tension, extraction fraction, and oxygen diffusibility as a function of axial segmentation of cortical capillary bed ($z$ direction), calculated for baseline condition of human brain. Abscissa: Axial segments of capillary bed ($z$). Ordinate: Capillary oxygen extraction fraction ($E_{O_2}(z)$), tension ($P_{O_2}^{cap}(z)$), and diffusibility ($L(z)$) as functions of axial segment. The variables were calculated with (5.7–5.9) under normal conditions of whole human brain, as listed in Tables 5.1 and 5.2: Normal CMR$_{O_2}$ (141 µmol hg$^{-1}$ min$^{-1}$ at the baseline), normal blood flow ($F$, 43 ml hg$^{-1}$ min$^{-1}$), normal arterial oxygen concentration (8.7 mM), normal $P_{50}^{cap}$, the half-saturation tension of oxygen (35 mmHg when corrected for the Bohr shift), and normal $\overline{h}$, the Hill coefficient (3.5 when corrected for Bohr shift). $P_{O_2}^{cap}(0)$ was assumed to be equal to the arterial oxygen tension $P_{O_2}^{art}$ (114 mmHg)

the capillary bed is then (Vafaee and Gjedde 2000),

$$\overline{P_{O_2}^{cap}} = \overline{P_{50}^{cap}} \sqrt[\overline{h}]{\frac{2}{\overline{E}_{O_2}} - 1}, \tag{5.7}$$

where $\overline{P_{50}^{cap}}$ is the average capillary half-saturation tension of oxygen binding to hemoglobin,[2] and $\overline{h}$ is the average value of the Hill coefficient in the capillary.

[2] As an alternative to the assumption of arterial hemoglobin saturation of unity, it is again possible to define an apparent average half-saturation tension that corrects for the degree of desaturation of hemoglobin that exists prior to the entry into the capillary. In the compartmental model, in agreement with (5.4), $\overline{P_{50}^{app}}$ is given by the expression:

$$\overline{P_{50}^{app}} = \frac{P_{50}^{cap}}{\sqrt[\overline{h}]{\frac{2 - S_{O_2}^{art} - S_{O_2}^{ven}}{S_{O_2}^{art} - S_{O_2}^{ven}}}},$$

**Table 5.2** Average properties of human whole-brain and cerebral cortex

| Variable [unit] | Whole-brain | Cerebral cortex |
|---|---|---|
| $CMR_{glc}$ [$\mu$mol g$^{-1}$ min$^{-1}$] | 0.25 | 0.30 |
| $CMR_{O_2}$ [$\mu$mol g$^{-1}$ min$^{-1}$] | 1.40 | 1.60 |
| CBF [ml g$^{-1}$ min$^{-1}$] | 0.43 | 0.50 |
| Oxygen extraction fraction ($E_{O_2}$) [ratio] | 0.39 | 0.38 |
| Oxygen diffusibility ($L$) [$\mu$mol g$^{-1}$ min$^{-1}$ mmHg$^{-1}$] | 0.032 | 0.041 |
| Cytochrome oxidase activity ($J_{max}$) [umol g$^{-1}$ min$^{-1}$] | 1.59 | 1.70 |

Modified from Kuwabara et al. (1992), Vafaee et al. (1999), and Gjedde et al. (2005a)

Substituting $\overline{L} = L(0.5)$, $\overline{P}^{cap}_{O_2} = P^{cap}_{O_2}(0.5)$, and $\overline{P}^{mit}_{O_2} = P^{mit}_{O_2}(0.5)$ in (5.6) yields,

$$\overline{L} = \frac{\overline{J}_{O_2}}{\overline{P}^{cap}_{O_2} - \overline{P}^{mit}_{O_2}}, \tag{5.8}$$

where $\overline{P}^{mit}_{O_2}$ is the tissue's minimum oxygen tension and $\overline{L}$ is the mean oxygen diffusibility, equal to $L(z)$ at $z = 0.5$. The magnitude of $\overline{L}$ depends on the unknown value of $\overline{P}^{mit}_{O_2}$, the oxygen tension at cytochrome $c$ oxidase. Equations (5.7) and (5.8) can be used to estimate $\overline{L}$ by identifying an extraction fraction at which the oxygen reserve of the mitochondria has been exhausted and calculating the corresponding value of $\overline{L}$ by means of (5.8). It follows from (5.8) that the oxygen consumption is,

$$\overline{J}_{O_2} = \overline{L}\left[P^{cap}_{50}\sqrt[\overline{h}]{\frac{2}{\overline{E}_{O_2}} - 1} - \overline{P}^{mit}_{O_2}\right], \tag{5.9}$$

where the average value of $h$ is $\overline{h}$. Solutions to these equations are shown in Figs. 5.1 and 5.2 for normal values of human brain.

## 5.3 Oxygen in Brain Tissue

### *5.3.1 Cytochrome Oxidation*

The solutions to the equations earlier depend on the magnitude of the mitochondrial oxygen tension in (5.8) and (5.9). This tension, together with the affinity of cytochrome $c$ oxidase toward oxygen, the maximum reaction rate of each mitochondrion, and the number of mitochondria in a given segment of tissue, determines

which has the value 1.02 for an extraction of 40% at an arterial oxygen tension of 114 mmHg (standard). At a tension of 52 mmHg the denominator has the value 1.16 at an extraction of 40% and 1.12 at 60% extraction.

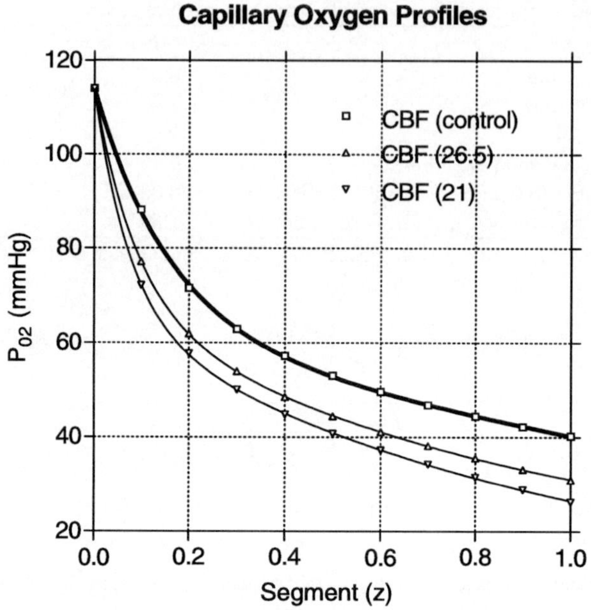


**Fig. 5.2** Capillary oxygen profiles. Abscissa: Axial segment of capillary bed ($z$). Ordinate: Capillary oxygen tensions ($P_{O_2}^{cap}(z)$) as function of axial segment, calculated as explained in Fig. 5.1 for baseline (CBF 43 ml hg$^{-1}$ min$^{-1}$), and two oligemia (low blood flow) states (CBF 26.5 and 21 ml hg$^{-1}$ min$^{-1}$), chosen to be just at and just below limit of adequate oxygen delivery. Numerical solution of the three equations with the three unknowns $E_{O_2}(z)$, $P_{O_2}^{cap}(z)$, and $x_o(z)$ yielded the capillary oxygen tension profile for insufficient oxygen supply

the oxygen consumption according to the simplest formulation of the Michaelis–Menten equation (Gnaiger et al. 1995, 1998; Guilivi 1998):

$$J_{O_2} = \frac{J_{max}\, P_{O_2}^{mit}}{P_{50}^{cyt} + P_{O_2}^{mit}}, \tag{5.10}$$

where $P_{50}^{cyt}$ is the apparent half-saturation tension of the oxygen reaction with cytochrome oxidase, $P_{O_2}^{mit}$ the mitochondrial oxygen tension, and $J_{max}$ is the maximum velocity of the cytochrome oxidase reaction. This equation is applicable to all mitochondria but it is not known to which extent the magnitude of each of the parameters of the equation, as well as the resulting oxygen consumption, apply generally to the tissue as a whole. Therefore, it is necessary to operate with the concept of an average mitochondrion. In vivo, the affinity is considerably lower (an order of magnitude) than the inherent affinity observed under in vitro circumstances. The expected saturation of the enzyme, therefore, is not as pronounced in vivo as in vitro. The standard properties of this enzyme are given in Table 5.1.

The actual flux of oxygen in each tissue volume element depends on the regional oxygen consumption, which often is assumed to be the same per unit tissue mass or volume everywhere but may in fact vary considerably from site to site. The flux is the difference between the average rate of oxygen consumption of the tissue segment and the regionally (but not locally) invariant rate of oxygen delivery to each unit of tissue. The mitochondrial oxygen tension in turn links the oxygen consumption with the oxygen delivery. The mitochondrial oxygen tension represents the balance between the diffusion of oxygen to the site of metabolism and the reaction of oxygen with cytochrome *c* oxidase, according to the simple Michaelis–Menten expression. The local variability of mitochondrial oxygen tension in brain tissue is therefore the key to the understanding of oxygen homeostasis.

### 5.3.2 Mitochondrial Oxygen Tension

#### Distributed Model of Tissue and Mitochondrial Oxygen

In the classic model of nonuniformly distributed oxygen, the mitochondria are evenly scattered in the tissue but the oxygen tension varies spatially as oxygen is consumed. This situation is reminiscent of the conditions existing in the Krogh cylinder, in which the local mitochondrial oxygen tension varies with the distance of the mitochondria from the microvessels, while the oxygen consumption rate is assumed to be the same for each mitochondrion, because affinity or maximum reaction rate or both are adjusted to maintain the same oxygen consumption rate in every tissue element. This situation is consistent with an equation of the form:

$$j_{O_2}(z,x) = -K\,A(z)\,\frac{\partial p_{O_2}(z,x)}{\partial x} = (1-x)\,J_{O_2}(z), \tag{5.11}$$

where $j_{O_2}(z,x)$ is the flux of oxygen from the capillary segment extending between 0 and $z$ to its surrounding tissue segment extending to a fractional distance, $x$, from the microvessels, $J_{O_2}(z)$ the regionally invariant oxygen consumption of each unit of tissue mass or volume, $K$ Krogh's diffusion coefficient for a unit surface area, and $A(z)$ is the magnitude of that area. The sites of oxygen consumption generate an extravascularly directed gradient of oxygen tension from the microvascular compartment to the tissue mass or volume fraction, $x$, served by the microvascular compartment. Assuming $J_{O_2}(z) = z\overline{J}_{O_2}$, and $A(z) = z\overline{A}$, dependence on $z$ disappears hence,

$$\frac{d\mathrm{P}_{O_2}(x)}{\mathrm{d}x} = -\left(\frac{1-x}{K\,\overline{A}}\right)\overline{J}_{O_2}, \tag{5.12}$$

where $\overline{A}$ is the diffusion area of a segmental surface. Integration with respect to the tissue mass or volume fraction $x$ yields the pressure as a function of the tissue mass or volume fraction served by this segment of the capillary,

$$P_{O_2}(x) = P_{O_2}(1) + \left[\frac{\overline{J}_{O_2}}{2\,K\,\overline{A}}\right](1-x)^2\,, \tag{5.13}$$

where $P_{O_2}(1)$ is the minimum oxygen tension in the tissue ($x = 1$). For evenly distributed (i.e., equally abundant more and less "distant") sites of oxygen consumption in tissue the relationship between the Krogh's coefficient $K$ and the weighted average oxygen diffusibility is,

$$\overline{L} = 2\, K\, \overline{A} \tag{5.14}$$

which follows from (5.8) and (5.13), with $P_{O_2}(0) = \overline{P}^{\mathrm{cap}}_{O_2}$, $P_{O_2}(1) = \overline{P}^{\mathrm{mit}}_{O_2}$, and $x = 0$. Substitution of $\overline{L}/2$ for $K\,\overline{A}$, and using (5.8) to eliminate $\overline{P}^{\mathrm{mit}}_{O_2}$ yields the weighted average oxygen tension as a function of the weighted average distance from the capillary,

$$P_{O_2}(x) = \overline{P}^{\mathrm{cap}}_{O_2} - \frac{\overline{J}_{O_2}}{\overline{L}}\left[2x - x^2\right], \tag{5.15}$$

where $\overline{L}$ is the weighted average oxygen diffusibility defined in (5.5) and $\overline{J}_{O_2}$ is the average rate of oxygen consumption of the entire tissue. This solution is shown in Fig. 5.4 for the cases of one normal and two low levels of blood flow to the human cerebral cortex.

For $x = 1$, (5.15) yields the minimum mitochondrial oxygen tension reached at the end of the diffusion path,

$$\overline{P}^{\mathrm{mit}}_{O_2} = \overline{P}^{\mathrm{cap}}_{O_2} - \frac{\overline{J}_{O_2}}{\overline{L}} \tag{5.16}$$

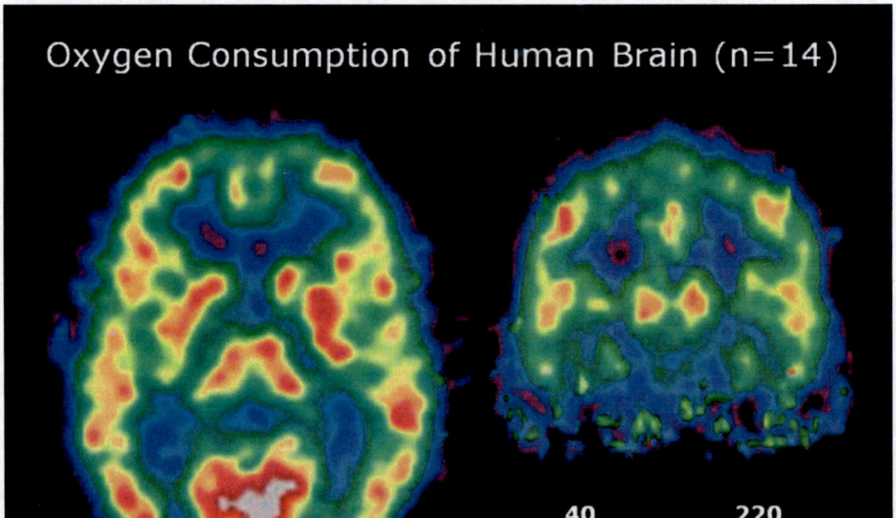


**Fig. 5.3** The case of two compartments: Map of average cerebral oxygen consumption rates of 14 healthy adult humans, determined by the method of Ohta et al. (1992) according to (1.32). Courtesy of Christopher Bailey, PhD, Aarhus PET Center

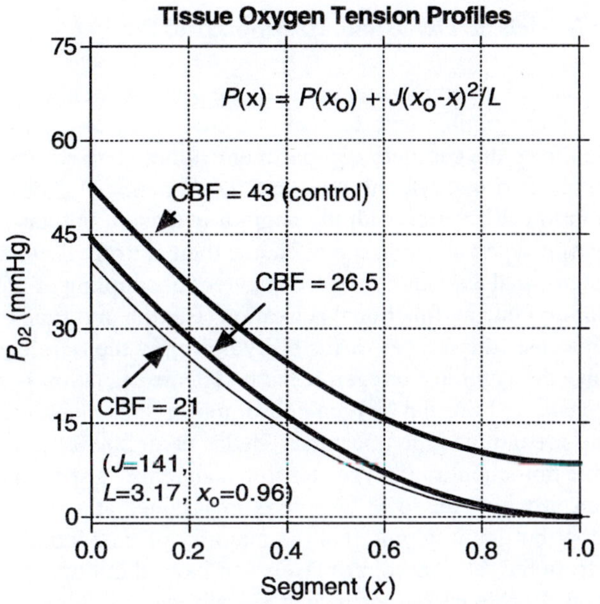


**Fig. 5.4** Tissue oxygen tensions of distributed tissue model. Abscissa: Tissue volume fraction ($x$). Ordinate: Oxygen tension profiles ($P_{O_2}(x)$) as a function of tissue volume fraction for the three cases of Fig. 5.2, using the equations indicated in the figure, identical to (5.15) and (5.24) for $x$ less than unity. The normal end-path oxygen tension was assumed to be 8.5 mmHg, as listed in Table 5.1

which is identical to the evenly distributed oxygen tension associated with the linear diffusion case earlier. Integration and normalization of (5.15) yields the weighted tissue average of the oxygen tension,

$$\overline{P}^{\text{br}}_{O_2} = \overline{P}^{\text{cap}}_{O_2} - \frac{2\,\overline{J}_{O_2}}{3\,\overline{L}}, \tag{5.17}$$

where $\overline{P}^{\text{br}}_{O_2}$ is the weighted average tissue tension.

Equation (5.15) is in fact a special solution to a more general formulation in which the local mitochondrial oxygen tension varies with the distance from the microvessels, depending on the local oxygen consumption rate, which in turn varies with the oxygen tension according to the Michaelis–Menten equation for cytochrome c oxidase, provided the enzyme has the same affinity and maximum reaction rate everywhere. In that case the diffusion equation assumes the form:

$$K\,\overline{A}\,\frac{d^2 P_{O_2}(x)}{dx^2} = J_{\max}\left(\frac{P_{O_2}(x)}{P_{50}^{\text{cyt}} + P_{O_2}(x)}\right), \tag{5.18}$$

which equation in principle is soluble only by numerical means. When $P_{O_2}(x) \gg P_{50}^{\text{cyt}}$, the equation reduces to the special case of spatially uniform oxygen consumption discussed earlier. Whether this happens depending on the tension and on the affinity.

## Compartment Models of Tissue and Mitochondrial Oxygen

In the simplest case of a compartment model, the local mitochondrial oxygen tensions are uniformly negligible with respect to the capillary oxygen tension and the oxygen diffusion from the capillary compartment, although they would remain significant with respect to the cytochrome c oxidase reaction. In this case the local oxygen consumption still varies with the negligible tension but has little influence on the delivery of oxygen to the tissue, which is then entirely flow-limited. This is the basis for the original explanation of the apparent uncoupling of blood flow from oxygen consumption during functional activation (Buxton and Frank 1997; Gjedde 1996a, b). In this case, the oxygen diffusibility is simply the ratio between oxygen consumption and the capillary oxygen tension. This explanation is not consistent with the maintenance of normal oxygen consumption rates also during substantial flow decreases, for example in hypocapnia (Gjedde et al. 2005a).

The negligible mitochondrial oxygen tension is in reality a special case of a more general situation in which the local tension is maintained at a constant magnitude with uniform distribution, implying that the majority of mitochondria reside at the same distance from the relevant microvessels and have about the same rate of oxygen consumption. In this case, the oxygen is delivered by the simplest diffusion relation established by the tension difference between the microvessels and the mitochondria. The fixed distance from the relevant microvessels and the mitochondria represents a diffusion barrier, of which the tension at the far end is given by:

$$\overline{P}_{O_2}^{mit} = \overline{P}_{O_2}^{cap} - \frac{\overline{J}_{O_2}}{\overline{L}}, \tag{5.19}$$

where $\overline{P}_{O_2}^{mit}$ is the tension of the uniformly distributed mitochondrial oxygen. The resulting compartmental model is shown in Fig. 5.5. The gradient associated with the interface between the capillary and mitochondrial compartments allows the approximation of an average tissue oxygen diffusibility according to (5.16).

It is of course possible that distributed and compartmental oxygen states coexist in the tissue. The local mitochondrial oxygen tension, affinity, maximum reaction rate, and distance from microvessels, may vary unpredictably for all mitochondria, the density of which may also vary to maintain the appropriate oxygen consumption for each tissue element. However, because every arrangement discussed earlier yields the same or similar expressions of the average oxygen diffusibility, it appears generally and for most practical purposes valid to estimate the lowest oxygen tension in the tissue (which may also be the tension of the uniformly distributed oxygen, of course) by the formula:

$$\overline{P}_{O_2}^{mit} = \overline{P}_{O_2}^{cap} - \frac{\overline{J}_{O_2}}{\overline{L}} \tag{5.20}$$

in which $\overline{P}_{O_2}^{mit}$ is either the uniform tissue oxygen tension in the linear case or the minimum tissue oxygen tension in the nonlinear case.

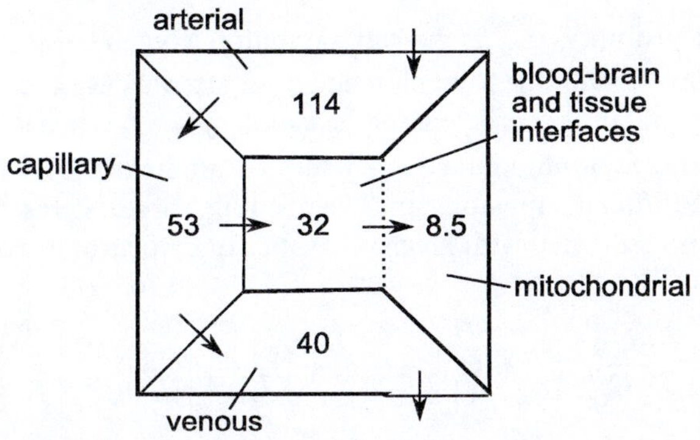


**Fig. 5.5** Combined compartment model of oxygen tensions in brain. The compartments include arterial, capillary, venous, and mitochondrial spaces. The interface between the capillary and mitochondrial compartments is a diffusion barrier, the exact position of which is not known with certainty, although it may be dominated by the capillary endothelium. The numbers refer to normal oxygen tensions in units of mmHg, calculated from (5.7) for capillary oxygen tension, (5.5) for venous oxygen tension ($z = 1$), and (5.9) for mitochondrial oxygen tension ($x = 1$). The oxygen tension of the average tissue compartment or capillary-mitochondrial diffusion interface is a simple linear average. Note that the term capillary bed is used for the entire portion of the vascular bed that interacts with the tissue. This portion may include elements of arterial microvessels (Duling and Berne 1970)

The average tissue oxygen tension, on the other hand, depends on the actual oxygen distribution but generally must observe the following inequalities:

$$\overline{P}_{O_2}^{cap} - \frac{2\,\overline{J}_{O_2}}{3\,\overline{L}} \leq \overline{P}_{O_2}^{br} \leq \overline{P}_{O_2}^{cap} - \frac{\overline{J}_{O_2}}{2\,\overline{L}}. \tag{5.21}$$

## 5.4 Flow-Metabolism Coupling of Oxygen

The expressions for the blood–brain transfer and metabolism of oxygen must be combined to obtain a general expression of the flow-metabolism coupling of oxygen. In this coupling, the mitochondrial oxygen tension represents the balance between the diffusion of oxygen to the site of metabolism and the reaction of oxygen with cytochrome *c* oxidase, known to follow the simple Michaelis–Menten expression given earlier. Using (5.1) with $z = 0.5$ to solve (5.9) for cerebral blood flow yields:

$$F = \frac{\overline{J}_{O_2}}{2\,C_{O_2}^{art}} \left( 1 + \left[ \frac{\overline{J}_{O_2} + \overline{L}\,P_{O_2}^{mit}}{\overline{L}\,\overline{P}_{50}^{cap}} \right]^{\overline{h}} \right), \tag{5.22}$$

where $F$ is the blood flow to the tissue (per unit weight or volume), $\overline{J}_{O_2}$ the average oxygen metabolism of the tissue, $C_{O_2}^{art}$ the arterial oxygen concentration,

$\overline{L}$ the oxygen diffusibility, $\overline{P}^{\mathrm{cap}}_{50}$ is the half-saturation oxygen tension of hemoglobin in capillaries, corrected for the Bohr shift and prior arterial desaturation, and $\overline{h}$ is the weighted average of Hill's coefficient for the capillary bed.[3] When the two equations involving the uniform or minimum oxygen tension are linked (5.10 and 5.22), $\overline{P}^{\mathrm{mit}}_{\mathrm{O_2}}$ can be eliminated from the relationship. The resulting formula prescribes the rate of blood flow that is associated with given properties of cytochrome c oxidase,

$$F = \frac{\overline{J}_{\mathrm{O_2}}}{2\,C^{\mathrm{art}}_{\mathrm{O_2}}}\left(1 + \left(\left[1 + \frac{\overline{L}\,P^{\mathrm{cyt}}_{50}}{J_{\mathrm{max}} - \overline{J}_{\mathrm{O_2}}}\right]\left[\frac{\overline{J}_{\mathrm{O_2}}}{\overline{L}\,P^{\mathrm{cap}}_{50}}\right]\right)^{\overline{h}}\right), \tag{5.23}$$

where all the parameters in principle are variables under homeostatic control. The variables have all been defined earlier. The equation describes a mechanism of nonlinear flow-metabolism coupling, cast in terms of the nominally independent variables or constants $F$, $C^{\mathrm{art}}_{\mathrm{O_2}}$, $\overline{L}$, $P^{\mathrm{cyt}}_{50}$, $J_{\mathrm{max}}$, $P^{\mathrm{cap}}_{50}$, and $\overline{h}$, and the nominally dependent variable $\overline{J}_{\mathrm{O_2}}$. If independent variables other than flow remain constant, oxygen consumption must follow flow nonlinearly. As oxygen metabolism is known not to depend rigidly on blood flow, for example in hypocapnia, it is safe to conclude that several independent variables undergo simultaneous change. One possible adjustment is the compensation for low oxygen saturation of arterial blood suggested by the substitution of $\overline{P}^{\mathrm{app}}_{50}$ for $\overline{P}^{\mathrm{cap}}_{50}$ as argued earlier.[4]

In principle, the relationship between oxygen consumption and blood flow described by the flow-metabolism couple is flow-limited. It is shown in Fig. 5.6 for normal values of the independent variables and constants for human brain. Assuming these parameters to be invariant, the relationship is highly nonlinear and fixed. The changes of oxygen consumption, relative to baseline, on the other hand, and the accompanying changes of blood flow, also relative to baseline, are linearly linked, as shown in theory in Fig. 5.7, and exemplified in Figs. 5.13 and 5.14, indicating that the assumption of invariant variables and constants other than blood flow and oxygen consumption, is valid under these conditions.

It is increasingly clear, however, that the independent variables as well as the constants must undergo substantial changes in some circumstances. Independent

---

[3] Adjustment of $\overline{L}$ may compensate for the degree of hypoxemic desaturation of arterial blood, provided the product $\overline{L}\,\overline{P}^{\mathrm{app}}_{50}$ can be considered a constant. This is the case if the magnitude of $L$ is adjusted to compensate for the decline of $\overline{P}^{\mathrm{app}}_{50}$ by desaturation, by a process of recruitment. Equation (5.4) shows that the combined desaturation prior to capillary entry and a commensurate adjustment of the blood–brain oxygen diffusion capacity together could serve to render the baseline magnitude of the product $LP^{\mathrm{app}}_{50}$ valid for all degrees of desaturation.

[4] The complete flow-metabolism equation shows that the effect of arterial hypoxemia can also be counteracted by maintaining a constant $P^{\mathrm{mit}}_{\mathrm{O_2}}/P^{\mathrm{cyt}}_{50}$ ratio as $P^{\mathrm{mit}}_{\mathrm{O_2}}$ declines. It is of interest that experimental observations conform less well to the equation when it is correctly modified for the prior arterial desaturation of hemoglobin. This suggests either that reserves of oxygen are available in the tissue for use by the mitochondria, or that "recruitment" may reset $\overline{L}$ to compensate for the prior arterial desaturation of hemoglobin.

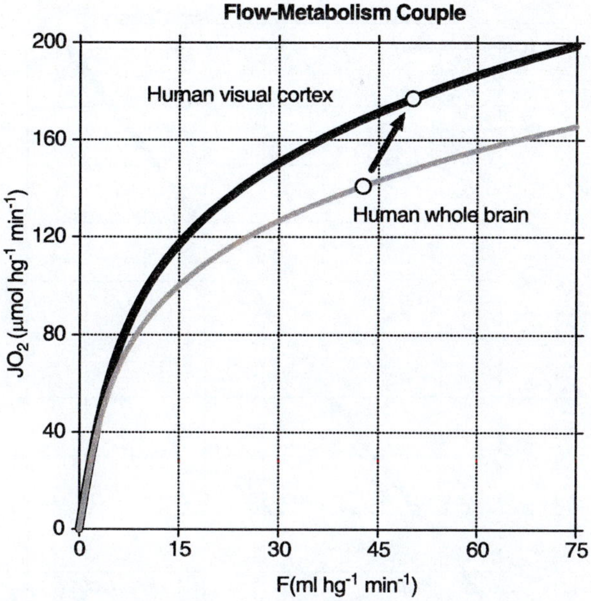


**Fig. 5.6** Flow-metabolism couple predicted by (5.22) and (5.23) for human whole-brain and visual cortex. Parameters used to generate curves are given in Tables 5.1 and 5.2 for whole human brain and human cerebral cortex. Abscissa: Blood flow (ml $hg^{-1}$ $min^{-1}$). Ordinate: Oxygen consumption (μmol $hg^{-1}$ $min^{-1}$). Points indicate normal values listed in Table 5.2 for the two tissues. Change from whole-brain to cortex is associated with slight decline of the oxygen extraction fraction ($E_{O_2}$ declines from 0.39 to 0.38)

variables or constants that may undergo simultaneous adjustment include the maximum reaction rate or affinity, or both, of cytochrome oxidase. There is evidence that they change in the same direction during activation of cytochrome oxidase in vitro, while there is other evidence that the affinity of cytochrome oxidase is subject to inhibition by nitric oxide which is likely to play a major role in flow adjustment as well (Gjedde et al. 2005a).

## 5.5 Limits to Oxygen Supply

Oxygen consumption is secured by an adequate blood flow and the possible adjustments of independent variables in (5.23). However, these adjustments eventually become exhausted when the flow declines to a certain threshold, below which the compulsory flow-limitation of oxygen consumption intervenes. At that threshold it is no longer possible for the circulation to maintain normal oxygen metabolism of all parts of the tissue.

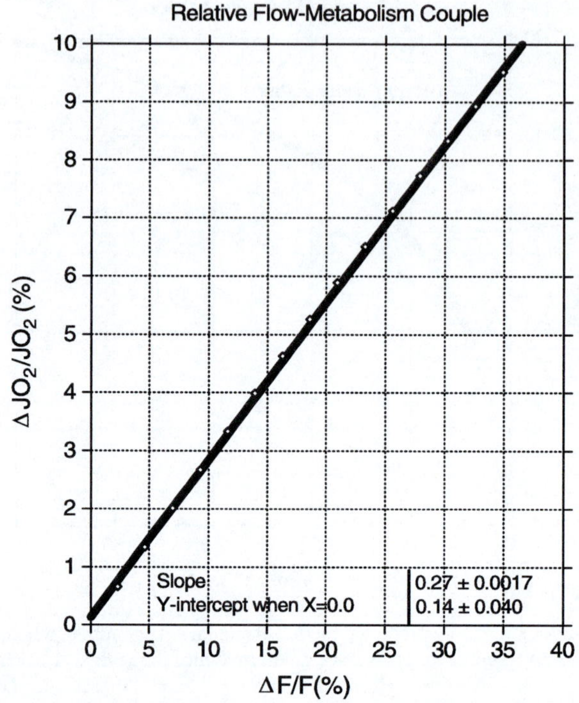


**Fig. 5.7** Changes of oxygen consumption and blood flow, relative to respective baselines for human whole-brain, calculated from human whole-brain curve shown in Fig. 5.6. Abscissa: Change of blood flow, relative to baseline (%). Ordinate: Change of oxygen consumption, relative to baseline (%). Note approximate linearity, slope of which is close to reciprocal of magnitude of average capillary Hill coefficient ($1/\overline{h}$), as indicated by (5.22) and (5.23)

The imbalance may be due to insufficient capillary driving pressure or increased maximum reaction rate or both, and the effect of the imbalance is different for the different arrangements and regulations of mitochondria and cytochrome oxidase discussed earlier. However, in all cases, the oxygen tension will decline until no mismatch exists between the metabolism and the delivery, but the needs of tissue may not be satisfied in part or in whole.

### *5.5.1 Distributed Model of Insufficient Oxygen Delivery*

The issue of when the needs of the tissue are not satisfied can be resolved by consideration of the tissue oxygen distribution. Equations (5.11–5.15) apply to the situation in which the capillary oxygen supply is not sufficient to supply the entire tissue at the necessary rate. When that happens, the segmental fraction $x$ served with oxygen will not reach unity ($x < 1$).

In the case of the nonuniform oxygen distribution, the effect depends on the distribution of metabolism. It is possible that some parts of the tissue exhaust the oxygen supply, leaving other parts without oxygen. The exhaustion of the oxygen supply by a given fraction ($x$) of the tissue mass or volume can be dealt with by introduction of a limit, $x_o$, at which the oxygen supply is exhausted ($x_o < 1$), and performing an analysis similar to that for (5.11–5.15), but starting from the assumption that $j_{O_2}(z,x) = [x_o(z) - x]\overline{J}_{O_2}$ rather than (5.11). An equation similar to (5.15) will result, though diffusibility which varies with $z$ will replace the average diffusibility, $\overline{L}$.

$$P_{O_2}(x,z) = P_{O_2}^{cap}(z) - \frac{2x - x^2}{L(z)}\overline{J}_{O_2}, \tag{5.24}$$

where $\overline{J}_{O_2}$ is the normal oxygen consumption of the tissue, excluding the part that is not supplied with oxygen (equal to $1 - x_o$), and $x_o$ is that fraction, able to consume oxygen at the normal rate. The oxygen tensions accompanying the depletion of the oxygen supply are given by (5.24), which yields the value of $x$ ($x_o$) at which the tension reaches zero, when the supply does not match the demand. The actual oxygen consumption average of the tissue is then $J'_{O_2} = x_o\,\overline{J}_{O_2}$.

In this case, (5.1) is no longer valid as written but must be modified to account for the nonlinear rise of the extraction fraction caused by the exhaustion of the oxygen supply. Equation (5.1) modifies to:

$$E_{O_2}(z) = \frac{z\,\overline{J}_{O_2}}{F\,C_{O_2}^{art}} \int_o^z x_o(z)\,\mathrm{d}z, \tag{5.25}$$

where as above $\overline{J}_{O_2}$ is the normal oxygen consumption per unit mass or volume of tissue. To solve (5.25) for $x_o(z)$, it is necessary to establish the capillary oxygen tension as a function of the extraction fraction reached in segment ($z$) serving the volume fraction $x$ according to (5.5),

$$P_{O_2}^{cap}(z) = P_{50}^{app}(z)\ \sqrt[h(z)]{\frac{1}{E_{O_2}(z)} - 1} \tag{5.26}$$

as well as the fraction of the tissue supplied with oxygen at that tension. It follows from evaluating (5.24) at $x = x_o$, the point at which $\overline{P}_{O_2}^{mit} = 0$, that

$$x_o(z) = \sqrt{\frac{L(z)\,P_{O_2}^{cap}(z)}{\overline{J}_{O_2}}}, \tag{5.27}$$

where $L(z)$ and $\overline{J}_{O_2}$ are assumed to be known from the normal cases of adequate oxygen supply. Thus, when $L(z)$ and $\overline{J}_{O_2}$ are known, numerical solution of the three (5.25–5.27) with the three unknowns $E_{O_2}(z)$, $P_{O_2}^{cap}(z)$, and $x_o(z)$ yields both the capillary oxygen tension profile for a case of insufficient oxygen supply, and the resulting oxygen consumption $J'_{O_2}(z) = x_o(z)\,\overline{J}_{O_2}$.

The solution can be further illustrated by consideration of the flux capacity by differentiation of (5.24), which yields a description of the oxygen metabolism in each segment of the tissue ($x$) at each segment of the capillary bed ($z$),

$$j_{O_2}(z, x) = (x_o(z) - x)\,\overline{J}_{O_2} = \frac{L(z)\,P_{O_2}^{cap}(z)}{x_o(z)} - x\,\overline{J}_{O_2} \tag{5.28}$$

which is a line of slope $-\overline{J}_{O_2}$ and ordinate intercept $J'_{O_2}(z)$. The ordinate intercept is the actual oxygen flux from that part of the capillary bed, equal to the product of the capillary oxygen pressure and oxygen diffusibility, relative to the tissue fraction served at that point ($x_o(z) \leq 1$).

The linearly declining flux capacity of (5.28) is illustrated in Fig. 5.8 for normal human brain as well as for lowered blood flow rates, including a blood flow rate just below the ischemic threshold. For values of $z$ at which $x_o < 1$, $E_{O_2}(z)$ fails to rise linearly and the compartmental simplification is then possible only by numerical solution of (5.25–5.27).

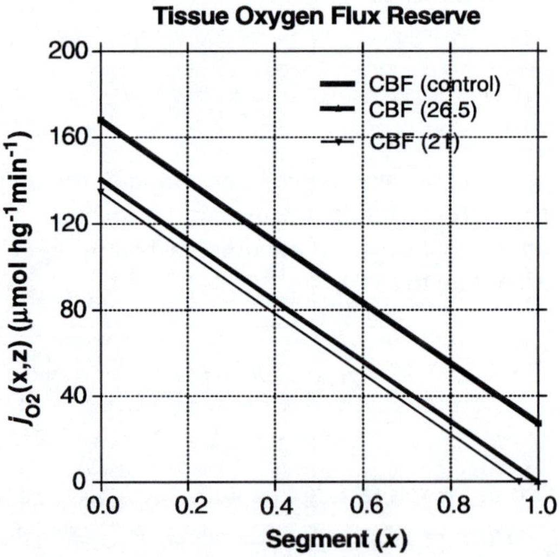


**Fig. 5.8** Flux-capacitor lines for distributed model of tissue oxygen, predicted from normal values shown in Tables 5.1 and 5.2, and in conditions of reduced blood flow as indicated in the graph. Dependent variable was calculated by means of (5.11) and (5.28). Abscissa: Tissue segment ($x$) served by average capillary compartment ($z = 0.5$). Ordinate: Oxygen flux as function of $x$. Flux reserve at normal blood flow is consistent with minimum mitochondrial oxygen tension of 8.5 mmHg. At the lowest blood flow rate, only 96% of tissue ($x_o = 0.96$) can consume oxygen at the normal rate

### 5.5.2 *Compartment Model of Insufficient Oxygen Delivery*

In the case of the uniform oxygen distribution, the reaction rate will decline in every mitochondrion until the consumption again matches the supply, as dictated by the flow. The extraction of oxygen at which this happens can be evaluated by means of (5.9) by setting $\overline{P}_{O_2}^{mit}$ equal to zero and solving for $\overline{E}_{O_2}$, assuming diffusibility and hemoglobin saturability to remain constant. The threshold and the declining oxygen consumption below the threshold can also be illustrated as a contraction of the tissue fraction served by the capillary compartment. This is shown in Fig. 5.9 as a shortening of the capillary segmental fraction $z$ served by the capillary compartment, beginning at the blood flow rate at which $x$ declines below unity before the end of the capillary bed has been reached (Fig. 5.10).

The actual oxygen flux is less than the arterial supply of oxygen because it depends on the average oxygen tension in the capillary, which in turn depends on the extraction of oxygen. The discrepancy between the arterial delivery of oxygen and the actually available oxygen supply is a result of the inherent diffusion-limitation of oxygen delivery to brain tissue (Gjedde et al. 1991; Kassissia et al. 1995), as expressed in the magnitude of $\overline{L}$. Thus, while oxygen consumption is flow limited, oxygen delivery is diffusion limited.

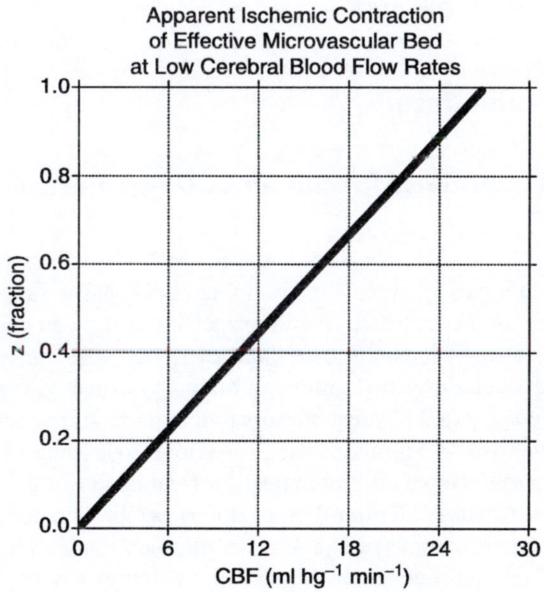


**Fig. 5.9** Prediction from (5.1) of apparent ischemic contraction of effective capillary bed serving brain tissue, using standard values listed in Tables 5.1 and 5.2. Abscissa: Cerebral blood flow in units of ml hg$^{-1}$ min$^{-1}$. Ordinate: Capillary bed segments serving entire tissue segment ($x_o = 1$). Effective capillary bed starts contracting at blood flow rates below 27 ml hg$^{-1}$ min$^{-1}$

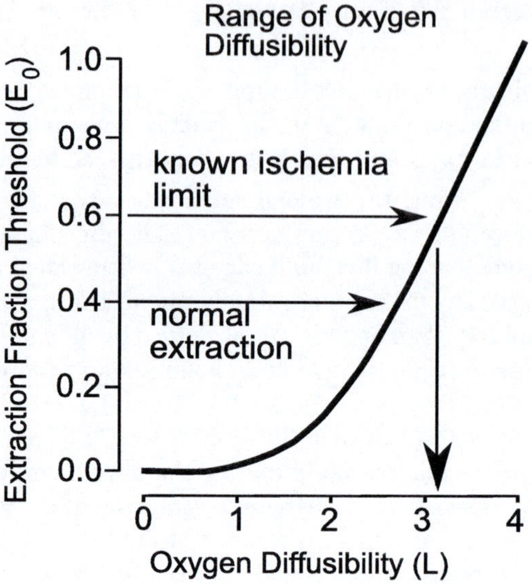


**Fig. 5.10** Estimation of whole-brain oxygen diffusibility ($\overline{L}$) from (5.7) and (5.8), using standard values listed in Tables 5.1 and 5.2. Abscissa: Possible magnitudes of oxygen diffusibility in units of μmol hg$^{-1}$ min$^{-1}$ mmHg$^{-1}$. Ordinate: Predicted oxygen extraction fraction threshold as a function of possible oxygen diffusibility. Known ischemic oxygen extraction limit ($E_o$) of 60% is consistent with magnitude of $\overline{L}$ of 3.17 μmol hg$^{-1}$ min$^{-1}$ mmHg$^{-1}$

## 5.6 Experimental Results

### *5.6.1 Brain Tissue and Mitochondrial Oxygen Tensions*

With values listed in Tables 5.1 and 5.2, the inequality given as (5.21) yields the prediction that average oxygen tensions in human brain tissue range from 20 to 30 mmHg, depending on the distribution and properties of oxygen consumption sites. Measurements of oxygen tensions date back many decades for animal brain but are not particularly well described in human brain. As expected from the treatment leading to the inequality (5.21), most authors find that values in human whole-brain tissue in vivo range from 100 mmHg close to arterioles to less than 30 mmHg further from the microvessels. However, the majority of values lie in the range of 20–40 mmHg (Zauner et al. 2002; Hemphill et al. 2005), which is completely consistent with the tissue compartmental average of 30 mmHg summarized in Fig. 5.5.

Mitochondrial oxygen tension in human brain tissue in vivo can only be inferred indirectly from the oxygen binding properties of hemoglobin and the oxygen metabolizing properties of cytochrome c oxidase. They have been summarized in Table 5.1. Using these values and the ischemic threshold of oxygen extraction of 60% (see later), Gjedde et al. (2005a) employed (5.19) to estimate an average mitochondrial oxygen tension of 8.5 mmHg in human brain.

### 5.6.2 Flow-Metabolism Coupling

Equation (5.23) is the quantitative expression of the flow-metabolism couple in compartmental kinetics. The equation considers only steady-state, although it could possibly be used to predict some transient events, provided the rate constants of the transients were of the appropriate magnitude. The foundations of the equation is shown in Figs. 5.1, 5.2, and 5.4. Figure 5.1 shows that the oxygen extraction as a function of capillary segment rises linearly.

Fundamentally, similar models of nonlinear flow-metabolism coupling have been formulated by several authors (Weibel 1984; Gjedde 1996b; Buxton and Frank 1997; Vafaee and Gjedde 2000; Aubert and Costalat 2002). The inverse relation between oxygen extraction fraction and oxygen tension of the capillary bed is central to these revisions of the conventional coupling of flow to metabolism, which was assumed to maintain a normal oxygen extraction fraction. The revisions arose from the insight that an inverse relation is necessary because every increase of the extraction fraction represents a reduction of the partial pressure of oxygen in the capillary compartment.

The formulation of the steady-state flow-metabolism couple is consistent with evidence obtained in visual cortex activation (Vafaee et al. 1999; Gjedde and Marrett 2001), motor cortex activation (Kastrup et al. 2002; Vafaee and Gjedde 2004), and hypothermia (Sakoh and Gjedde 2003) as shown in Figs. 5.11–5.13.

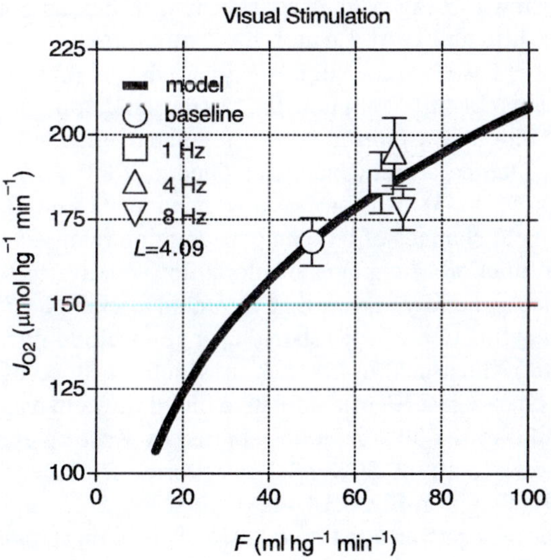


**Fig. 5.11** Flow-metabolism coupling during stimulation of visual cortex of human brain, measured with positron emission tomography of labeled oxygen uptake. Curve shows segment of cortical flow-metabolism couple illustrated in Fig. 5.6 for cerebral cortex. Abscissa: Cortical blood flow in units of ml hg$^{-1}$ min$^{-1}$. Ordinate: Oxygen consumption in human visual cortex in units of μmol hg$^{-1}$ min$^{-1}$ (modified from Vafaee et al. 1999)

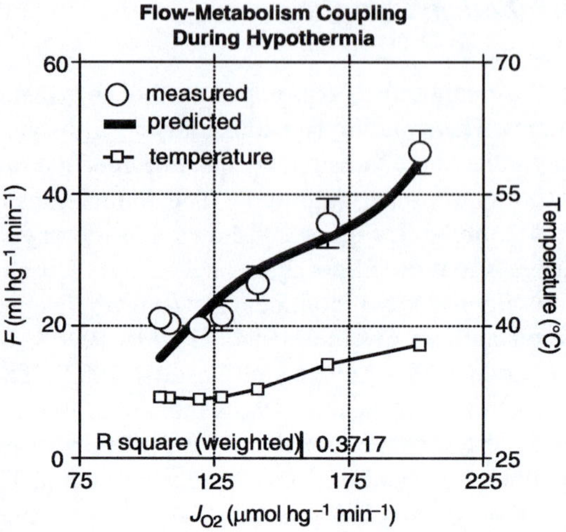


**Fig. 5.12** Flow-metabolism coupling during hypothermia, calculated for porcine cortical tissue with oxygen diffusibility of 5 μmol hg$^{-1}$ min mmHg$^{-1}$. Abscissa: Oxygen consumption in units of μmol hg$^{-1}$ min$^{-1}$. Ordinate: Cortical blood flow in units of ml hg$^{-1}$ min$^{-1}$ and cerebral core temperatures in units of °C. Note that axes have been inverted from Fig. 5.11 (modified from Sakoh and Gjedde 2003)

The evidence shown in Fig. 5.11 is consistent with human cortical gray matter having an oxygen diffusibility of 4 μmol hg$^{-1}$ min mmHg$^{-1}$, while the evidence shown in Fig. 5.12 is based on the slightly higher whole-brain oxygen diffusibility of approximately 5 μmol hg$^{-1}$ min mmHg$^{-1}$ in the smaller pig brain. The evidence also confirms that blood flow rates are down-regulated when the energy demand of neurons falls at low temperatures (Sakoh and Gjedde 2003).

The formulation of (5.23) also consistent with evidence from functional MRI experiments which yield changes of oxygen consumption relative to baseline oxygen consumption as a function of changes of blood flow relative to the baseline blood flow. This evidence generally shows that the relative change of the oxygen consumption is a linear function of the relative change of blood flow. Equation (5.23) makes the prediction illustrated in Fig. 5.7, in which the slope of the relative oxygen consumption change is 27% of the relative blood flow change for whole human brain. This is similar to the 30% relation reported by Kastrup et al. (2002), shown in Fig. 5.13, but lower than the 50% relation reported for human visual cortex by Hoge et al. (1999), shown in Fig. 5.14. As implied by (5.22) and (5.23), the slope is a measure of the Hill power $h$, as it equals $h^{-1}$. It is important to keep in mind that a linear relation between changes of oxygen consumption and blood flow with a slope different from unity, relative to the respective baselines, is proof of a nonlinear relation between the absolute values of the respective variables.

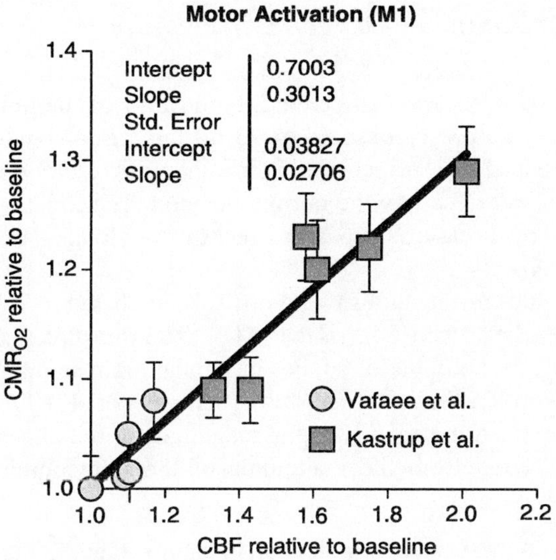


**Fig. 5.13** Relative changes of blood flow and oxygen consumption in human motor cortex, measured by functional MRI (Kastrup et al.) and positron emission tomography (Vafaee et al.). Abscissa: Change of blood flow, relative to baseline in fractions. Ordinate: Change of oxygen consumption in human motor cortex activation by finger motion, relative to baseline in fractions (modified from Kastrup et al. 2002 and Vafaee and Gjedde 2004)

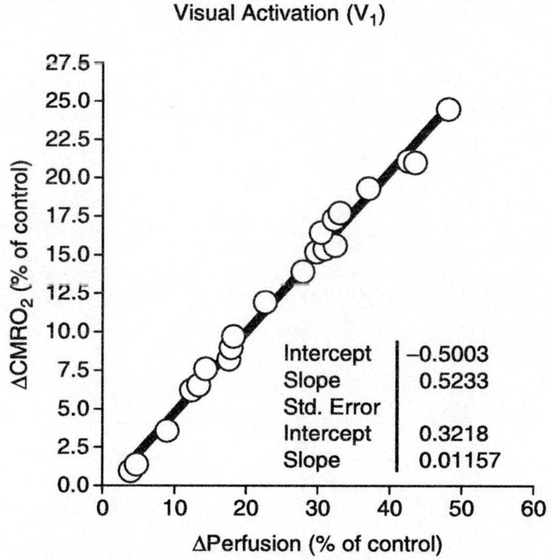


**Fig. 5.14** Relative changes of blood flow and oxygen consumption during stimulation of human visual cortex, measured by functional MRI. Abscissa: Change of blood flow, relative to baseline in percentage. Ordinate: Change of oxygen consumption in human motor cortex activation by photic stimulation of visual cortex, relative to baseline in percentage (modified from Hoge et al. 1999)

### 5.6.3 *Ischemic Limits of Oxygen Diffusibility*

Figures 5.2, 5.4, and 5.9 show that cerebral blood flow to the human brain limits the oxygen delivery below a threshold of 25 ml $hg^{-1}$ $min^{-1}$ when brain energy metabolism proceeds at the normal rate. Below this rate of blood flow, the oxygen supply is not sufficient to satisfy the normal demand. The limit is lower when the demand is lowered by disease, as appears to be the case with hepatic encephalopathy (Gjedde et al. 2010b).

For standard values of metabolism and diffusibility, listed in Tables 5.1 and 5.2, the threshold is reached when $\overline{E}_{O_2}$ rises to 60%. The threshold extraction of 60% has been confirmed in a number of studies of human brain summarized by Gjedde et al. (2005a, b) as well as in an animal study (Scheufler et al. 2002).

The threshold extraction depends on the magnitude of the diffusibility $\overline{L}$, which in turn can be inferred from the magnitude of the mitochondrial oxygen tension $\overline{P}_{O_2}^{mit}$.

Using the well-known ischemic limit as an indication that $\overline{P}_{O_2}^{mit}$ is negligible, (5.7) and (5.8) give rise to a relationship between $\overline{L}$ and the ischemic threshold shown in Fig. 5.10, according to which the magnitude of $\overline{L}$ is close to 3 μmol $hg^{-1}$ $min^{-1}$ $mmHg^{-1}$ (3.17 μmol $hg^{-1}$ $min^{-1}$ $mmHg^{-1}$ according to Gjedde et al. 2005a).

# Chapter 6
# Blood–Brain Glucose Transfer*

## 6.1 Brief History

Under normal circumstances, a mixture of $\alpha$- and $\beta$-D-glucopyranose (D-glucose) is the only fuel of brain energy metabolism (Pardridge 1983). The breakdown of D-glucose is regulated by complex mechanisms that influence the activities of phosphofructokinase and hexokinase. The metabolism depends heavily on the glucose concentration of the brain intracellular and interstitial fluids. As the tissue glucose concentration, in turn, depends on the glucose phosphorylation rate, the blood–brain transfer assumes a pivotal role in the supply of glucose to the brain.

The interface between the vascular and extravascular spaces in brain is the *barrière hèmato-encéphalique*, originally so named by Stern and Gautier (1921) because of its uncertain location. In a much quoted passage, August Krogh (1946) suggested that the endothelial lining of the brain vasculature may function as an extended cell membrane. The suggestion raised the question of whether the paracellular route past the cerebral capillary endothelium, i.e., the transport between the endothelial cells, is restricted to such an extent that it unmasks the properties of endothelial cell membranes in general, or whether the brain capillary endothelial membranes themselves have unusual properties that render them particularly restrictive to some classes of compounds, and specifically permeable to others.

The concept of facilitated diffusion evolved in parallel with the study of the transport of electrolytes and polar nonelectrolytes across cellular membranes. The nature of the specific glucose-transporting properties of all membranes remained a mystery until the discovery of insulin heralded the study of facilitated diffusion. By 1940, many workers speculated that insulin acted to facilitate the transfer of glucose across muscle cell membranes by interaction with a specific "property" of the membrane (Lundsgaard 1939; Levine et al. 1949). In 1948, LeFèvre showed that glucose efflux from human erythrocytes was a function of the intracellular

* Adapted from Gjedde (1992) Blood–brain glucose transfer. In: *Physiology and Pharmacology of the Blood–Brain Barrier*, Chap. 6a: *Handbook of Experimental Pharmacology*, MWB Bradbury, ed. Springer-Verlag, Berlin Heidelberg 1992, pp. 65–115, with permission from Springer-Verlag, Berlin Heidelberg.

A. Gjedde et al., *Neurokinetics: The Dynamics of Neurobiology In Vivo*, 
DOI 10.1007/978-1-4419-7409-9_6, 

glucose concentration in the manner predicted by the Michaelis–Menten equation (Michaelis and Menten 1913). Although it was apparent that human erythrocytes must possess these "specific" properties in abundance, the relation to insulin, if any, remained obscure.

In an abstract, Crone (1960) reported that the blood–brain transfer of D-glucose appeared to be much faster than predicted from the properties of the physically similar fructose. He also noted that the rate of transfer appeared to change in inverse proportion to the D-glucose concentration, as expected of substances subject to facilitated diffusion (Crone 1965). In the years between 1960 and 1980, the transfer was shown to obey the Michaelis–Menten criteria of saturability, competitive inhibition, and stereospecificity. Since 1980, bofth insulin-sensitive and insulin-insensitive glucose transporters have been isolated, identified, sequenced, and cloned. Although a mechanism of insulin action has been proposed, it is uncertain how the insulin-insensitive glucose transport is regulated, if at all.

## 6.2 Brain Endothelial Glucose Transporter

### *6.2.1 Molecular Biology*

Glucose transporters are grouped on the basis of their sodium-dependence and hence on the energy-requirements of the transport. The sodium-dependent glucose transfer through epithelial brush-border membranes is driven by hydrolysis of ATP. The inward transport concentrates glucose sufficiently to allow this nutrient to cross the basolateral membrane by passive, facilitated diffusion. Of the putative family of passive, facilitative glucose transporter genes, the six members listed in Table 6.1 have been identified (Kayano et al. 1988, 1990). Gerhart et al. (1989) and Pardridge et al. (1990a, b) showed that the glucose transporter of the cerebral capillary endothelium is of the GLUT1 isoform. Bagley et al. (1989) found evidence of GLUT1 in neuropil, as well as in endothelium, but the contribution may be quantitatively insignificant.

Kasahara and Hinkle (1976, 1977) first purified the transporter in human erythrocyte membrane, using reconstitution of D-glucose transport in erythro-

**Table 6.1** Mammalian glucose transporter genes

| Isoform | Insulin effect | Tissue or membrane | $K_t$ [mM] |
|---|---|---|---|
| GLUT1 | − | HepG2/erythrocyte/BBB | 1 |
| glut-2 | − | Hepatocytes, $\beta$-cells, basolateral membranes | 20 |
| glut-3 | − | Fetal skeletal muscle cells/brain tissue | <1 |
| glut-4 | + | Muscle cells/adipocytes | 5 |
| glut-5 | | Brush-border membranes | 1–2 |
| glut-6 | | Non-expressed pseudo-gene | |

cyte ghosts as a criterion. Cytochalasin B inhibits transport of hexoses across cell membranes (Kletzien et al. 1972) by specific binding to a protein component of the membrane involved in glucose transport with an affinity constant of 100 nM (Lin and Spudich 1974). When cytochalasin B was found to be a potent noncompetitive inhibitor of glucose transport in human erythrocytes by Taverna and Langdon (1973), photoaffinity labeling with this mold metabolite, and purification, identified a 54,000 g mol$^{-1}$ integral component of human erythrocyte membranes (Baldwin et al. 1982; Carter-Su et al. 1982) and of isolated rat, porcine, and bovine brain microvessels (Baldwin et al. 1984; Dick et al. 1984) with the properties of a glucose transporter. Dick et al. (1984) also found that the microvessel density of transporters was 20-fold higher than that of the particulate fraction of cerebral cortex.

Using an antibody raised against the human erythrocyte transporter, Mueckler et al. (1985) cloned a transporter cDNA from the human hepatoma HepG2 line and presented the structure shown in Fig. 6.1. Birnbaum et al. (1986) isolated a cDNA encoding a rat brain glucose transporter with an amino acid sequence 98% identical to that of the HepG2 protein (Matthaei et al. 1987a, b; Fukumoto et al. 1988). Subsequently, the gene was shown to be expressed also in peripheral nerve and choroid plexus epithelium and to reside on chromosome 1 with 35,000 bases, 10 exons and 9 introns (Froehner et al. 1988; Shows et al. 1987; Baldwin and Lienhard 1989; Gerhart et al. 1989).

The GLUT1 transporter is expressed in most tissues (Pilch 1990), particularly in cultured cells and during fetal life (Asano et al. 1988; Fukumoto et al. 1988; Werner et al. 1989). Postnatally, the GLUT1 expression declines in most tissues other than

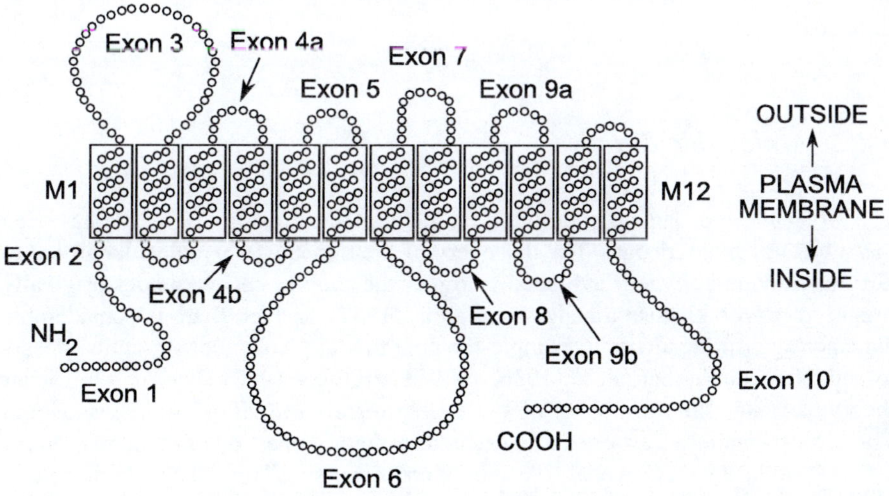


**Fig. 6.1** Blood–brain glucose transporter. The model is based on the exon-intron organization of ancestral facilitative glucose-transporter gene. *Circles* indicate amino acids encoded by each of 12 exons in the ancestral gene. The membrane-spanning $\alpha$-helices are numbered M1–M2. From Bell et al. (1990)

placenta, kidney, and brain (Bell et al. 1990), although oncogenic transformation causes reexpression of the gene (Thorens et al. 1988). The GLUT1 protein belongs to the glucose-regulated family of stress-inducible proteins (Wertheimer et al. 1991). In brain, the GLUT1 transporter is confined to the capillary endothelium. In other endothelia, it is only expressed in the testes in which the capillary endothelium is similarly restrictive as in brain (Harik et al. 1990).

In brain, the GLUT1 transporter gene is expressed at the blood–brain barrier (Boado and Pardridge 1990; Pardridge et al. 1990a, b) and in glial membranes, while the GLUT3 gene is expressed in neuronal membranes. GLUT3 also encodes for a glucose-transporter in fetal muscle and fat cells, and both GLUT1 and GLUT3 encode glucose transporters with comparatively low transport Michaelis constants ($K_t \sim 1$ mM).

GLUT2 encodes the glucose transporter of hepatocytes, the basolateral membrane of small intestine epithelial cells, the distal tubule cells of the kidney, and the $\beta$-cells of the islets of Langerhans. This gene resides on chromosome 3 in humans and is 52% homologous with GLUT1 (Thorens et al. 1988; Fukumoto et al. 1989). GLUT2 expresses a transporter with a $K_t$ of 20 mM which is difficult to reveal in vivo in the presence of higher-affinity transporters, also because cytochalasin-B binds to GLUT2 with comparatively low affinity (Axelrod and Pilch 1983).

GLUT4, the gene of the insulin-sensitive glucose transporter of muscle and fat cells, has been localized to chromosome 17 in humans and appears to be 65% homologous with GLUT1 (James et al. 1988, 1989; Kayano et al. 1988; Birnbaum 1989; Charron et al. 1989). In view of the discussion of an effect of insulin on blood–brain glucose transfer, it is interesting that there is evidence against the presence of the insulin-sensitive glucose transporter in endothelial membranes (Slot et al. 1990).

### *6.2.2 Molecular Kinetics*

The molecular mechanism underlying the facilitated diffusion of glucose has been the subject of much debate. The discovery of the transporter's molecular structure did not provide sufficient evidence to reject the mobile carrier model originally proposed for the glucose transfer by Widdas (1952). In fact, there is considerable evidence against the alternative single-file pore that the molecular structure appears to support (Cunningham et al. 1986, 1989; Carruthers 1990). The mobile carrier theory has taken the form either of an "alternating conformation" model, shown in Fig. 6.2, in which the carrier can be reached from only one side of the membrane at a time (Widdas 1952; Oka et al. 1989) or of an "fixed-sites" model, in which the carrier can be accessed and occupied from both sides of the membrane simultaneously (Baker and Widdas 1973; Helgerson and Carruthers 1989; Naftalin 1988). The latter model departs from the mobile carrier concept by allowing several transport sites to be exposed to both sides of the membrane simultaneously. Cunningham et al. (1989) tested the alternating-conformer model by simulation of two membranes

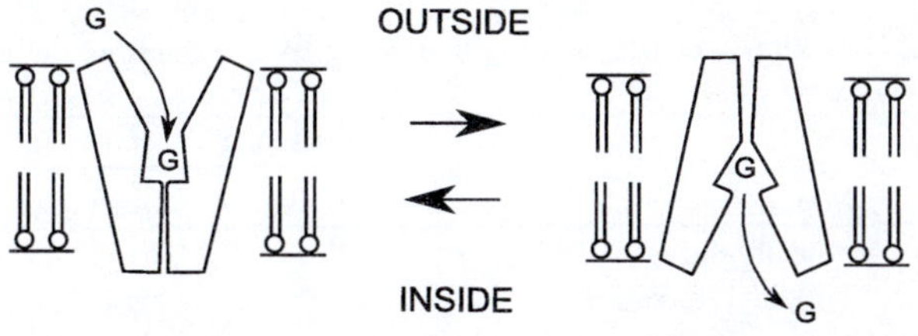


**Fig. 6.2** Blood–brain glucose transporter. Schematic representation of the "alternating conformation" model of the glucose transporter. (One cannot exclude the possibility that the inner and outer glucose-binding site may be composed of completely different domains.) From Oka et al. (1990)

in series, separated by a minute, well-mixed endothelial space. Using the nomenclature proposed by Stein (1986), and kinetic constants estimated by Lowe and Walmsley (1986), the simulation confirmed the experimental estimates of an apparent Michaelis constant ($K_t$) of 8 mM for the transendothelial transport when the half-saturation concentration of the carrier, i.e., the affinity measured in the absence of glucose on the "trans"-side of the membrane (zero-trans-condition) was 7 mM. Again, this value is considerably higher than expected for the GLUT1 transporter.

Studies of the binding of cytochalasin B in isolated brain microvessels have yielded densities of glucose transporters between 1 and 10 pmol mg$^{-1}$ endothelial cells (Dick et al. 1984; Baldwin et al. 1985; Dick and Harik 1986; Matthaei et al. 1986; Kasanicki et al. 1987; Harik et al. 1988; Kalaria et al. 1988; Morin et al. 1988; Pardridge et al. 1990a, b; Mooradian and Morin 1991), 1 mg being the approximate quantity of endothelial cells in 1 g of brain tissue (Gjedde 1983). Cunningham et al. (1989) calculated a transporter quantity of 40 pmol necessary to generate the known maximum transendothelial flux of 2–4 μmol min$^{-1}$. The significance of this discrepancy is not clear. It may indicate that cytochalasin binding is incomplete (Morin et al. 1988).

### *6.2.3 Structural Requirements of Glucose Transport*

Oldendorf (1971), Betz et al. (1975), and Cremer et al. (1983) studied a series of hexoses and other competitors for blood–brain glucose transport to determine the degree of its specificity. In addition, deoxy- and fluorodeoxyglucose have been particularly well studied because of their use in measurements of brain glucose metabolism (Sokoloff et al. 1977; Phelps et al. 1979). As shown in Table 6.2, the relative affinities for the substrates compare well with those observed in erythrocytes and other organs. The affinities indicate that the hydroxyl groups of carbon-atoms 3 and 4 are particularly important for identification of the molecule by the transporter. There is evidence that the orientation is preserved during transport such that the molecule enters the transporter from the extracellular space with C6 facing forward and from the intracellular space with C1 facing forward (Baldwin and Lienhard 1981).

**Table 6.2** Inhibitory constants of glucose transport inhibitors, relative to michaelis constant of glucose

| | | Inhibitory constants (mM) | |
|---|---|---|---|
| Carbon | Inhibitor | BBB[a] | Other membranes[b] |
| C1 | 1-deoxyglucose | | 10 |
| | 1-methylglucose | | high |
| C2 | 2-deoxyglucose | 0.7 | |
| | 2-fluoro-2-deoxyglucose | 0.6 | |
| | 2,2-dichloro-2-deoxyglucose | | 1 |
| | 2-*O*-methylglucose | | 15 |
| C3 | 3-fluoro-3-deoxyglucose | | 1 |
| | 3-*O*-methylglucose | 1.1 | 1 |
| | mannose | 1.7 | |
| | 3-deoxyglucose | | 10 |
| C4 | 4-O-propylglucose | | 1 |
| | galactose | 2.7 | 13 |
| C6 | 6-fluoro-6-deoxyglucose | | 0.2 |
| | 6-chloro-6-deoxyglucose | 0.3 | |
| | 6-chloro-6-deoxymannose | 0.4 | |
| | 6-tosyl-6-deoxygalactose | 0.7 | |
| | 6-chloro-6-deoxygalactose | 0.9 | |
| | 6-deoxyglucose | | 1 |
| | xylose | Infinity | Infinity |
| | fucose | infinity | infinity |

[a] From Pardridge and Oldendorf (1975), Betz et al. (1975), Cremer and Cunningham (1979), Fuglsang et al. (1986), Crane et al. (1983), Pardridge (1983), Gjedde 1984b

[b] From Carruthers (1990)

## 6.3 Theory of Blood–Brain Glucose Transfer

The study of glucose transport across the blood–brain barrier has been complicated by several kinetic problems that include:

- The flow-dependent relationship between clearance and permeability.
- The time-dependent relationship between "first pass" and net extraction, i.e., the problem of bidirectional flux.
- The concentration-dependent relationship between apparent permeability and maximal transport.

These issues have never been dealt with in their entirety in a single publication, although important elements have been discussed by Lund-Andersen (1979), Cunningham and Cremer (1981), Gjedde (1982, 1983), Lassen and Gjedde (1983), Pardridge (1983), and Cunningham et al. (1986, 1989).

### 6.3.1 *Apparent Permeability and Flux*

The quantity of glucose transported from blood to brain is a function of an apparent endothelial permeability ($P$, in units of cm $s^{-1}$ or nm $s^{-1}$), the area of endothelial surface available for transport ($S$, in $cm^2$ $g^{-1}$ or $ml^{-1}$), and the concentrations of glucose on the two sides of the interface. The apparent permeability is measurable only as a clearance which is not directly proportional to the permeability-surface area product because tracer must escape from the capillary during transit.

A second assumption is made that the net rate of tracer transfer across the endothelium equals the difference between the unidirectional rates of transfer. From the definition of transport by passive diffusion, it follows that

$$\Delta j(t) = P_1 S \frac{\bar{c}_c(t)}{\alpha} - P_2 S \frac{m_e(t)}{V_d} \tag{6.1}$$

in which $\Delta j(t)$ is the net flux across the membrane, $\bar{c}_c(t)$ the weighted average concentration of the tracer in capillary blood or plasma, $P_1$ the apparent luminal permeability, $P_2$ the permeability to tracer returning from the brain to blood, $V_d$ the solvent (water) volume of distribution in brain, $\alpha$ the solubility of the tracer in blood or plasma relative to water, and $m_e(t)$ is the quantity (mass) of exchangeable tracer glucose in brain.

**Specific Activity** The average capillary tracer concentration $\bar{c}_c(t)$ varies both with the arterial tracer concentration $c_a(t)$ and the concentration of exchangeable tracer glucose in the brain interstitial fluid. This exchange causes the specific activity to fall from the arterial to the venous end of the capillary. Crone (1963), Johnson and Wilson (1966), Gjedde et al. (1980, 1982), and Christensen et al. (1982) derived solutions for the tracer concentration in the venous effluent, $c_v(t)$. When, for simplicity, $P = P_1 S/\alpha$, and $p = P_2 S/V_d$, the solution is,

$$c_v(t) = c_a(t) e^{-P/F} + m_e(t) \frac{p}{P} \left(1 - e^{-P/F}\right) \tag{6.2}$$

where $F$ is the blood flow. The equation indicates that the specific activity of tracer glucose will decline from the arterial to the venous end of the capillary and that this decline will be time-dependent as shown in Fig. 6.3 from Lassen and Gjedde (1983). By Fick's perfusion principle, $\Delta j(t) = F[c_a(t) - c_v(t)]$. Using (6.2) to eliminate $c_v(t)$, the net exchange is given by:

$$\Delta j(t) = P \left( \frac{1 - e^{-P/F}}{P/F} \right) c_a(t) - p \left( \frac{1 - e^{-P/F}}{P/F} \right) m_e(t), \tag{6.3}$$

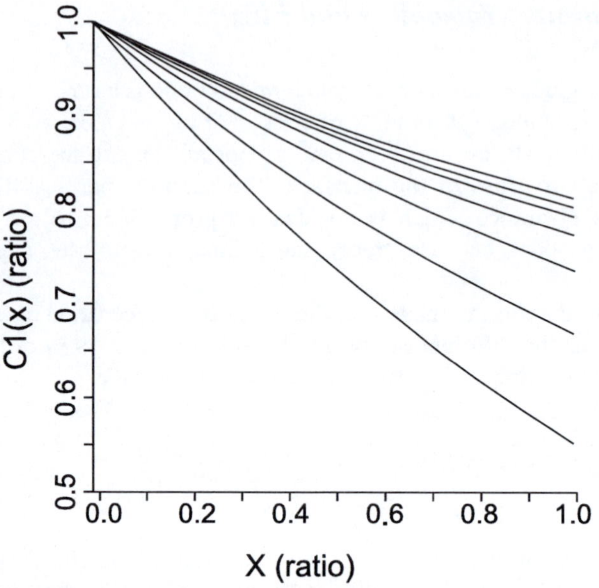


**Fig. 6.3** Effect of capillary passage on specific activity of glucose tracer. Concentration profile in capillary of tracer glucose for $T_{max} = 400$ and $CMR = 70$ μmol hg$^{-1}$ min$^{-1}$, $C_a = 9$ mM, $K_t = 7$ mM, and F = 40 ml hg$^{-1}$ min$^{-1}$ (7,8). The *lines (from the bottom)* represent the times '0', 1, 2, 3, 4, 5, and 45 min. From Lassen and Gjedde (1983)

where the term $1-e^{-P/F}$ is the extraction fraction, $E(o)$, for unidirectional transfer of tracer. Equation (6.3) yields definitions for $K_1$, $k_2$, and $V_e$,

$$K_1 \equiv P\left[\frac{1-e^{-P/F}}{P/F}\right] = F\left(1-e^{-P/F}\right) = E(o)F \tag{6.4}$$

and

$$k_2 \equiv p\left[\frac{1-e^{-P/F}}{P/F}\right] = \frac{F}{V_e}\left(1-e^{-P/F}\right) = \frac{E(o)F}{V_e}, \tag{6.5}$$

where $K_1$ is the initial ("unidirectional") clearance of tracer glucose, and $k_2$ the fractional clearance of the brain interstitial fluid. From these definitions, it follows that $K_1/k_2 = P/p$, defined as the partition volume $V_e$.

Assuming that the total gain of radioactivity in the brain equals the net sum of tracer glucose transferred across the capillary endothelium at all times, the gain of tracer per unit time is given by:

$$\frac{dm(t)}{dt} = K_1 c_a(t) - k_2 m_e(t), \tag{6.6}$$

where $m(t)$ is the content of tracer glucose or tracer-derived molecules in brain. Equation (6.6) can be solved directly only when either (1) all extravascular tracer glucose is exchangeable for the duration of the study, i.e., $m(t) = m_e(t) + V_o c_a(t)$ or (2) no extravascular tracer is exchangeable, i.e., $m_e(t) \simeq 0$. In the former case,

$$\frac{dm(t)}{dt} = (K_1 + k_2 V_o) c_a(t) - k_2 M(t), \tag{6.7}$$

where $V_o$ is a vascular volume of distribution, rather than the vascular volume itself. In the latter case,

$$\frac{dm(t)}{dt} = K_1 c_a(t). \tag{6.8}$$

Note that by definition, $m(t)$ is never zero.

**Initial vs. Steady-State Influx** According to (6.1), the tracer flux is unidirectional when

$$\Delta j(t) \cong \frac{P_1 S}{\alpha} \bar{c}_c(t) = P \bar{c}_c(t). \tag{6.9}$$

It is convenient to refer to this flux as the "influx," $\vec{j}(t)$. Initially when $m_e(t) = 0$, it follows from (6.2) that,

$$\bar{c}_c(t) = \left[ \frac{1 - e^{-P/F}}{P/F} \right] c_a(t). \tag{6.10}$$

Hence, initially,

$$\vec{j}(t) = K_1 c_a(t). \tag{6.11}$$

Using the approximation $\bar{c}_c(t) \cong [c_a(t) + c_v(t)]/2$, to eliminate $c_v(t)$ from Fick's perfusion equation, $\Delta j(t) = F[c_a(t) - c_v(t)]$, yields $\bar{c}_c(t) = c_a(t) - \frac{\Delta j}{2F}$, where $\Delta j$ is the net transfer of native glucose across the endothelium. Substituting the last expression in (6.9) implies that at steady-state,

$$\vec{j} = P c_a - \Delta j \frac{P}{2F}, \tag{6.12}$$

where $\vec{j}$ is the influx and $c_a$ the arterial concentration, of native glucose. Equation (6.12) reduces to $\vec{j} \cong P c_a$ for $\Delta J \ll F c_a$. Steady-state rates of influx must be calculated from values of $P$ obtained under unidirectional circumstances (6.7 or 6.8) and calculated by means of (6.12). In the past, experimental conditions were often far from unidirectional, and erroneous steady-state fluxes were calculated with (6.11).

The unidirectionality of transfer can be evaluated by solving (6.2) for the extraction fraction at time $t$, as defined by Gjedde (1982), and where $E(t)$ is used for a time-varying quantity for consistency with the reference.

$$E(t) \equiv \frac{c_a(t) - c_v(t)}{c_a(t)} = \left(1 - e^{-P/F}\right)\left(1 - \frac{m_e(t)}{V_e c_a(t)}\right) \tag{6.13}$$

according to which $E(t) = E(o)$ for $m_e(t) = 0$. Hence, "first pass" extraction means negligible tissue content, allowing (6.6) to be solved for $m_e(t) \cong 0$ (6.8) (Crone 1963).

The fraction of tracer extracted *in* (rather than *at*) the time $T$, is

$$\tilde{E}(T) \equiv 1 - \left[ \int_0^T c_v(t)\mathrm{d}t \Big/ \int_0^T c_a(t)\mathrm{d}t \right] = \frac{m_e(T)}{\left[ F \int_0^T c_a(t)\mathrm{d}t \right]}. \tag{6.14}$$

Since $F[c_a(t) - c_v(t)]$ is the rate at which tracer is extracted from the capillary bed, its integral from 0 to $T$ is the amount of tracer in tissue at $T$, which accounts for the numerator, earlier. Similarly, the integral of $Fc_a(t)$ is the amount supplied, and accounts for the denominator. Integrating (6.2) from 0 to $T$, using $p/P = 1/V_e$, and $e^{-P/F} = 1 - E(o)$, yields:

$$\int_0^T c_v(t)\mathrm{d}t = (1 - E(o)) \int_0^T c_a(t)\mathrm{d}t + \frac{E(o)}{V_e} \int_0^T m_e(t)\mathrm{d}t.$$

Dividing by $\int_0^T c_a(t)\mathrm{d}t$, and substituting in (6.14), implies that, for low values of $T$, (Gjedde 1982),

$$\frac{\tilde{E}(T)}{E(o)} = 1 - E(T)\frac{F \int_0^T m_e(t)\mathrm{d}t}{V_e m_e(T)} \cong 1 - E(T)\frac{FT}{2V_e}, \tag{6.15}$$

where the approximation $\frac{1}{T} \int_0^T m_e(t)\mathrm{d}t \cong \frac{m_e(T)}{2}$ was used to get the second equality. This describes the relative backflux in the period after the injection. If the measured extraction is 40% in 20 s, if flow is 0.5 ml $\mathrm{min}^{-1}$, and if the partition volume is 0.5 ml, backflux is 7% in 20 s. In 3 min, it would be 60%. For longer periods, however, the equation underestimates the backflux.

### 6.3.2 *Facilitated Diffusion*

#### Michaelis–Menten Equation

Facilitation of diffusion must involve ligand recognition as the initial step. The criteria of facilitation are therefore the same as for pure receptor–ligand interactions and include *saturability, stereospecificity*, and *competitive inhibition*, as well as evidence of the consequent counter transport.

Transport rates in vivo usually reveal a lower affinity between the hypothetical receptor and the substrate than described for pure receptor–ligand interactions (Gjedde et al. 1986). The reason for the low affinity is plain from Michaelis' and

Menten's (1913) original steady-state solution of the enzyme kinetic equation, observed by LeFevre (1961) also to describe the transport of tracer glucose into red blood cells,

$$\vec{J} = \frac{T_{\max} C_a}{K_t + C_a}, \tag{6.16}$$

where $\vec{J}$ is the unidirectional flux, $K_t$ the Michaelis or half-saturation concentration, $T_{\max}$ the maximal transport rate, and $C_a$ is the external steady-state substrate concentration. The equation is the equilibrium solution of a differential equation describing the change per unit time of the quantity of substrate bound to the transporter. The change of bound substrate is the difference between substrate associating with, and dissociating from, the receptor. The speed of association reflects the likelihood of a ligand molecule joining the receptor from the vicinity of a site ($k_a$), multiplied by the number of molecules and the number of receptor sites, see Chap. 2, Sect. 2.2.4. The speed of dissociation reflects the combined likelihood of dissociation ($k_d$) by return to the substrate solution ($k_r$) and transport through the membrane ($k_t$), multiplied by the number of receptor–ligand complexes. For "true" receptors, the $k_d$ reflects only the rate of return to the solution and, hence, is much lower. For transporters, the $k_d$ reflects both processes. The maximal transport rate is $k_t B_{\max}$, where $B_{\max}$ is the total number of sites. The actual transport rate is $k_t B$, where $B$ is the number of receptor–ligand complexes. If $\vec{J}$ is substituted for $k_t B$, $T_{\max}$ for $k_t B_{\max}$, and $K_t$ for $(k_r + k_t)/(k_a)$, (6.16) is obtained.

### Michaelis–Menten Constants

Bachelard et al. (1973), Pardridge and Oldendorf (1975), and Gjedde (1980) defined the apparent permeability of the endothelium to substrates of facilitated diffusion on the basis of (6.16),

$$P = \frac{P_1 S}{\alpha} \equiv \frac{T_{\max}}{K_t + C_a}. \tag{6.17}$$

From this equation, it follows that simultaneous measurements of $P$ and $\vec{J}$ are linearly related. Thus, $\vec{J}$ is a linear function of $P$,

$$\vec{J} = -K_t P + T_{\max} \tag{6.18}$$

in which $T_{\max}$ is the ordinate intercept and $-K_t$ the slope. Equation (6.18) is identical to the Eadie–Hofstee (Eadie 1942; Hofstee 1952) or Scatchard (Scatchard 1949) plots of enzyme reaction velocity. An example of the use of (6.17) and (6.18) to evaluate $T_{\max}$ and $K_t$ is given in Fig. 6.4 (Crone 1985).

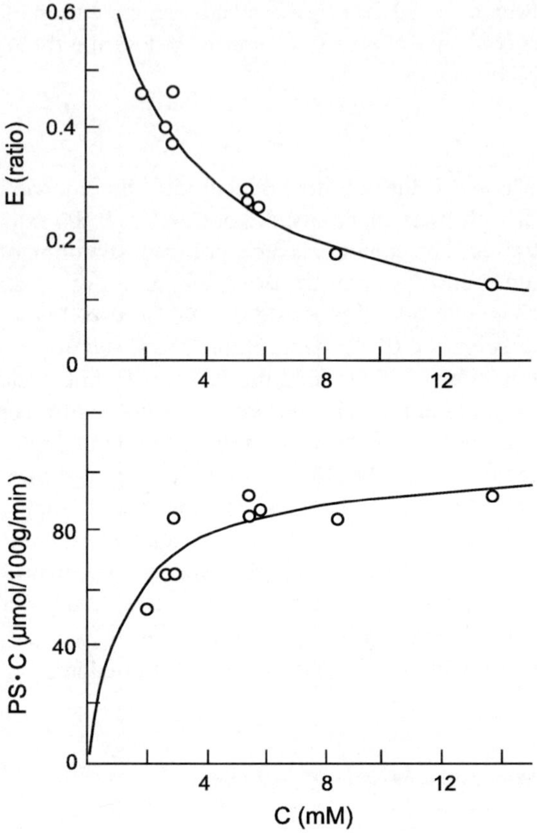


**Fig. 6.4** Blood–brain glucose transport in dog. The *upper part* of the figure shows the initial unidirectional extraction of labeled D-glucose in the dog's brain during a single passage. The abscissa shows the concentrations of glucose in the blood. The *lower part* shows the calculated unidirectional transport rate of various glucose concentrations. Modified from Crone (1965)

## Partition Volume

Assuming symmetrical transport, the fractional rate of transfer from the brain tissue to the circulation is given by,

$$p = \frac{P_2 S}{V_\mathrm{d}} \equiv \frac{\alpha T_{\mathrm{max}} / V_\mathrm{d}}{K_\mathrm{t} + C_e} \tag{6.19}$$

in which $K_\mathrm{t}$ is the half-saturation concentration of the tracer relative to whole-blood or plasma and $C_e$ is the concentration of native glucose relative to whole-blood or plasma (hence $\alpha$ and $V_\mathrm{d}$). Division yields the $P/p$ ratio for facilitated diffusion of a single substrate, defining the tracer partition volume,

$$V_\mathrm{e} = \frac{K_1}{k_2} = \frac{P}{p} = \frac{V_\mathrm{d}}{\alpha}\left[\frac{K_\mathrm{t} + C_e}{K_\mathrm{t} + C_\mathrm{a}}\right]. \tag{6.20}$$

Equation (6.20) predicts that the partition volume must be the same for native glucose and all glucose tracers with identical values of $T_{max}$. For glucose for which $\alpha \cong 1$, the equation reduces to,

$$V_e = \frac{V_d K_t + M_e}{K_t + C_a}, \tag{6.21}$$

where $M_e$ is the tissue content of native glucose. This equation was used by Gjedde (1983) to calculate the glucose content of brain tissue on the basis of measurements of $V_e$ and $C_a$ when $K_t$ and $V_d$ were known.

### 6.3.3 Multiple Membranes

The estimates of $K_t$ and $T_{max}$ according to (6.17) or (6.18) are empirical measurements with an uncertain relation to the properties of the individual membranes interposed between the circulation and the cytosol. Glucose and glucose tracers must pass both membranes of the capillary endothelium as well as the membranes of glial and neuronal cells before they can react with metabolic enzymes. At least four resistances may be important (i.e., both membranes of the endothelium, the cell membrane, and the hexokinase reaction), and probably one or more unstirred layers. The half-time of endothelial equilibration is of the order of 0.1 s (Gjedde and Christensen 1984); the half-time of extracellular fluid equilibration 0.5–2 min (Lund-Andersen and Kjeldsen 1977) depending on the relative resistances of cell membrane transfer and phosphorylation by hexokinase; and the half-time of total brain water equilibration, in the absence of $CO_2$ loss, close to 10 min (Gjedde 1982, Brøndsted and Gjedde 1988). It is a question whether the time of equilibration is defined entirely by the size of the water space, or whether the glial and neuronal cell membranes impose an additional delay, relative to the rate of diffusion in water. A problem is also whether the cell membranes impose a delay on the equilibration relative to the rate of metabolism which would tend to establish a steady-state concentration difference between the extra- and intravascular space.

#### Endothelial Membranes

The simple Michaelis–Menten kinetic analysis lumps the two membranes of the cerebral capillary endothelium. While the combination assumes that the endothelial transfer of glucose is so rapid, and the volume of the intraendothelial space so small, that the tracer in the endothelium achieves almost instant equilibration with the tracer in the capillary, the kinetic constants nonetheless depend on the individual constants of the two membranes.

The consequences of using a single Michaelis–Menten expression for the quantitation of transendothelial flux were examined by Pappenheimer and Setchell (1973) and Gjedde and Christensen (1984). The latter authors concluded that steady-state between the blood and the endothelial cells reached in 0.1 s. They formulated a model of two endothelial cell membranes with Michaelis–Menten properties and showed that the constants of the transendothelial transport are simple functions of the properties of the individual membranes when transport across the endothelium is rapid ($P > 10$ nm s$^{-}$1),

$$T_{\max} = \frac{J^a_{\max} J^b_{\max}}{J^a_{\max} + J^b_{\max} + \frac{\Delta J (K^b_t - K^a_t)}{K^a_t}} \tag{6.22}$$

and

$$K_t = \frac{J^a_{\max} K^b_t + J^b_{\max} K^a_t + \Delta J (K^b_t - K^a_t)}{J^a_{\max} + J^b_{\max} + \frac{\Delta J (K^b_t - K^a_t)}{K^a_t}}, \tag{6.23}$$

where $a$ and $b$ refer to the individual membranes and $J_{\max}$ and $K_t$ refer to their Michaelis constants. For symmetrical transport, i.e., $K^b_t = K^a_t$ and $J^b_{\max} = J^a_{\max}$, (6.22) and (6.23) can be used to establish both the ratio between $T_{\max}$ and $\Delta J$ and the value of $K_t$ by nonlinear regression analysis. The $T_{\max}/\Delta J$ ratio is 4 in awake rats (Gjedde and Christensen 1984).

In the same analysis, the Michaelis–Menten constants determined in tracer experiments described the facilitated diffusion in the steady-state only when the two membranes had identical kinetic properties (Gjedde and Christensen 1984). By examination of published relationships between blood and brain glucose, the authors discovered no discrepancy between tracer and steady-state measurements and hence concluded that the two membranes appeared to have identical glucose transport properties. A more detailed discussion of the molecular implications of the two-membrane endothelium has been presented by Cunningham et al. (1989).

### Glial and Neuronal Cell Membranes

Almost all models of blood–brain glucose transfer have assumed that the transfer across cell membranes is so rapid that glucose concentrations are equal in extra- and intracellular fluid.

Raichle et al. (1975) used coincident detection of positron emission to determine net and unidirectional blood–brain glucose transfer in the anesthetized monkey. The uptake was followed every second for several minutes and yielded a glucose distribution space of no more than 15% of the brain water space. In a study using the double-indicator diffusion method, (see Table 6.3) Knudsen et al. (1990) also concluded that the initial glucose distribution in brain roughly equals the volume of the extracellular space. Using the model assuming nonmixing of tracer in the extravascular compartment (unstirred compartment) the latter authors attempted to determine the distribution of tracer glucose in the endothelium but the size of the space yielded by their three-compartment model was too large to be consistent with endothelial anatomy. Using a similar model, but now assuming complete mixing in the extravascular compartment, in order to determine the magnitude of initial distribution in the extracellular space, the authors obtained a value of approximately 15%

**Table 6.3** Methods of studying blood–brain glucose transfer

| Tissue sampling | | Arteriovenous sampling |
|---|---|---|
| Artificial perfusate | Native perfusion | |
| Intracarotid bolus (BUI) | Intravenous bolus (INT) | Intracarotid bolus (IND) |
| | | Intravenous bolus (IVB) |
| Intracarotid perfusion (ICP) | Intravenous infusion (INF) | |

of the water space. The experiments may not have lasted long enough to determine steady-state distribution in the tissue, and they were too long to determine endothelial distribution, despite the high temporal resolution of 1 s sampling. However, they do suggest the influence of some resistance at cellular membranes.

In contrast to the high time-resolution studies, that revealed a possible resistance at the cell membrane, low time-resolution have suggested that cell membrane transport is not rate-limiting in the steady-state. The studies of Lund-Andersen and Kjeldsen (1976, 1977) concluded that glucose tracer equilibration between the extra- and intracellular compartments is immeasurably fast in vitro, rendering diffusion in the extracellular space rate-limiting for transport into the intracellular space. The minimum value for the coefficient of membrane transfer was at least an order of magnitude greater than the phosphorylation coefficient of 2-deoxyglucose.

Neither has the question of the relationship between cell membrane transport of glucose and the cellular metabolic rate for this nutrient been answered by steady-state methods. Measurements of brain glucose concentrations in rodents indicate that the space in which native glucose is dissolved in the steady-state must be much greater than the extracellular space if a concentration gradient is to exist between plasma and tissue glucose. The steady-state distribution of 3-*O*-methylglucose, a nonmetabolizable hexose, averages 60% of total brain water in both rodents and humans, and has been shown to approach 100% of the brain water volume in hyperglycemic states (Gjedde 1983; Brooks et al. 1986a, b), indicating saturation of the luminal glucose transporter and distribution of 3-*O*-methylglucose in the entire water volume in the brain.

The results of a study of Diemer et al. (1985) also favor distribution of glucose in the entire water phase of brain tissue because it yielded glial and neuronal cell membrane permeabilities to glucose close to those observed for the blood–brain barrier per unit surface area, indicating a much greater total permeability because of the much greater cellular surface (5000:1).

## 6.4 Evidence of Blood–Brain Glucose Transfer

It has been extraordinarily difficult to obtain consistently accurate and precise measurements of blood–brain glucose transport in any species. In many studies, the change of glucose transport under abnormal circumstances is less than the variation observed in normal values between different studies. For this reason, some disagreement still remains about normal values in rodents and humans.

### 6.4.1 Methods

The final common path for all measurements of glucose transport the determination of $K_1$, the unidirectional clearance, either by direct measurement or as the product, of an extraction fraction and blood or plasma flow. Most newer studies derived the PS product and hence the flux $(P_1 S C_a/\alpha)$ from estimates of $K_1$ according to (6.18).

#### Operational Equations

**Clearance** Direct measurements of clearance can be made only by sampling arterial perfusate and brain tissue simultaneously. The equation underlying the calculation of $K_1$ is one of several solutions to (6.7),

$$m(T) = (K_1 + k_2 V_o)e^{-k_2 T} \int_0^T c_a(t)e^{k_2 t} dt. \tag{6.24}$$

Equation (6.24) can be solved by nonlinear regression analysis of at least three separate observations. However, statistically, multilinear regression analysis may be preferable to nonlinear regression. By integration of (6.7), Blomqvist (1984) derived a sum of integrals,

$$m(T) = K_1 \left(1 + \frac{V_o}{V_e}\right) \int_0^T c_a(t)dt - \left(\frac{K_1}{V_e}\right) \int_0^T m(t)dt, \tag{6.25}$$

where $c_a(t)$ is the arterial concentration, $V_e$ the partition volume, and $V_o$ is the initial volume of distribution. If the magnitude of one or more of the parameters is independently known, the number of required observations can be reduced proportionately. In experiments with a single observation of $m(T)$, (6.24) can be solved directly for $K_1$ when $k_2$ is known,

$$K_1 = \frac{m(T)}{\int_0^T c_a(t)e^{k_2(T-t)}dt} - V_o k_2. \tag{6.26}$$

For short periods of circulation, when $k_2$ is unknown, this equation reduces to the solution of (6.8) first introduced by Schaefer et al. (1976) and Blasberg et al. (1978),

$$\lim_{T \to 0} K_1 \cong \frac{m(T)}{(1 + \frac{V_o}{V_e}) \int_0^T c_a(t)dt} = \frac{m(T) - V_o c_a(T)}{\int_0^T c_a(t)dt}. \tag{6.27}$$

Equation (6.25) is soluble by multilinear regression only when a sufficient number of measurements are available. If each measurement is a separate experiment, (6.25) can be normalized to obtain comparable measures of virtual volumes of distribution in each experiment,

$$V(T) = \frac{m(T)}{c_a(T)} = K_1 \left[ \left(1 + \frac{V_o}{V_e}\right) \frac{\int_0^T c_a(t)dt}{c_a(T)} - \left(\frac{1}{V_e}\right) \frac{\int_0^T m(t)dt}{c_a(T)} \right]. \tag{6.28}$$

When $V_e$ is sufficiently large, the equation shrinks to the normalized solution of (6.14),

$$V(T) = K_1 \Theta(T) + V_o, \tag{6.29}$$

where $\Theta(T)$ is the normalized integral $\int_0^T c_a(t)dt/c_a(T)$ which has unit of time and equals time when the tracer concentration in plasma is constant. $V(T)$ is a linear function of $\Theta(T)$ with a slope of $K_1$ when the $\int_0^T m(t)dt/V_e$ ratio remains negligible compared to $\int_0^T c_a(t)dt$. Equation (6.29) underlies the Gjedde–Patlak plot (Gjedde 1981b, 1982; Patlak et al. 1983). Continued linearity of this plot signifies absent backflux. Examples are shown in Fig. 6.5 for a series of glucose analogs entering neonatal rat brain (Fuglsang et al. 1986).

**Extraction Fractions** Extraction fractions can be measured directly only by sampling arterial and venous perfusate of the brain simultaneously, using (6.13) to calculate $E(t)$, or (6.14) to calculate $E(T)$. Extraction fractions can also be calculated indirectly as the ratio between clearance and perfusion rate.

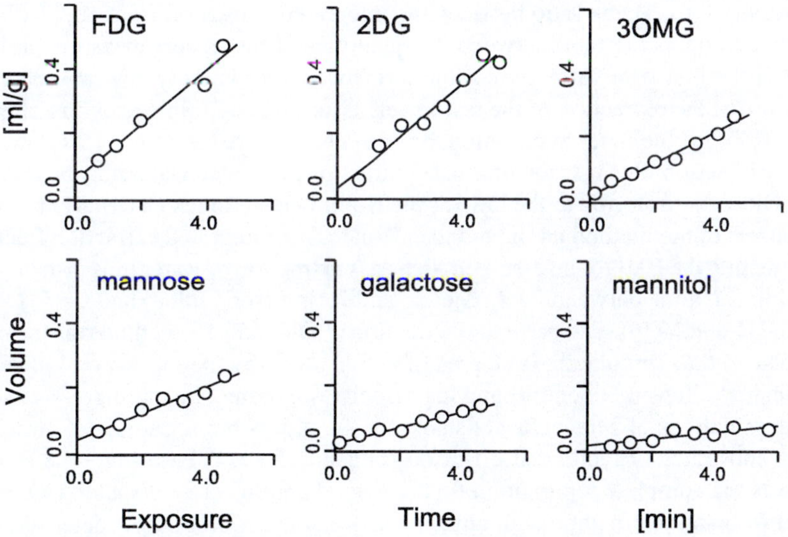


**Fig. 6.5** Apparent distribution volume in brain of five glucose analogs and mannitol (a plasma marker) as function of time. The five hexoses include fluorodeoxyglucose (FDG), deoxyglucose (2DG), and 3-*O*-methylglucose (3OMG). Abscissae: blood integrals, normalized against concentration, weighted exposure time (minutes); ordinates: brain/blood radioactivity ratios. Each point is one observation. Each graph represents one litter. From Fuglsang et al. (1986)

### Experimental Procedures

The methods applied to living tissue can roughly be divided into two groups, including in the first group all methods requiring in vivo or ex vivo sampling of tissue, and in the second group the methods that require sampling of arterial and cerebral venous blood. In the former group, the methods can be further subdivided into two groups, one including the methods in which the brain is perfused by native blood, the other including the methods that employ an artificial perfusate (Table 6.3). In the table, the abbreviations correspond to the descriptions later.

**IND** The indicator diffusion method (Crone 1963; Lassen et al. 1971) requires that a mixture of test and reference substance be sampled in cerebral venous blood after intracarotid injection. The method yields an estimate of $E(t)$. The reference substance must remain in the vasculature to provide the measure of $c_a(t)$ in (6.13). The time resolution of the method is excellent but the blood–brain transfer values have an uncertain relation to individual brain regions or even the brain as a whole. The intracarotid injection also posed ethical problems in normal volunteers.

**IVB** For the reasons indicated earlier, as well as to take capillary heterogeneity and backflux into account, the indicator diffusion method was modified for intravenous, rather than arterial, bolus injection in rats (Sawada et al. 1989) and humans (Knudsen et al. 1990).

**BUI** In 1970, Oldendorf published a modified method of intracarotid bolus injection of a mixture of a test tracer and a reference tracer, usually labeled water. According to (6.14), the ratio between the integrated extraction fraction, $[E(T)]$ of two tracers equals the ratio between the quantities of the tracers measured in brain tissue. If the first-pass extraction of the reference tracer is unity, the ratio becomes a measure of the extraction of the test tracer, assuming negligible backflux of either tracer. To fulfill the latter assumption, brain contents were measured 15 s (later 5 s) after the injection. Because the first-pass extraction of labeled water may be less than unity, Oldendorf referred to the ratio as the Brain Uptake Index (BUI). Many subsequent users of the method attempted to calculate true integrated extraction fractions by correcting the BUI for an assumed first-pass extraction of water of less than unity, choosing fractions between 0.43 (Gjedde and Rasmussen 1980a) and 0.85 (Oldendorf and Braun 1976). The particular attraction of Oldendorf's method was the claim that the cerebral vasculature is cleared of native blood by the intracarotid injection. This feature allowed Oldendorf and his associates to complete an impressive series of determinations of Michaelis constants for the blood–brain transfer of almost all known substrates of facilitated diffusion (Oldendorf 1971; Pardridge 1983). Only later was the complete separation of native blood and injectate disputed (Pardridge et al. 1985) and the method is in any case unsuitable for blood flow determination, necessary for calculation of PS-products.

**INT** Several authors used Saperstein's (1958) indicator fractionation principle to devise an intravenous bolus injection version of Oldendorf's method in which extraction fractions could be measured by comparing brain uptake of two different tracers and clearances could be measured by comparing the brain uptake with the integrated arterial concentration. This method, now commonly known as the integral method, appeared in different forms, one designed for blood flow measurements (Schaefer et al. 1976; Van Uitert and Levy 1978; Gjedde et al. 1980) but with no provision for backflux of highly diffusible tracers, and another designed to yield blood–brain transfer rates (Bachelard et al. 1973; Daniel et al. 1978; Gjedde et al. 1980). The intravenous bolus injection method obtains the tracer delivery by mechanical integration of the arterial concentration and subsequent calculation of the kinetic circulation time $\Theta$ on the basis of the integral and the final concentration (Gjedde 1981b, 1982; Patlak et al. 1983). The advantage of this method is the feasibility of measuring blood flow simultaneously. In the later experiments, tracer glucose was administered as a rapid intravenous bolus and the experiment terminated after 10 or 20 s to avoid backflux (Gjedde et al. 1980).

**INF** In the earlier experiments, the tracer often circulated for a minute or more. The glucose tracer was infused continuously in a manner calculated to yield a constant arterial level of tracer glucose (Daniel et al. 1978) according to (6.29) in which $\Theta$ is then real time.

**MET** In a variation of the integral method based on the glucose metabolism method of Sokoloff et al. (1977), Cunningham and Cremer (1981) determined glucose clearance ($K_1$) and metabolism simultaneously by neurochemical separation of glucose tracer and metabolites in brain. In a later treatment, the authors extended the method to include calculation of the PS-product from the clearance (Hargreaves et al. 1986).

**ICP** The method of isolating and independently perfusing the vasculature of the brain via the intracarotid artery provided a combination of the intracarotid and intravenous administration methods, by means of which the rate of perfusion and composition of the perfusate can both be controlled and glucose transfer determined, either by bolus administration of the tracer in the perfusate and venous rather than tissue sampling (Betz et al. 1973) or by continuous infusion of the tracer and tissue sampling at the end of the experiment (Takasato et al. 1984; Zlokovic et al. 1986). Despite the controlled infusion, the method may require that the investigators measure perfusate flow simultaneously. The advantage is the control of perfusate composition.

### The Blood–Brain Barrier In Vitro

Microvessels have been separated from other components of brain tissue and studied in vitro (Betz et al. 1979; Joo 1985). Glucose and glucose analog "uptake" has been measured with labeled tracers, and the number of glucose binding sites has been estimated with cytochalasin B (Betz et al. 1983). As discussed earlier, isolated microvessels with a surface-volume ratio at least 1,000-fold higher in vitro than in vivo are not suited for transport studies that last more than a few seconds. For this reason, the results of the uptake studies (e.g., Kolber et al. 1979) throw little light on the blood–brain transfer of hexoses (Bradbury 1985). Cultured endothelial cells from the brain provide an alternative to in vivo studies of transendothelial flux.

## 6.4.2 Normal Values in Awake Subjects

Originally, the observations with the tissue sampling method were made after a minute or more, without correction for backflux (Bachelard et al. 1973; Daniel et al. 1978). In addition, glucose transfer rates were calculated from the clearance rather than the PS-product, and the subjects were invariably anesthetized. For these reasons, only values obtained in awake subjects after about 1980 can reasonably be argued to represent normal values, as listed in Tables 6.4–6.6.

### Permeability and Flux in Normoglycemia

The first demonstration of facilitated diffusion of glucose across the blood–brain barrier was provided by Crone (1960) who used the indicator diffusion method. The result is illustrated in Fig. 6.4, redrawn from Crone (1965) by Crone (1985).

**Table 6.4** Blood–brain glucose transport in awake subjects

| Reference | Species | $C_a$ | PS | PSx$C_a$ | BUI/E | $K_1$ | $K_1$x$C_a$ |
|---|---|---|---|---|---|---|---|
| Sokoloff et al. (1977) | Rat | 9 | | | | 19 | 168 |
| Nemoto (1978) | Rat | 5 | | | | 10 | 48 |
| Gjedde and Rasmussen (1980a) | Rat | 9 | | | | 18 | 160 |
| Gjedde and Rasmussen (1980b) | Rat | 8 | | | | 19 | 160 |
| Gjedde and Rasmussen (1980b) | Rat | 9 | | | | 13 | 114 |
| Cremer et al. (1981) | Rat | 10 | 17 | 170 | | | |
| Braun et al. (1985) | Rat | [10] | 11 | 108 | | | |
| Crane et al. (1985) | Rat | 8 | 14 | 127 | | | |
| LaManna and Harik (1985) | Rat | 10 | 17 | 170 | | | |
| Hargreaves et al. (1986b) | Rat | 10 | 18 | 180 | | | |
| LaManna and Harik (1986) | Rat | 11 | 17 | 186 | | | |
| Planas and Cunningham (1987) | Rat | 10 | 16 | 158 | | | |
| Pelligrino et al. (1990a) | Rat | 9 | | | | 15 | 136 |
| Pelligrino et al. (1990b) | Rat | 9 | 16 | 144 | | | |
| Hertz et al. (1981) | Man | 5 | | | 14 | 9 | 14 |
| Hertz and Paulson (1982) | Man | 5 | 7 | 36 | 15 | 8 | |
| Blomqvist et al. (1985) | Man | 5 | | | | 6 | 30 |
| Brooks et al. (1986a) | Man | 5 | | | | 6 | 30 |
| Brooks et al. (1986b) | Man | 4 | | | | 5 | 22 |
| Feinendegen et al. (1986) | Man | 4 | | | | 16 | 67 |
| Gutniak et al. (1990) | Man | 6 | | | | 6 | 39 |

**Table 6.5** Michaelis constants of blood–brain glucose transfer in awake subjects

| Reference | Species | $T_{max}$ | $K_t$ |
|---|---|---|---|
| Nemoto et al. (1978) | Rat | 1.9 | 12 |
| Gjedde and Rasmussen (1980a) | Rat | 4.1 | 8 |
| Gjedde and Rasmussen (1980b) | Rat | 4.1 | 9 |
| Braun et al. (1985) | Rat | 1.6 | 8 |
| Crane et al. (1985) | Rat | 1.9 | 8 |
| LaManna and Harik (1985) | Rat | 2.5 | 5 |
| LaManna and Harik (1986) | Rat | 2.5 | 5 |
| Hargreaves et al. (1986a) | Rat | 4.1 | 8 |
| Brooks et al. (1986a, b) | Man | 0.4 | 4 |
| Feinendegen et al. (1986) | Man | 2.0 | 4 |
| Ribeiro (1991) | Man | 0.8 | 2 |
| Blomqvist et al. (1991) | Man | 0.6 | 4 |

The graph shows a Michaelis–Menten-type relationship between the unidirectional transport of glucose from the circulation to brain tissue and the plasma glucose concentration at which the transport occurs. The flux of glucose was calculated from the fractions of extracted glucose shown in the upper graph, indicating a maximal glucose transport rate close to 90 μmol $hg^{-1}$ $min^{-1}$ and a Michaelis half-saturation constant close to 1 mM.

**Average Gray Matter Values** From Table 6.3, the following conclusions can be drawn: In the awake adult rat, the gray matter blood–brain glucose transfer is close to 200 μmol $hg^{-1}$ $min^{-1}$ at a glucose concentration of about 8 mM, indicating a $P_1S$-product of approximately 25 ml $hg^{-1}$ $min^{-1}$, depending on the region of the brain. The glucose influx is therefore approximately twice as high as the net utilization of 100 μmol $hg^{-1}$ $min^{-1}$ (Sokoloff et al. 1977). In resting humans, the influx, measured with glucose, 3-*O*-methylglucose, or fluorodeoxyglucose, is close to 50 μmol $hg^{-1}$ $min^{-1}$ at 4 mM, or approximately twice the net consumption as well, corresponding to a PS-product of 12.5 ml $hg^{-1}$ $min^{-1}$ and a value of $K_1$ of 10 ml $hg^{-1}$ $min^{-1}$, as listed in Table 6.5. Measurements of blood–brain fluorodeoxyglucose transfer are particularly abundant because of the use of fluorodeoxyglucose in positron tomography.

**Regional Variation** In vivo, the regional variation of blood–brain glucose transport generally follows the distribution of blood flow and glucose consumption (Raichle et al. 1975; Cremer et al. 1983; Hawkins et al. 1983; Gjedde and Diemer 1985a,b; Planas and Cunningham 1987; Gjedde et al. 1990). In vitro the distribution of glucose transporters frequently has been shown to be different from that of in vivo transport and metabolism (Tucker and Cunningham 1988; Bagley et al. 1989; Dick and Harik 1986), perhaps indicating regional differences of glucose transporter reserves. Somewhat inappropriately, this phenomenon has been termed the flow-metabolism couple (Roy and Sherrington 1890). More correctly, the couple refers

**Table 6.6** Blood–brain clearance of fluorodeoxyglucose in man

| Reference | $K_1^*$ (ml hg$^{-1}$ min$^{-1}$) |
|---|---|
| Phelps et al. (1979) | 10 |
| Friedland et al. (1983a, b)[a] | 13 |
| Hawkins et al. (1983a, b)[a]: | |
| Young adults | 8 |
| Older subjects | 10 |
| Kato et al. (1984) | 15 |
| Gjedde et al. (1985) | 10 |
| Reivich et al. (1985) | |
| $k_3$-model | 11 |
| $k_4$-model | 10 |
| Evans et al. (1986) | 8 |
| Hawkins et al. (1986) | 4 |
| Lammertsma et al. (1987) | 6 and 7 |
| Redies et al. (1989) | 8 |
| Eastman et al. (1990) | 10 |
| Kuwabara et al. (1990) | 9 |
| Shapiro et al. (1990) | 14 |
| Jagust et al. (1991) | 16 |
| Kuwabara and Gjedde (1991) | 9 |

[a] No corrections for the radioactivity within vessels in tissue

to the preservation of the relationship between the three variables during functional changes. They will be considered later.

## Michaelis–Menten Constants

The maximal transport capacity of the cerebral capillary endothelium has no strong interspecies variation, although it tends to be inversely proportional to the size of the subject, as shown in Table 6.6. In rats, the estimates of $T_{max}$ range from 40 μmol hg$^{-1}$ min$^{-1}$, observed by Pardridge et al. (1982) during heavy pentobarbital anesthesia, to 600 μmol hg$^{-1}$ min-1, observed by Gjedde and Diemer (1985a, b) in fine structures of the brain, with a median between 200 and 400 μmol hg$^{-1}$ min$^{-1}$, depending on the duration of the experiment, the crudeness of tissue sampling (manual vs. autoradiography), the absence or presence of anesthesia, the stage of postnatal development, and the method of calculation (i.e., permeability-surface-area-product or clearance).

The estimates of the half-saturation concentration, $K_t$, vary from 1 to 12 mM, with a definite median at 8 mM in rats, (Fig. 6.6) and 4 mM in humans (Table 6.5). Thus, it is reasonable to conclude that the maximal transport is close to 400 μmol hg$^{-1}$ min$^{-1}$ in the awake restrained rat and 100 μmol hg$^{-1}$ min$^{-1}$ in the resting awake human. The maximally possible apparent permeability-surface area product, equal to the $T_{max}/K_t$ ratio, is 50 ml hg$^{-1}$ min$^{-1}$ in awake restrained rats and 25 ml hg$^{-1}$ min$^{-1}$ in awake, resting humans. Estimates of the endothelial surface area

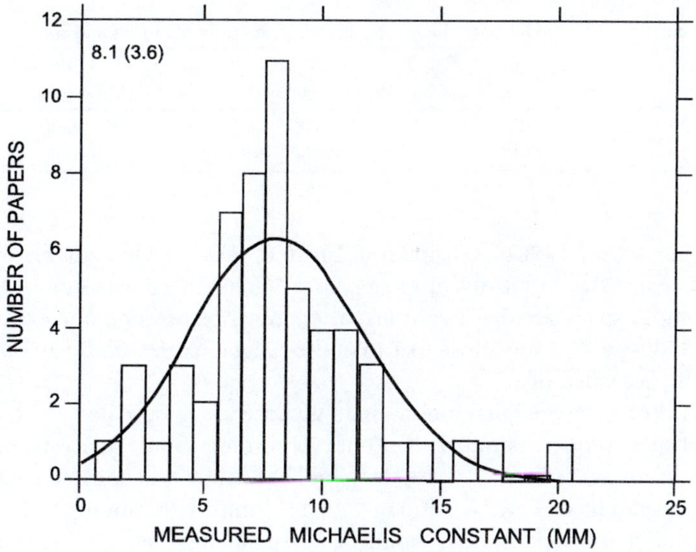


**Fig. 6.6** Frequency of measured $K_t$-values for blood–brain glucose transfer in rat, using a variety of techniques and anesthetic states

average 50 $cm^2\ g^{-1}$ in humans (Hunziker et al. 1979) and 100 $cm^2\ g^{-1}$ in rats (Gjedde and Diemer 1985b). These values yield identical apparent maximal permeabilities in the two species of approximately 800 nm $s^{-1}$ (Table 6.7). Part of the variation of $T_{max}$ may be due to a formal flaw in the logic underlying $T_{max}$ and $K_t$ measurements. By lowering glucose concentrations with insulin or raising them with glucose, it cannot be assumed that the conditions of blood–brain transfer remain the same; blood flow or metabolic changes may alter the kinetic properties of the cerebral capillary endothelium and hence invalidate the use of the Michaelis–Menten equation.

In studies in which transfer rates in both directions, as well as the glucose concentrations on both sides of the endothelium, are known, it is possible to calculate apparent values for $T_{max}$ and $K_t$ for a single physiological state. These calculations assume symmetrical transport and have been attempted only in a few cases (Ribeiro et al. 1991; Blomqvist et al. 1991). For reasons that are not understood, these measurements tend to be lower than those obtained at different arterial plasma concentrations.

## Non-Saturable Glucose Transfer

In many studies of glucose transfer, the data appeared to support the presence of a nonsaturable diffusion component of glucose transport, ranging in magnitude from 0.5 to 5 ml $hg^{-1}\ min^{-1}$. It is interesting that in vitro studies of glucose transport by the GLUT1 transporter indicate an intrinsic Michaelis half-saturation constant

**Table 6.7** Fundamental blood–brain glucose transport characteristics in rat and man

| Species | $T_{max}$ μmol hg$^{-1}$ min$^{-1}$ | $K_t$ mM | $P_{max}S$ ml hg$^{-1}$ min | $S$ cm$^2$ g$^{-1}$ | $P_{max}$ nm s$^{-1}$ | $C_a$ mM | PS $C_a$ μmol hg$^{-1}$min$^{-1}$ |
|---|---|---|---|---|---|---|---|
| Rat | 400 | 8 | 50 | 100 | 800 | 8 | 200 |
| Man | 100 | 4 | 25 | 50 | 800 | 4 | 50 |

of 1 mM (Carruthers 1990). As shown in Fig. 6.6, most in vivo studies of blood–brain glucose transfer in rat reveal an apparent Michaelis constant close to 8 mM. It is tempting to speculate that two transporters must be present, one expressed by GLUT1, yielding a $K_t$ value close to 1 mM, the second expressed by an other gene, yielding a higher value of $K_t$.

Gjedde (1981b) showed that blood–brain glucose transport data can be analyzed for two transport processes rather than one. According to this analysis, one transporter has a $K_t$ of 1 mM and a $T_{max}$ of 160 μmol hg$^{-1}$ min$^{-1}$ in anesthetized rats, the second transporter has a $T_{max}/K_t$ ratio of 5 ml hg$^{-1}$ min$^{-1}$. Assuming a $K_t$ of 20 mM for the second transporter, the $T_{max}$ would be 100 μmol hg$^{-1}$ min$^{-1}$. When analyzed for a single transporter, the two transporters yield an average $K_t$ close to 8 mM at commonly employed glucose concentrations. The second transporter may be responsible for the frequent observations of "nonsaturable" glucose transport believed to be specific because L-glucose is subject to no comparable transport. According to Gjedde (1981b), the low affinity transporter is responsible for the majority of the net transport at high concentrations of glucose in plasma and hence would be the one expected to be sensitive to consistently elevated glucose concentrations.

## Metabolism-Flux Ratio and the Lumped Constant

The observation that glucose influx normally exceeds glucose consumption by a factor of two is reflected not only in the brain glucose content but also in certain more specialized indicators, such as the metabolism index $[k_3/(k_2 + k_3)]$ or the "lumped constant" (Sokoloff et al. 1977). The lumped constant is the net rate of deoxyglucose transfer, relative to the net glucose transfer. It depends exclusively on the relation between glucose demand and supply and can therefore be calculated from the ratio between unidirectional and net glucose transfer. In turn, when the lumped constant is known, the ratio between glucose demand and supply, i.e., the metabolism-flux couple, can be determined. The observation that lumped constants are similar in a wide variety of species confirms the uniform relationship between unidirectional and net glucose transfer (Gjedde 1987).

At half the unidirectional flux, the brain interstitial glucose concentration is one third of the $K_t$, or 2.7 mM in rats and 1.3 mM in humans. This is generally the range observed in awake rats and calculated in humans. Several studies of the relationship between brain and plasma glucose as a function of the plasma glucose concentration were analyzed by Gjedde and Christensen (1984)) who concluded that the relationship supported symmetrical blood–brain glucose transfer with a $T_{max}/\Delta J$ ratio of

3–5, and a $K_t$ value of 7 mM. The question is whether the glucose concentration of the brain tissue changes under normal circumstances, i.e., of how finely tuned the regulation of glucose supply is under changing physiological conditions, and of how sensitive it is to pathological influences.

### *6.4.3 Acute Changes of Glucose Transport*

The factors that influence glucose transfer acutely are included in (6.12) and (6.20). They include blood flow ($F$) which may increase the average capillary glucose concentration, albeit almost negligibly; whole-blood solubility ($\alpha$) which may increase the apparent permeability by lowering the distribution in red blood cells; the endothelial surface available for facilitated diffusion ($S$) which in theory may increase by recruitment of reserve capillaries; and the apparent permeability ($P$) which may be increased by recruitment of transporters and by changes of transporter affinity, for example by changes of ATP concentration (Carruthers 1990). According to this theory, ATP regulates the affinity of the glucose transporter, causing the affinity to decline when ATP concentrations are low (Jacquez 1984; Carruthers 1986; Helgerson et al. 1989).

#### Activation and Deactivation

**Anesthesia** The interest in the effect of anesthetic agents on blood–brain barrier characteristics has been twofold, focused on one hand on the consequences of the ability of these agents to depress brain function in general, and on the other on the consequences of specific effects of the agents on membrane properties, if any. These factors are not easily separated in vivo. Most investigators of the blood–brain transfer of glucose in animals used anesthetics, of course. Only in some studies did the animal, usually a rat, remain awake during the application of a noninvasive procedure but these studies were often designed for other purposes. As a result, the number of studies in which the differences between the awake and the anesthetized states were specifically recorded is comparatively few.

Before 1980, Betz et al. (1973) in dogs, Horton (1973) in mice, and Nemoto et al. (1978) in rats, observed no consistent changes of the blood–brain glucose transfer in pentobarbital or halothane anesthesia. The significance of these findings is difficult to assess because opposite changes of blood flow and metabolism were expected with anesthetic agents like halothane, and because the methods ignored the effect of blood flow changes on the measured unidirectional clearance (i.e., as distinct from changes of the apparent PS-product or Michaelis constants of the transport). The absent change of glucose transfer was consistent with measurements of brain glucose content which generally went up as exemplified in Fig. 6.7, adapted from Mayman et al. (1964). The applications of (6.22) and (6.23) to the findings of Mayman et al. (1964), revealed that the ratio between $T_{max}$ (the maximal transport) and $\Delta J$ (the net transport) increased from 2.5 to 5 in phenobarbitone anesthesia, suggesting a decline of metabolism in the absence of a decline of transport.

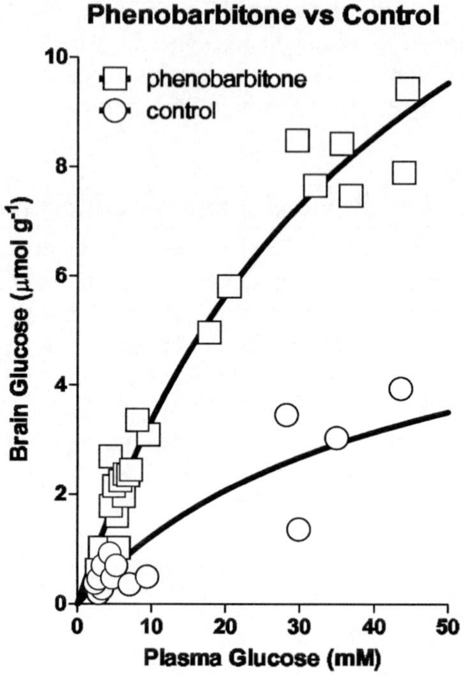


**Fig. 6.7** The relation of brain glucose to serum glucose in hyperglycaemia. Phenobarbitone (250 mg/kg i.p.) was given 1 h before sacrifice. At the same time 25 ml/kg of 1 M-glucose was injected i.p. into those mice with serum glucose concentrations above 10 m-moles/l. The points represent experimentally determined values. For simplicity, only the particular control values which were determined concurrently with the phenobarbital values are shown. The curves represent values calculated for a reversible, carrier-mediated glucose transport system described in the text. From Mayman et al. (1964)

After 1980, several studies, listed in Table 6.8, showed that pentobarbital, which consistently reduces both brain metabolism and blood flow, causes blood–brain glucose transfer to decline as well (Gjedde and Rasmussen 1980b; Crane et al. 1985; Hargreaves et al. 1986; LaManna and Harik 1986). The decline varied from 25% to 75% of the control value observed in awake rats, depending on the depth of anesthesia and the condition of the awake rats. The estimates of $T_{max}$ declined approximately 50% and the estimates of $K_t$ did not change significantly. In contrast to the earlier studies, these studies suggested that blood–brain glucose transfer declines in proportion to the decline of blood flow and metabolism observed during pentobarbital anesthesia. Also, Sokoloff et al. (1977) found a small but insignificant decline of the lumped constant in pentobarbital anesthesia, indicating parallel declines of transport and metabolism. Together, the early and the late studies show that both metabolism and flux decline in anesthesia but the metabolism perhaps more so than the flux. However, earlier measurements of increased brain glucose contents in anesthesia may have been influenced by postmortem breakdown of glucose, likely to affect rapidly metabolizing brains more than slowly metabolizing brains (Gjedde 1984b).

**Activation** There are few studies of glucose transport changes during normal physiological activation of the brain and the metabolism-flux couple has rarely been tested by comparing two states of different functional activation of the same region or regions of the brain. The reason is that physiological excitation of neurons is difficult to study in rats, and humans; in humans because physiological excitation may

**Table 6.8** Blood-brain glucose transfer in pentobarbital anesthetized rats

| | Flux (Concentration) [μmol hg$^{-1}$ min$^{-1}$ (mM)] | | $T_{max}$ ($K_t$) | |
|---|---|---|---|---|
| Authors | Awake | Pentobarbital | Awake | Pentobarbital |
| Gjedde and Rasmussen (1980a, b) | 160 (8) | 114 (9) | 410 (9) | 240 (5) |
| Crane et al. (1985) | | | | |
| Fed | 127 (8) | 49 (8) | 190 (8) | 100 (10) |
| 2-day fast | 167 (6) | 50 (7) | 180 (7) | 70 (8) |
| Hargreaves et al. (1986) | 174 (6) | 95 (6) | – | – |
| LaManna and Harik (1986) | 186 (11) | 144 (12) | 250 (5) | 210 (4) |

not be maintained without attenuation for more than a few minutes, in rats because a condition comparable to that of the resting human is likely to be replaced with a state of excitation due to physical restraint. The sedated rat may actually be a more correct parallel to the relaxed, resting human. Also, one result of excitation may be increased blood glucose. Systemic excitation may cause blood glucose to rise and hence elevate blood–brain glucose transfer without changing the transport characteristics. Despite these limitations, in the only study of its kind, Gjedde et al. (1991) observed a 15% increase of both glucose influx and metabolism during somatosensory stimulation in humans, apparently due to increased capillary diffusion capacity (recruitment) rather than increased blood flow.

In a few studies, attempts were made to enhance neuronal activation further by less physiological means. Cremer et al. (1981, 1983) compared glucose transport and consumption in brain of rats administered pyrethroid insecticides. The insecticide caused the transport and metabolism of glucose to increase by 50–100% in most regions suggesting an increase of the capillary diffusion capacity for glucose, as shown in Fig. 6.8. Calculated on the basis of a constant $K_t$, the $T_{max}$ increased somewhat less. In rats first subjected to an overnight fast, during which the plasma glucose concentration declined by 50%, the glucose clearance but not the net flux increased. In this group, the brain glucose content declined, accompanied by a marked increase of $T_{max}$ from 337 μmol hg$^{-1}$ min$^{-1}$ in the cerebellum of control rats to more than 700 μmol hg$^{-1}$ min$^{-1}$ in fasted rats in which a strong tremor was evident (Cremer et al. 1983). The decline of brain glucose content signifies an increase of the lumped constant and hence an insufficient increase of glucose transport. The studies suggest that some capillaries may be recruited to allow increased glucose transport in conditions in which glucose consumption is increased but strong evidence for an actual increase of the total number of perfused capillaries is lacking. Less heterogeneity of capillary perfusion velocities may be sufficient to account for these results.

**Seizures** Generalized seizures in young and adult mammals are generally believed to be accompanied by substantial increases of oxygen and glucose metabolism and blood flow but there are few reliable reports that directly or indirectly reveal the relationship between unidirectional and net blood–brain glucose transfer in this condition of massive departure from steady-state. A few studies indicate that the

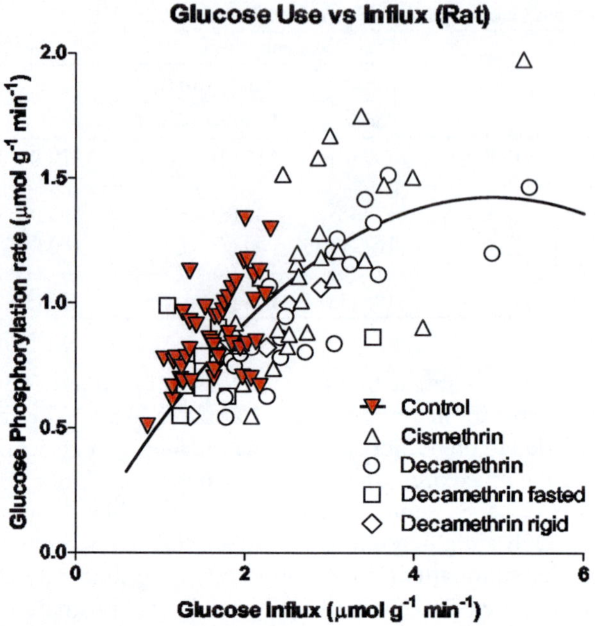


**Fig. 6.8** The relationship between rate of glucose influx ($K_1C_a$) and rate of glucose phosphorylation ($KC_a$) during excitation of rat brain with insecticide. Abscissa: rate of glucose influx (μmol $g^{-1}$ $min^{-1}$. From Cremer et al. (1981)

lumped constant rises significantly in certain types of seizures, suggesting isolated increases of metabolism in these conditions. Gjedde and Diemer (1985a, b) used 3-*O*-methylglucose to calculate a 50% increase of rat hippocampal lumped constant in kainic-acid-induced seizures, while Fujikawa et al. (1989) used measurements of brain glucose content to calculate lumped constant increases of more than 100% in newborn marmosets given bicuculline. Also, Cremer et al. (1988) noted an increase of the distribution of 2-deoxyglucose relative to glucose in rats given kainic acid, a change that signified an increase of the lumped constant. It is evident that glucose transport fails to keep pace with glucose metabolism during increases of the latter that exceed 20–50% of normal levels. These findings alone suggest that the reserve of capillaries available for recruitment, if any, cannot be substantial.

### Carbon Dioxide and Spreading Depression

Although the regional variation of glucose transport from blood to brain appears to be correlated to differences of blood flow and metabolism, primary blood flow changes do not necessarily lead to changes of blood–brain glucose transfer.

**Hypercapnia** Carbon dioxide is a strong stimulant of blood flow change. As it is unlikely that carbon dioxide is the agent responsible for the changes of blood flow

**Table 6.9** Effect of carbon dioxide on blood–brain glucose transfer in rat

| Group | $P_aCO_2$ (mmHg) | Plasma glucose (mM) | PS ml $hg^{-1}$ $min^{-1}$ | E (%) | (n) |
|---|---|---|---|---|---|
| Hypocapnia | 25±1.0 | 11.8±0.6 | 17.0±3.2 | 44±6 | (5) |
| Normocapnia | 38±1.5 | 12.2±1.6 | 23.4±3.0 | 44±5 | (7) |
| Hypercapnia | 67±2.0 | 12.7±0.2 | 20.8±3.4 | 12±3 | (4) |

From Gjedde and Lauritzen (1985)

observed during neuronal excitation, it is not a priori certain that carbon-dioxide-induced changes of blood flow must change blood–brain glucose transfer. In fact, in rabbits, in 1975, Berson et al. measured a decline of glucose extraction during massive hypercapnia (100 mmHg). Low extraction was also observed in two more recent studies of glucose transport during hypercapnia (Hertz and Paulson 1982; Gjedde and Lauritzen 1985). The low extraction fractions show that carbon-dioxide-induced blood flow changes are not accompanied by parallel changes of glucose transport capacity, i.e., of the number of capillary glucose transporters. Originally, the calculation of blood–brain glucose flux on the basis of simultaneous measurements of blood flow suggested increased glucose transport (Hertz and Paulson 1982) but later re-analysis of the same data (Knudsen et al. 1991) revealed no increase of blood–brain glucose flux. As shown in Table 6.9, the same conclusion was drawn by Gjedde and Lauritzen (1985) for rats.

**Hypocapnia** In hypocapnia, leading to low blood flow, it is the expectation that the first-pass glucose extraction must go up, as also observed by Hertz and Paulson (1982). In rat, Gjedde and Lauritzen (1985) observed no significant change of glucose transport, while in man, the transport appeared to go up. The explanation for the latter change is not clear.

**Spreading Depression** Gjedde et al. (1981) measured the changes of glucose transfer during Leao's spreading cortical depression. The condition is characterized by massively elevated potassium concentrations and transiently elevated blood flow. No change of glucose transfer, measured as the *PS*-product, occurred during the wave of increased blood flow, confirming the conclusion drawn earlier that the blood flow changes per se cause no change of blood–brain glucose transfer indicative of recruitment.

### Anoxia and Ischemia

The question of blood–brain glucose transfer modulation in anoxic or ischemic states has been particularly difficult to answer because the marked changes of blood flow, metabolism, and glucose concentrations on both sides of the blood–brain barrier may affect blood–brain glucose transfer without changing the fundamental characteristics of the transport.

**Anoxia** Anoxia causes cerebral blood flow to increase. In the absence of other changes, the increased blood flow lowers the extraction of glucose. A reduction of the first-pass extraction indicates a reduced flux-flow couple and hence insufficient recruitment. Agnew and Crone (1967) and Berson et al. (1975) observed the anoxic decline of glucose extraction in rabbits. To circumvent the problem of increased blood flow, Betz et al. (1974) and Kintner et al. (1980) mechanically perfused the head of dogs without oxygen, adding radioactive glucose to the perfusate to measure blood–brain glucose transfer. The transfer was calculated as the product of the measured extraction fraction and the rate of outflow through a venous catheter, relative to the weight of the brain. Calculated on this basis, the unidirectional glucose influx appeared to decline by 50% within 15 min of the onset of anoxic perfusion. The reason for the decline is not clear but may be associated with loss of ATP from brain tissue and endothelial cells. It is also possible that the blood flow rate was underestimated by the venous outflow method; unless venous blood is collected in toto, outflow is in reality a pressure measurement, reflecting the low resistance during anoxia. During reoxygenation, glucose transport slowly returned to normal.

**Ischemia** During complete ischemia, it is impossible to measure glucose transport because of the lack of perfusion. Studies have therefore focused on the changes of glucose transport after ischemia. In the postischemic phase, Betz et al. (1983), in gerbils, and Gjedde et al. (1990), in living humans after an episode of stroke, reported significantly reduced glucose transfer, related to a reduction of the capillary diffusion capacity, as expressed by a calculated density of 'standardized' capillaries. Although the latter result suggested that the change is not specifically related to glucose transport but to general changes of the microvasculature, part of the decline may also be attributed to low transport affinity, according to Gjedde and Siemkowicz (1978) who reported an increase of $K_t$ from 6 to 12 mM, again perhaps as a result of ATP loss from endothelial cells.

The separate changes of blood flow and glucose transport observed after ischemia, and during spreading depression and hypo- or hypercapnia, indicate that blood flow and glucose transport are regulated by different mechanisms in the vascular bed, possibly in different sectors of the vasculature. According to this possibility, blood flow regulation may be an arteriolar function, glucose transport regulation a capillary function.

## *6.4.4 Chronic Changes*

### Pre- and Postnatal Development

Under normal circumstances, the only known permanent change of blood–brain glucose transfer occurs during pre- and postnatal development. GLUT1 seems to be ubiquitous in fetal tissues but declines after birth. Thus, erythrocyte GLUT1 transporters are lost after birth in many species (Widdas 1955; Goodwin 1956; Jacquez 1984). The signal for the continued expression of the GLUT1 transporter in brain capillary endothelium has not been identified but it is possible that the increasing tightness of the blood–brain barrier in postnatal life, and the decline of glucose concentration in brain extracellular space relative to plasma, maintains the expression of the GLUT1 transporters, perhaps by influence of a humoral agent released by glial foot processes (Maxwell et al. 1989).

Several studies show an increase of blood–brain glucose transfer in parallel with the postnatal increase of glucose consumption, with little change of the lumped constant, indicating constant relations between unidirectional and net glucose transfer (Fuglsang et al. 1986; Dyve and Gjedde 1991). Cremer et al. (1976) noted an increase of Oldendorf's Brain Uptake Index of first-pass extraction from 19% to 32% between 18 days and 9 weeks of age in rats. Daniel et al. (1978) noted at least a doubling of the blood–brain clearance between 2 and 9 weeks of age in rats and, interestingly, a decline toward neonatal values in senescent (64 weeks old) rats. Cremer et al. (1979) described a 40% increase of both blood–brain glucose transfer and $T_{max}$ between the ages of 2.5 weeks and adulthood in rats. Cornford and Cornford (1986) noted a several-fold increase of blood–brain glucose transfer between 1 day old and 28-day old rabbits, the result of an almost 20-fold increase of the $T_{max}$ of the glucose transporter, in contrast to two earlier studies in which the Brain Uptake Index of glucose changed very little during postnatal development of rabbits (Braun et al. 1980; Cornford et al. 1982).

Cremer et al. (1979) and Cornford and Cornford (1986) observed very little increase of $K_t$ with age, in contrast to Fuglsang et al. (1986) who calculated a much higher value of $K_t$ in newborn rats than in adult rats. Both Cremer et al. and Cornford and Cornford determined the so-called nonsaturable component of glucose transport separately. If it is correct that the nonsaturable component is in reality a less-high-affinity transporter then it is possible that the GLUT1 transporter may be not only maintained but also specifically induced during development, in response to the declining ratio between glucose in brain and glucose in blood. This speculation predicts that the GLUT1 transport system is particularly sensitive to changes of the extracellular glucose concentration.

### Dementia

There is evidence of a decline of glucose transport in cerebral cortex in Alzheimer's disease in parallel with the decline of functional activity and metabolism, related both to the cortical atrophy and to the decline of functional activity of the surviving tissue (Kalaria and Harik 1989; Jagust et al. 1991). It is unlikely that this phenomenon is related to any direct mechanism of regulation of glucose transport. It is more likely that the decline represents an acceleration of the normal decline in senescence.

## Hypo- and Hyperglycemia

The Michaelis–Menten nature of the blood–brain glucose transfer dictates that glucose clearance ($K_1$) and apparent permeability (PS) must increase when blood glucose declines, or decline when blood glucose increases. Despite continued rediscovery of this aspect of the Michaelis–Menten equation, the real question is whether additional, specific changes occur when the state of abnormal glucose concentrations persists. Glucopenia caused an upregulation of glucose transport in cultured fibroblasts (Ullrey et al. 1975; Ullrey and Kalckar 1981) and low extracellular glucose caused glucose transporter numbers to increase in chicken embryo fibroblasts (Pessin et al. 1982; Kletzien and Perdue 1985). The concept of glucose-starvation enhancement of sugar transport in animal cell cultures (GSE, Germinario et al. 1982) is by now so entrenched that the question is not whether it occurs but whether it is a feature of all glucose transporters and whether it occurs in vivo. In the rat L6 cell line of skeletal muscle, basal glucose transport is subserved by GLUT1 type transporters (Walker et al. 1989). By elevating glucose in the medium from 5 to 15 mM, glucose transport was reduced 30% by depletion of GLUT1 transporter protein and cytochalasin-B binding sites (Klip and Paquet 1990). Thus, in vitro, the GLUT1 transporter appears to be down-regulated (Klip et al. 1991). However, in vivo, the situation may be different. In mildly diabetic rats, GLUT4 transporters in muscle cells but not GLUT1 were diminished when measured by cytochalasin B binding (Ramlal et al. 1989) but mRNA levels were not reduced, suggesting a translational block (Klip and Paquet 1990). The significance of this difference between in vivo and in vitro conditions is not clear.

**Hyperglycemia** There are no confirmed inborn or idiopathic abnormalities of blood–brain glucose transfer in vivo. Both Wyke (1959) and Harris and Prout (1970) published case reports suggesting the existence of an idiopathic or acquired state of relative cerebral hypoglycemia, i.e., a condition in which the blood glucose level is normal, yet blood–brain glucose transfer appears to be abnormal. Similar observations were made by DeFronzo et al. (1977, 1980) and Lilavivathane et al. (1979) in diabetics in which the blood sugar level was normalized rapidly.

The discussion of the pre- and postnatal development of glucose transport supported the prediction that elevated glucose in brain may inhibit the expression of the GLUT1 transporter. Observations of this kind led Gjedde and Crone (1981) to examine the blood–brain glucose transfer in rats with maintained hyperglycemia to test whether prolonged exposure of the blood–brain barrier to elevated glucose inhibited the transport. The results, obtained with the intravenous bolus injection method and tissue sampling, indicated a down-regulation of the blood–brain glucose transfer in the hyperglycemic state. The finding can be criticized because only two (glucose clearance and blood flow) of the three variables required to calculate blood–brain glucose transfer (i.e., glucose clearance, blood flow, and blood volume) were simultaneously measured.

The result was initially confirmed by other studies, both directly with another method (McCall et al. 1982, using intracarotid bolus injection and tissue sampling) and indirectly by determination of the number of cytochalasin B binding sites in isolated cerebral microvessels in vitro (Matthaei et al. 1986). At the same time, Brooks et al. (1986a, b) found no difference of unidirectional blood–brain glucose transfer between normal humans and diabetics, and Duckrow (1988) and Harik and LaManna (1988), in separate reports, found no change of glucose transfer in chronically diabetic rats. Likewise, Harik et al. (1988) reported an increase, rather than a decrease, of cytochalasin B binding sites in isolated brain microvessels, and in erythrocytes from diabetics (Harik et al. 1991a, b). However, Harik and LaManna (1988) also observed a significantly reduced L-glucose space in the chronically hyperglycemic rats.

To address the issue, using the carotid perfusion method first introduced by Takasato et al. (1984), Pardridge et al. (1990a) controlled most of the factors that affect glucose transport. In rats, 1 week after the elevation of a plasma glucose from 5.8 to 24.6 mM, at a perfusate glucose concentration of 10 mM, the PS-product of glucose transport had declined from 28 to 16 ml $hg^{-1}$ $min^{-1}$. The rate of tissue perfusion also declined in these rats, despite identical inflow rates in the two groups. The same group of investigators also found an increase of mRNA, confirming a translational block in this condition (Choi et al. 1989). Using the BUI method, also Mooradian and Morin (1991) noted a significant decline of 3-*O*-methylglucose uptake from 42% to 33%, 8 days after induction of hyperglycemia. Acutely hyperglycemic animal, were not affected.

Using the method designed to yield glucose metabolic rate, first introduced by Cunningham and Cremer (1981), Pelligrino et al. (1990a) observed an increase of the unidirectional glucose clearance ($K_1$) in rats hyperglycemic at 33 mM for 6–8 weeks with acutely normalized glucose levels; from an average of 16 ml $hg^{-1}$ $min^{-1}$ (8.5 mM glucose) in normal rats to 35 ml $hg^{-1}$ $min^{-1}$ (7.9 mM glucose) in the chronically hyperglycemic rats. Glucose concentration in brain and glucose consumption were also elevated. Although glucose PS-product, glucose influx, or blood or plasma flow, were not measured, the glucose concentration and metabolism measurements are difficult to reconcile with a reduction of glucose transport. The duration of hyperglycemia may play a role; the original observation by Gjedde and Crone (1981) was made 3 weeks after the induction of hyperglycemia and predicted a modest decline of the glucose PS product from 21 to 17 ml $hg^{-1}$ $min^{-1}$ at a plasma glucose concentration of 10 mM.

**Insulin** Originally, a fourth factor was not considered in these studies, i.e., the insulin level in chronic hyperglycemia, but for apparently good reasons: Crone (1965), Buschiazzo et al. (1970), Betz et al. (1973, 1979), and Daniel et al. (1975b) had concluded that insulin had no measurable effect on unidirectional blood–brain glucose transfer. Although some of these studies can be criticized on methodological grounds (Bradbury 1985; Hertz 1986), the conclusion was later supported by the finding that the glucose transporter of the cerebral capillary endothelium is of the insulin-insensitive kind, and that the insulin-sensitive glucose transporter is not expressed in endothelial cells (Slot et al. 1990).

In 1981, Hertz et al. (1981) measured unidirectional blood–brain glucose transfer in humans with the intracarotid bolus injection method and arteriovenous sampling. Insulin at a 100-fold elevated concentration reduced the unidirectional blood–brain glucose transfer by 50%. The study appeared to confirm that blood–brain glucose transfer may be low in the absence of massively elevated insulin levels. The finding received support from in vitro measurements of glucose transfer into, and cytochalasin B binding to, isolated cerebral microvessels (Djuričić et al. 1983; Matthaei et al. 1986). Both studies have been criticized, the former because glucose transfer into isolated microvessels cannot be unidirectional (Bradbury 1985), the latter by Harik et al. (1988) who observed a 30% increase of cytochalasin B binding in hyperglycemia and suggested that sonication of tissue in the experiments of Matthaei et al. in part abolished specific cytochalasin B binding.

In 1987, Namba et al. reported no effect of insulin on the unidirectional blood–brain transfer of 3-O-methyl-glucose, a nonmetabolizable glucose analog. As the intracarotid bolus injection method in humans does not distinguish between glucose transport in different brain regions, it is possible that the effect on glucose transport was caused by uptake into regions with a specific sensitivity to glucose (Reith et al. 1987). When the problem was reinvestigated with a regional method (PET) at a more physiological insulin level, no change of the $K_1$ of fluorodeoxyglucose was seen (Hasselbalch et al. 1991). Concluding from 10 years of study of unidirectional blood–brain glucose transfer in prolonged hyperglycemia and insulin-deficient states, it appears unlikely that insulin has a significant effect at physiological levels but likely that prolonged hyperglycemia may reduce blood–brain glucose transfer under some circumstances.

**Hypoglycemia** It is of interest to observe the change of blood–brain glucose transfer under the opposite circumstance of prolonged hypoglycemia and potentially glucose-deficient states like inanition. Contrary to the original report by Daniel et al. (1971), inanition with ketoacidosis was shown to stimulate blood–brain transfer of ketone bodies (Gjedde and Crone 1975; Moore et al. 1976) and glucose (Christensen et al. 1981) while hypoglycemia stimulated glucose transfer (McCall et al. 1986; Pelligrino et al. 1990b). If confirmed, the studies of Pelligrino et al. (1990a, b) indicate somewhat implausibly that both hypo- and hyperglycemia stimulate unidirectional blood–brain glucose transfer. The stimulation of glucose transfer during inanition was not confirmed by Crane et al. (1985) in rats, or Redies et al. (1989) in obese humans undergoing therapeutic weight loss. In conclusion, it appears that sustained hypoglycemia may stimulate blood–brain glucose transfer, whereas inanition, with or without ketoacidosis, probably has little effect. For the reasons discussed earlier, it is not likely that insulin is responsible for the stimulation of unidirectional blood–brain glucose transfer during sustained insulin-induced hypoglycemia.

# Chapter 7
# Metabolism of Glucose*

## 7.1 Basic Principles of Metabolism

The metabolism of the different organs of the body is specialized in several ways. For example, under normal circumstances, adult brain energy metabolism is fueled almost exclusively by glucose. The work performed by the brain consists of the transfer of ions across cellular and mitochondrial membranes. As the extrusion of sodium ions from the intracellular space in a single minute replenishes the total extracellular content of sodium ions, the work is considerable. While both the oxygen reduction to water and the oxidative phosphorylation of ADP appear to be driven by the mitochondrial ADP/ATP ratio, in collaboration with the cytochrome oxidase activity, glucose consumption is driven by the phosphofructokinase activity outside the mitochondrion. Although well-correlated in the steady-state, these activities become uncoupled in the absence of steady-state.

Other organs, notably the liver, produce rather than use glucose under certain circumstances, and yet others use glucose preferentially only during certain kinds of activity. Cardiac and skeletal muscle use several fuels. During fasting and inactivity when insulin levels are low, these muscle types depend on free fatty acids as fuels.

This chapter describes the interaction between tracer glucose uptake and phosphorylation; derives the equations for the use of glucose analogs to calculate tissue glucose phosphorylation; and estimates the value of the "lumped constant" as a function of the glucose phosphorylation rate, the plasma glucose concentration, the maximal transport capacity of the capillary endothelium, and the blood flow. Kinetic models have been introduced or discussed by Sokoloff et al. (1977), Phelps et al. (1979), Gjedde (1982), Lassen and Gjedde (1983), Lammertsma et al. (1987), and Redies and Gjedde (1989). Gjedde (1987) has presented a review of these models; Redies and Gjedde (1989) provide practical guidelines.

* Adapted from Gjedde A (1995b) Glucose metabolism. In: *Principles of Nuclear Medicine*, 2nd edition, eds Wagner HNJr, Szabo Z, Buchanan JW. Saunders, Philadelphia, pp. 54–71, with permission from Saunders, Philadelphia.

A. Gjedde et al., *Neurokinetics: The Dynamics of Neurobiology In Vivo*, 211
DOI 10.1007/978-1-4419-7409-9_7, © Springer Science+Business Media, LLC 2011

### 7.1.1 Glycolysis

A kinetic model must be based on the known biochemical characteristics of the intermediary metabolism of glucose. Glucose crosses the endothelium of capillaries by facilitated diffusion through cell membranes (transcellular), and by bulk flow of water between cells (paracellular). In brain, the latter route is prevented by the osmotic forces generated by the very high reflection coefficients of the blood–brain barrier toward hydrophilic solutes. Outside the brain, the paracellular route is of sufficient capacity to allow paracapillary circulation. Thus, in brain, the interstitial glucose concentration is much less than that of blood plasma, while outside brain, the interstitial glucose level equals that of blood plasma.

From the interstitial fluid, glucose enters cells by means of a second transporter which may not be the same as the one that facilitates the diffusion across endothelial cells. In muscle and fat cells, this transporter is the insulin-sensitive glut4 transporter (see Gjedde 1992). In the absence of insulin, it has such low capacity that the glucose concentration in the cytosol is low. In the presence of insulin, the transport increases to allow glucose in the cytosol to approach the concentration of glucose in the interstitium. In the brain, the glut1 and glut3 glucose transporters are insensitive to insulin but normally so active that glucose inside nervous tissue cells has the same concentration as glucose in the interstitium.

In the cytoplasm, a series of 11 enzymes break down glucose in two stages, stages I and II of glycolysis. The enzymes are regulated by the concentration of precursors, substrates, and regulatory factors, rather than by structural constraints. Glucose is first phosphorylated by gluco- or hexokinase to glucose-6-phosphate. Hexokinase is more widely distributed than glucokinase and most cells use hexokinase to phosphorylate glucose and other hexoses. Glucokinase predominates in the liver and phosphorylates only glucose. Hexokinase is the only enzyme in brain and muscles. Both enzymes catalyze irreversible reactions. The affinity of hexokinase to glucose is so high that the enzyme is saturated in brain cells, and in muscle and fat cells under insulin stimulation. This renders the rate of glucose metabolism essentially equal to the maximal velocity of the hexokinase reaction ($J_{max}^{Hase}$). Several isoenzymes of hexokinase exist. Glucokinase has a low affinity for glucose but high capacity, causing it to operate only when the glucose concentration in the interstitium of the liver (and hence in plasma) is high.

Glucose 6-phosphate occupies a pivotal position in the metabolism of glucose, providing a gateway to five different paths. In most tissues, a major fraction of glucose-6-phosphate is either dephosphorylated by glucose 6-phosphatase or isomerized to fructose-6-phosphate by glucose phosphate isomerase. The dephosphorylation is prominent in the liver but normally negligible in brain where the enzyme is sparse and further sequestered in the cisterns of the endoplasmic reticulum without the necessary glucose 6-phosphate translocase that probably exists in liver (Fishman and Karnovsky 1986). This difference between the brain and the liver is illustrated in Fig. 7.1 which compares the accumulation of the glucose tracer fluorodeoxyglucose in normal liver and brain and in liver of patients with *von Gierke*'s Type I Glycogen Storage Disease in which the liver glucose 6-phosphatase is absent.

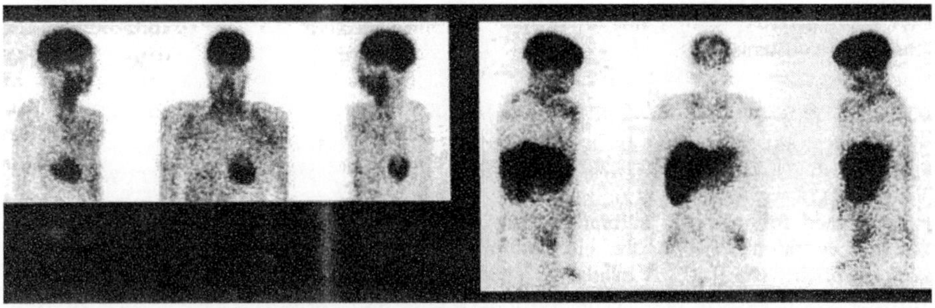

**Fig. 7.1** Uptake of fluorodeoxyglucose in body of normal volunteer and patient with von Gierke's disease. Note absent uptake in liver in the normal volunteer and trapping in brain and liver in the patient, indicative of absent glucose-6-phosphatase. From Phelps et al. with permission

In stage I of glycolysis, glucose is phosphorylated twice and ultimately cleaved into two molecules of glyceraldehyde 3-phosphate. In stage II of glycolysis, glyceraldehyde 3-phosphate is oxidized and dephosphorylated to pyruvate. In the last step of glycolysis, pyruvate is reduced to lactate by lactate dehydrogenase to a degree dictated by the specific isoenzymes of this enzyme. In muscle, the isoenzyme $M_4$ has such high capacity and affinity to pyruvate that the pyruvate-lactate ratio is shifted significantly toward lactate in the steady-state. In heart, the isoenzyme $H_4$ has such low capacity and affinity that the pyruvate-lactate ratio is low. In brain, the intermediate isoenzymes $M_2H_2$ and $MH_3$ result in an intermediate pyruvate-lactate ratio. Regional differences exist in the brain, as in muscle, some regions having a preponderance of the heart enzyme, others having a preponderance of the muscle enzyme.

The isoenzyme influences the lactate content. Thus, the heart enzyme maintains a fairly low lactate level for a given pyruvate level. The low level results in a low rate of lactate transport out of the cells. The muscle enzyme, on the other hand, maintains a high level of lactate for a given level of puryvate and hence stimulates lactate transport out of the cell with a consequent loss of glucose-derived carbon skeletons.

In the nonsteady-state, the isoenzyme also influences the rate at which the pyruvate level may rise during increases of the rate of glycolysis. The high pyruvate-lactate ratio maintained by the heart enzyme causes the pyruvate level to rise rapidly for a given increase of the rate of pyruvate synthesis, thus enhancing the entry of pyruvate into alternate pathways. The low pyruvate-lactate ratio maintained by the muscle enzyme causes the pyruvate level to rise slowly and delays the entry of pyruvate into other pathways. It also causes lactate levels to rise so much that the transport out of the cell keeps pace with the increased rate of glycolysis. When that is the case, the fraction of glucose molecules leaving the tissue as lactate increases significantly. This is particularly important in regions of the brain with a preponderance of the muscle-type enzyme. Net, the cell has phosphorylated one molecule of ADP for each molecule of lactate generated.

### *7.1.2 Oxidative Phosphorylation*

In the presence of oxygen, mitochondria phosphorylate ADP by transport of hydrogen ions across the inner membrane. The hydrogen ions are generated by the breakdown of pyruvate to carbon dioxide in the series of reactions known as the Krebs', citric acid, or tricarboxylic acid, cycle in the mitochondrion. The hydrogen ion flux is driven by the cytochrome oxidase's reduction of molecular oxygen to ionic oxygen ($O^{2-}$) which combine with hydrogen ions to form water. In the combined process of glycolytic synthesis and mitochondrial breakdown of two molecules of pyruvate, 19 molecules of ADP are phosphorylated net. The oxidative phosphorylation is driven by the breakdown of ATP in the cytosol and fueled by the breakdown of glucose. The ATP breakdown is driven by cellular work such as ion pumping in muscles and brain.

The oxidative phosphorylation requires a steady supply of ADP, produced in the cytosol by the breakdown of ATP during ion pumping. Insufficient diffusion of ADP to the mitochondria may inhibit oxidative phosphorylation and hence oxygen consumption during intense ion pumping.

### *7.1.3 Gluconeogenesis*

The metabolite glucose 6-phosphate plays a central role in the biosynthesis of mono-, di-, and polysaccharides, mucopolysaccharides, glycoproteins, nucleic acids, and 5-carbon sugars. Five enzymes determine the paths along which glucose 6-phosphate is metabolized to accomplish these syntheses. The relative activities of the enzymes depend on local regulatory factors in the particular organ. In brain, glucose phosphate isomerase is by far the most active of the enzymes. In liver, glucose 6-phosphatase is an active source of glucose.

To make glucose 6-phosphate from sources other than glucose, most organisms can reverse the glycolytic path from pyruvate. Using the same path, vertebrates convert lactate to glucose. This process mainly occurs in the liver which has the enzymes necessary to bypass the three irreversible reactions of the glycolytic chain. These irreversible steps include the phosphorylation of glucose to glucose 6-phosphate by gluco- or hexokinase, the phosphorylation of fructose 6-phosphate to fructose 1,6-diphosphate by phosphofructokinase-1, and the dephosphorylation of phosphoenolpyruvate to pyruvate by pyruvate kinase. The irreversible steps are the points of regulation of the rate of glycolysis. Phosphofructokinase-2 catalyzes the generation of fructose-2,6-phosphate, a product with the sole role of allosteric activation of phosphofructokinase-1.

To reverse glycolysis, equally irreversible reactions operate in the opposite direction, catalyzed by different enzymes under separate regulation. To reverse the pyruvate kinase step, pyruvate is converted to oxaloacetate by pyruvate carboxylase in mitochondria, and oxaloacetate is phosphorylated and decarboxylated to phosphoenolpyruvate in the cytosol. The role of oxaloacetate allows Krebs' cycle intermediates to contribute to gluconeogenesis by oxidation to oxaloacetate. The same is true of the amino acids that can be converted to Krebs' cycle intermediates.

To reverse the phosphofructokinase step, fructose 1,6-diphosphate is dephosphorylated by fructose diphosphatase. To reverse the gluco- and hexokinase step, glucose 6-phosphate is dephosphorylated by glucose 6-phosphatase. Glucose 6-phosphatase is a classic marker of the endoplasmic reticulum, to which glucose 6-phosphate may gain access by a translocase (Fishman and Karnovsky 1986). Brain possesses little pyruvate carboxylase, glucose 6-phosphatase, or glucose 6-phoshate translocase (Fishman and Karnovsky 1986; Pertsch et al. 1988; Schmidt et al. 1989). The lack of access to glucose 6-phosphatase prevents synthesis of glucose by brain.

### *7.1.4 Glycogenesis and Glycogenolysis*

In addition to being a substrate for glucose phosphate isomerase and glucose 6-phosphatase, glucose 6-phosphate is also the substrate of phosphoglucomutase which catalyzes the conversion to glucose 1-phosphate, the gateway to the synthesis of mono-, di-, and polysaccharides, as well as to glucuronic acid. The interaction of glucose 1-phosphate with UDPG-pyrophosphorylase leads to the formation of UDP-glucose which in turn is linked to an existing glycogen chain by the action of glycogen synthetase. In turn, glucose units are cleaved from the glycogen chain and phosphorylated to regenerate glucose 1-phosphate by means of glycogen phosphorylase.

### *7.1.5 Pentose-Phosphate Pathway*

The fifth enzyme acting on glucose 6-phosphate is glucose 6-phosphate dehydrogenase which catalyzes the synthesis of 6-phosphogluconolactone, an important precursor of nucleic acids and 5-carbon sugars. The synthesis regenerates NADPH by oxidation of glucose 6-phosphate. If produced in excess quantities, the ribose 5-phosphate intermediate can be reconverted to the glycolytic intermediates fructose 6-phosphate and glyceraldehyde 3-phosphate.

Deoxyglucose 6-phosphate may be a substrate for both phosphoglucomutase and glucose 6-phosphate dehydrogenase but both reactions are quantitatively insignificant in adult brain (Hakim and Moss 1972; Hakim et al. 1980; Dienel et al. 1992). Glucose 6-phosphate dehydrogenase can be abnormally active in tumors (Ikezaki et al. 1992) but does not directly generate metabolites that easily leave the brain.

## 7.2 Kinetics of Steady-State Glucose Metabolism

The net rate of native glucose transfer into tissue can be described in three different ways. First, it can be calculated from the arterio-venous difference and the plasma flow, according to the perfusion principle of Fick,

$$R = FC_{\mathrm{a}} - FC_{v} = KC_{\mathrm{a}} \tag{7.1}$$

in which $R$ is the net rate of glucose uptake, $F$ the plasma flow, and $K$ is the net clearance of glucose into tissue in the steady-state. These equations follow the convention that lower case symbols, with the exception of the relaxation constants $k_2$, $k_3$, and $k_4$, represent variables while upper case symbols represent physiological constants.

Second, in the steady-state, glucose consumption equals,

$$R = k_{\text{app}} M_e, \tag{7.2}$$

where $k_{\text{app}}$ is the net steady-state rate of glucose phosphorylation and $M_e$ is the tissue glucose content. The net steady-state rate of phosphorylation is the product of the unidirectional rate of phosphorylation and the isomerization fraction,

$$k_{\text{app}} \equiv \Phi k_3, \tag{7.3}$$

where $\Phi$ is the fraction of glucose that undergoes isomerization to fructose-6-phosphate. For glucose in the steady-state, the coefficient of phosphorylation by hexokinase can be defined as the ratio between the reaction velocity ($J$) and the steady-state tissue glucose content ($M_e$), as derived in Chap. 1,

$$k_3 \equiv \frac{J^{\text{Hase}}}{M_e} = \frac{J^{\text{Hase}}_{\text{max}}}{V_{\text{d}} K^{\text{Hase}}_m + M_e}, \tag{7.4}$$

where $J^{\text{Hase}}_{\text{max}}$ is the maximal rate of glucose phosphorylation by hexokinase, $V_{\text{d}}$ the volume of distribution in the organ, relative to the solubility in the reference solvent, and $K^{\text{Hase}}_m$ is the Michaelis half-saturation concentration of hexokinase toward glucose. The coefficient of dephosphosphorylation is defined in a similar manner, assuming the quantity of inhibitors to be negligible,

$$k_4 \equiv \frac{J^{\text{Pase}}_{\text{max}}}{V_{\text{d}} K^{\text{Pase}}_m + M_m}, \tag{7.5}$$

where $M_m$ is the tissue content of glucose-6-phosphate and $J^{\text{Pase}}_{\text{max}}$ is the maximal rate of dephosphorylation by glucose-6-phosphase. The volume $V_{\text{d}}$ assumes a special significance if the phosphatase is sequestered from the glycolytic enzymes in the cytosol, establishing a concentration difference between glucose-6-phosphate in the cytosol and glucose-6-phosphate at the phosphatase sites.

The magnitude of $\Phi$ is very close to unity. This prediction is based on the known tissue contents of glucose and glucose-6-phosphate which influence the magnitude of $\Phi$,

$$\Phi \equiv \frac{k_{\text{app}} M_e}{k_{\text{app}} M_e + k_4 M_m} \tag{7.6}$$

which is close to unity because $M_m \ll M_e$ for glucose, when $k_4 \ll k_{\text{app}}$. In this equation, $k_{\text{app}}$ is the net rate of native glucose phosphorylation in the steady-state. When $k_4$ is nil, $\Phi$ is unity, of course. As a rough calculation, when $M_e/M_m$ and $k_{\text{app}}/k_4$ are both 10, $\Phi$ is 0.99.

Third, in the steady-state, glucose consumption can be calculated from the difference between the arterial concentration ($C_{\text{a}}$) and the interstitial concentration of glucose ($C_e = M_e/V_{\text{d}}$), as discussed in Chaps. 1 and 2,

$$\Delta J_{\text{glc}} = K_1 C_{\text{a}} - k_2 M_e, \tag{7.7}$$

where $K_1$ is the unidirectional clearance of glucose by the organ in question and $k_2$ is the fractional clearance from the tissue. Equations (7.1), (7.2), and (7.7) combine to yield an expression for $K$, the net clearance of native glucose in the steady-state as a function of the individual transfer coefficients,

$$K = k_{\text{app}} \left[ \frac{K_1}{k_2 + k_{\text{app}}} \right] = EF, \tag{7.8}$$

where $E$ is the net extraction of glucose. If $\Phi$ is unity, i.e., in the absence of dephosphorylation, $k_{\text{app}} = k_3$.

## 7.3 Kinetics of Deoxyglucose Metabolism

When a tracer of glucose is introduced into the blood stream, the tracer is transported across the endothelial and cellular membranes and further subjected to metabolism by enzymes for which it retains affinity. For deoxyglucose and its fluorinated version (fluorodeoxyglucose), metabolism includes transport across the blood-tissue barrier and cellular membranes by the glucose transporter, phosphorylation by hexokinase, dephosphorylation by glucose-6-phosphatase, when present, but not isomerization by glucose phosphate isomerase.

Irreversible metabolism occurs when glucose-6-phosphatase activity is negligible. In brain, glucose-6-phosphate and glucose tracer 6-phosphate are not dephosphorylated to any significant degree to glucose or deoxyglucose.

For deoxyglucose in tracer quantity, the phosphorylation coefficient is defined in a manner similar to the definition of the phosphorylation constant of glucose,

$$k_3^* \equiv \frac{K_m^{\text{Hase}}}{K_m^{\text{Hase}*}} \left( \frac{J_{\max}^{\text{Hase}*}}{V_{\text{d}} K_m^{\text{Hase}} + M_e} \right) = \frac{j^{\text{Hase}*}}{m_{\text{e}}^*}, \tag{7.9}$$

where $j^{\text{Hase}*}$ is the variable hexokinase reaction velocity, $M_e$ the tissue content of glucose, $m_{\text{e}}^*$ the tissue content of deoxyglucose, and $J_{\max}^{\text{Hase}*}$ and $K_m^{\text{Hase}*}$ is the appropriate Michaelis–Menten constants of the tracer. The ratio between the rates of unidirectional phosphorylation of deoxyglucose and native glucose is the phosphorylation ratio,

$$\varphi \equiv \frac{k_3^*}{k_3} = \frac{J_{\max}^{\text{Hase}^*} K_m^{\text{Hase}}}{J_{\max}^{\text{Hase}} K_m^{\text{Hase}^*}} \tag{7.10}$$

which is used to define the lumped constant. If the tissue has the necessary phosphatase activity to dephosphorylate deoxyglucose, this enzymatic step can be assigned a rate constant,

$$k_4^* \equiv \frac{K_m^{\text{Pase}}}{K_m^{\text{Pase}^*}} \left( \frac{J_{\max}^{\text{Pase}^*}}{V_{\text{d}} K_m^{\text{Pase}} + M_m} \right) = \frac{j^{\text{Pase}^*}}{m_{\text{m}}^*}, \tag{7.11}$$

where tracer variables are indicated by an asterisk, and phosphatase constants and substrates indicated by the superscript Pase. Thus, $K_m^{\text{Pase}^*}$ is the Michaelis half-saturation constant of the phosphatase toward the tracer 6-phosphate, $k_3^*$ the rate of phosphorylation of the tracer, $k_4^*$ the rate of dephosphorylation, if any, and $M_m$ the quantity of glucose 6-phosphate in the tissue. Because some tracer metabolites may escape directly from the tissue, $m_z^*$ is a quantity of deoxyglucose metabolite (if any) that escapes directly from the tissue at the rate of $k_5^*$,

$$k_5^* = \frac{j^{\text{L}^*}}{m_z^*}, \tag{7.12}$$

where $j^{\text{L}^*}$ is the loss and $m_z^*$ is the mass of the tracer metabolite undergoing loss.

With these metabolism coefficients, the net rate of tracer transport across the capillary endothelium (in brain), or the cell membranes (outside brain), can be described,

$$\Delta j^* = K_1^* c_{\text{a}}^* - k_2^* m_{\text{e}}^* - k_5^* m_z^*. \tag{7.13}$$

The accumulation of tracer metabolites in the tissue can also be described by a differential equation:

$$\Delta j^* - \frac{\text{d}m_{\text{e}}^*}{\text{d}t} = k_3^* m_{\text{e}}^* - k_4^* m_{\text{m}}^* - k_5^* m_z^* \tag{7.14}$$

from which it follows by subtraction that

$$\frac{\text{d}m_{\text{e}}^*}{\text{d}t} = K_1^* c_{\text{a}}^* - (k_2^* + k_3^*) m_{\text{e}}^* + k_4^* m_{\text{m}}^*. \tag{7.15}$$

The solutions of these equations depend on the significance assigned a priori to the process described by $k_4$. In brain, it can be argued that the phosphatase activity is negligible. In other organs, the phosphatase activity may be too high to ignore. Therefore, it must be decided whether the organ, to which the deoxyglucose model is applied, has reversible phosphorylation or not. Either choice can be acceptable, depending on the purpose of the study.

**Table 7.1** Glucose tracer transport and phosphorylation ratios in brain

| Species | Tracer | $\tau$ | $\varphi$ | Author(s) |
|---|---|---|---|---|
| Rat | Deoxyglucose | 1.35 | 0.37 | Cunningham and Cremer (1981) |
| | | 1.75 | | Gjedde (1982) |
| | | 1.39 | 0.38 | Crane et al. (1983) |
| | | 1.52 | | Fuglsang et al. (1986) |
| | Fluorodeoxyglucose | 1.67 | 0.55 | Crane et al. (1983) |
| | | 1.65 | | Fuglsang et al. (1986) |
| Man | Fluorodeoxyglucose | 1.10 | 0.30 | Kuwabara et al. (1990)[a] |
| | | 1.10 | 0.33 | Kuwabara and Gjedde (1991)[b] |
| | | 1.48 | | Hasselbalch et al. (1996)[a] |
| | | | 0.39 | Hasselbalch et al. (1998)[a] |

[a] Model of irreversible metabolism
[b] Model of reversible metabolism

The transport coefficient is the ratio between the blood-tissue clearances of deoxyglucose and native glucose, under normal circumstances, and it approximates a constant, characteristic of symmetrical transport that obeys Michaelis–Menten kinetics,

$$\tau \equiv \frac{K_1^*}{K_1} = \frac{k_2^*}{k_2} \cong \frac{J_{\max}^{T*} K_m^T}{J_{\max}^T K_m^{T*}}, \tag{7.16}$$

where $J_{\max}^T$ and $K_m^T$ are the Michaelis–Menten constants of endothelial transport. This approximate constant will be used later to determine the lumped constant. Values of $\tau$ in brain are listed in Table 7.1.

### *7.3.1 Irreversible Metabolism*

The use of the metabolism coefficients defined earlier depends on the biochemistry of the tracer in question, particularly whether the metabolism of the tracer is reversible or irreversible.

Without dephosphorylation, deoxyglucose 6-phosphate and other metabolites accumulate irreversibly if $k_5^*$ is nil. This is the model of deoxyglucose metabolism proposed by Sokoloff et al. (1977). The model predicts that net tracer metabolism is directly proportional to the tracer quantity in the precursor pool $m_{\mathrm{e}}^*$,

$$\Delta j^* = k_3^* m_{\mathrm{e}}^* = K^* c_{\mathrm{a}}^*. \tag{7.17}$$

For $k_5^* = 0$, the combination of (7.13) and (7.17) yields,

$$K^* = k_3^* \left[ \frac{K_1^*}{k_2^* + k_3^*} \right] = E^* F \tag{7.18}$$

which states that the net steady-state of uptake is equal to the rate of phosphorylation multiplied by the apparent steady-state volume of distribution of unphosphorylated glucose tracer. By introducing a "nonmetabolism" index, $\rho^*$, defined as:

$$\rho^* = \frac{k_2^*}{k_2^* + k_3^*} = 1 - \frac{K^*}{K_1^*} \tag{7.19}$$

it is possible to further simplify (7.18),

$$K^* = E^* F = \rho^* k_3^* V_e, \tag{7.20}$$

where $V_e$ equals the $K_1^*/k_2^*$ and $K_1/k_2$ ratios (see Chaps. 1 and 2).

### *7.3.2 Lumped Constant*

Because a glucose tracer (such as deoxyglucose) has kinetic constants that are different from those of glucose, $K^*$ is not equal to $K$. The "lumped constant," $\Lambda$, converts the steady-state rate of tracer phosphorylation to rate of glucose phosphorylation. Although it is by definition an operational constant, it has been given several exact descriptions, of which the one of Sokoloff et al. (1977) is based on the assumption of irreversible metabolism. It follows from the definition that the interpretation of the lumped constant depends on the presence or absence of glucose 6-phosphatase.

In the absence of glucose-6-phosphatase, the lumped constant equals the ratio between the net clearance of the glucose tracer and glucose, equal to the ratio between the unidirectional rate of phosphorylation of the glucose tracer ($K^* = K_1^* k_3^*/[k_2^* + k_3^*]$) and the net clearance of glucose ($K = K_1 k_3/[k_2 + k_3]$). The unidirectional rate of phosphorylation reflects the fraction of glucose tracer which, once transported into tissue, is further phosphorylated. In the absence of dephosphorylation, the net and unidirectional rates are identical. Equations (7.1), (7.8), and (7.18) yield the definition of the lumped constant,

$$\Lambda \equiv \frac{E^*}{E} = \frac{K^*}{K} = \frac{K_1^* k_3^* (k_2 + k_3)}{K_1 k_3 (k_2^* + k_3^*)}. \tag{7.21}$$

Equation (7.21) is the basis for the determination of the lumped constant as the ratio between the steady-state extraction fractions. Equation (7.21) also shows that the lumped constant is the product of the ratio of phosphorylation rates and the ratio of the apparent distribution spaces in the tissue.

The lumped constant can *also* be assessed by combination and rearrangement of (7.10), (7.16), (7.19), and (7.21),

$$\Lambda = \tau - \rho^* (\tau - \varphi) \tag{7.22}$$

in which $\varphi$ is the $k_3^*/k_3$ ratio between the hexokinase affinities for the tracer and the native glucose, assumed to be approximately constant throughout the organ, $\tau$ the $K_1^*/K_1$ and $k_2^*/k_2$ ratios of the membrane transporter affinities for the glucose tracer and native glucose, also assumed to be constant throughout the organ, and $\rho^*$ the metabolism index, equal to the $k_2^*/(k_2^* + k_3^*)$ ratio which indicates the fraction of tracer that is not phosphorylated (metabolized). Kuwabara et al. (1990) (see later) showed that this relationship defines the value of the lumped constant from estimates of the net and unidirectional glucose tracer clearances when the values of $\tau$ and $\varphi$ are known,

$$\Lambda = \frac{K^*}{K} = \varphi + (\tau - \varphi)\frac{K^*}{K_1^*}. \tag{7.23}$$

This equation is the most useful expression of the lumped constant's dependence on the relationship between the net and unidirectional transport of the glucose tracer. The equation was used by Hasselbalch et al. (2001) to reevaluate the lumped constant for human cerebral metabolic rate for glucose with the constants reported in Table 7.1, compared to other values of the lumped constant listed in Table 7.3.

## *7.3.3 Reversible Metabolism*

### Effect of Dephosphorylation in Original Model

When labeled metabolites of deoxyglucose are lost only by dephosphorylation, the net uptake of the tracer is governed by (7.14) with $k_5^* = 0$ which, when divided by $m_\mathrm{e}^*$ takes the form,

$$\frac{\Delta j^*}{m_\mathrm{e}^*} - \frac{1}{m_\mathrm{e}^*}\frac{\mathrm{d}m_\mathrm{e}^*}{\mathrm{d}t} = k_3^* - k_4^*\frac{m_\mathrm{m}^*}{m_\mathrm{e}^*}.$$

Assuming the logarithmic derivative, $\frac{1}{m_\mathrm{e}^*}\frac{dm_\mathrm{e}^*}{dt} = \frac{d\ln(m_\mathrm{e}^*)}{dt}$ is negligible,

$$\frac{\Delta j^*}{m_\mathrm{e}^*} \cong k_3^* - k_4^*\frac{m_\mathrm{m}^*}{m_\mathrm{e}^*}. \tag{7.24}$$

Let

$$k_\mathrm{app}^*(t) \equiv k_3^* - k_4^*\frac{m_\mathrm{m}^*}{m_\mathrm{e}^*} \quad \text{and} \quad K_\mathrm{app}^*(t) \equiv k_\mathrm{app}^*(t)\frac{m_\mathrm{e}^*}{c_\mathrm{a}^*}. \tag{7.25}$$

With $k_5^* = 0$, the combination of (7.13) and (7.24) yields:

$$K_\mathrm{app}^*(t) = k_\mathrm{app}^*(t)\left[\frac{K_1^*}{k_2^* + k_\mathrm{app}^*(t)}\right] = E_\mathrm{app}^*(t)F. \tag{7.26}$$

By introducing an apparent metabolism index as a time variable,

$$\rho^*_{\text{app}}(t) = \frac{k_2^*}{k_2^* + k^*_{\text{app}}(t)} = 1 - \frac{K^*_{\text{app}}(t)}{K_1^*}. \tag{7.27}$$

Equation (7.26) can be simplified to,

$$K^*_{\text{app}}(t) = E^*_{\text{app}}(t)F = \rho^*_{\text{app}}(t)k^*_{\text{app}}(t)V_{\text{e}} \tag{7.28}$$

which as a steadily declining function ultimately reaches zero.

In the presence of significant phosphatase activity, the $k_3^*/k_4^*$ ratio is an index of the size of the metabolite pool, represented by $\sigma^*$,

$$\sigma^* = \frac{k_3^*}{k_4^*} = \left(\frac{J_{\text{max}}^{\text{Hase}*}}{J_{\text{max}}^{\text{Pase}*}}\right)\frac{K_m^{\text{Pase}*}\,K_m^{\text{Hase}}(K_m^{\text{Pase}}V_{\text{d}} + M_m)}{K_m^{\text{Hase}*}\,K_m^{\text{Pase}}(K_m^{\text{Hase}}V_{\text{d}} + M_e)}. \tag{7.29}$$

When $k_4^*$ is negligible, it follows that $k^*_{\text{app}}(t)$ equals $k_3^*$. In the brain, the metabolite pool for deoxyglucose is close to 10 (see Table 7.3). As the brain, for this reason, is often assigned a significant value of $k_4^*$, the $k^*_{\text{app}}(t)/k_3^*$ ratio is a variable, the magnitude of which decreases as a function of time,

$$\Phi^*(t) = \frac{k^*_{\text{app}}(t)}{k_3^*} = 1 - \frac{1}{\sigma^*}\left(\frac{m^*_{\text{m}}}{m^*_{\text{e}}}\right). \tag{7.30}$$

This relationship leads to a formula for the *apparent* phosphorylation ratio,

$$\varphi_{\text{app}}(t) \equiv \frac{k^*_{\text{app}}(t)}{k_{\text{app}}} = \frac{k_3^*}{k_{\text{app}}}\left(1 - \frac{1}{\sigma^*}\left[\frac{m^*_{\text{m}}}{m^*_{\text{e}}}\right]\right) = \frac{J_{\text{max}}^{\text{Hase}*}\,K_m^{\text{Hase}}}{\Phi J_{\text{max}}^{\text{Hase}}\,K_m^{\text{Hase}*}}\left(1 - \frac{1}{\sigma^*}\left[\frac{m^*_{\text{m}}}{m^*_{\text{e}}}\right]\right) \tag{7.31}$$

which shows that the phosphorylation ratio declines in a time-dependent fashion when the phosphorylated metabolite is subject to dephosphorylation.

The apparent lumped "constant" now is also a time-variable because the net extraction of the tracer continues to fall,

$$\Lambda_{\text{app}}(t) \equiv \frac{E^*_{\text{app}}(t)}{E} = \frac{K^*_{\text{app}}(t)}{K} = \frac{K_1^* k^*_{\text{app}}(t)\,(k_2 + k_3)}{K_1 k_3\left(k_2^* + k^*_{\text{app}}(t)\right)} = \tau - \rho^*_{\text{app}}(t)\left[\tau - \Phi\varphi_{\text{app}}(t)\right] \tag{7.32}$$

which shows that the "apparent" or measured lumped constant is a function of time when $k_4^*$ is not negligible. When $k_4^*$ is negligible, on the other hand, $\Lambda_{\text{app}}(t)$ equals $\Lambda$. The "true" and "apparent" lumped constants, therefore, are related by the equation:

$$\frac{K^*_{\text{app}}(t)}{\Lambda_{\text{app}}(t)} = \frac{K^*}{\Lambda} = K. \tag{7.33}$$

This relationship shows that the dephosphorylation has little influence on the relationship between $K^*$ and $\Lambda$. The net clearance of the tracer and the lumped constant must both be determined *as if* the phosphatase activity were nil. In other words, it is not correct to determine the lumped constant as the ratio between the fractions of net extraction at a given time and then proceed to correct the net tracer clearance for an assumed effect of dephosphorylation. Such a correction would yield values of the glucose metabolic rate that are erroneously high. Rather, the glucose phosphorylation rate must be calculated in the usual way, with *no correction* for loss of metabolites, because the variable lumped constant *already* incorporates this correction, such that the basic equation remains valid,

$$R = C_{\mathrm{a}} \frac{K^*_{\mathrm{app}}(t)}{\Lambda_{\mathrm{app}}(t)} = C_{\mathrm{a}} \frac{K^*}{\Lambda}. \tag{7.34}$$

### The Model Adapted to Dephosphorylation

When the deoxyglucose model is adapted to describe reversible metabolism, the dependence on time disappears and the definitions of $k_3$, $k_3^*$, $k_4$, $k_4^*$, and the lumped constant, assume their original meaning. The phosphorylation ratio is,

$$\frac{\varphi}{\Phi} = \frac{k_3^*}{k_{\mathrm{app}}} = \frac{J_{\max}^{\mathrm{Hase}^*} K_m^{\mathrm{Hase}}}{\Phi J_{\max}^{\mathrm{Hase}} K_m^{\mathrm{Hase}^*}} = \frac{\varphi_{\mathrm{app}}(t)}{\left(1 - \frac{1}{\sigma^*}\left[\frac{m_{\mathrm{m}}^*}{m_{\mathrm{e}}^*}\right]\right)}, \tag{7.35}$$

where $\varphi_{\mathrm{app}}(t)$ is the apparent phosphorylation ratio determined on the assumption of irreversible metabolism. The phosphorylation constant determined by Kuwabara and Gjedde (1991) for reversible metabolism of deoxyglucose in the human brain, averaged 110% of the apparent constant determined for irreversible metabolism, indicating that some dephosphorylation may indeed take place in the human brain. The two values, $\varphi$ and $\varphi_{\mathrm{app}}(t)$, estimated by Kuwabara and Gjedde (1991) for human brain, are listed in Table 7.1.

In the modified model, as a result of the reversibility of the metabolism, the true lumped constant cannot be determined as the ratio between the net extractions of tracer and glucose because there is no steady-state net extraction of the glucose tracer. The lumped constant can only be determined by means of (7.23), modified for dephoshorylation of glucose 6-phosphate,

$$\Lambda \equiv \frac{K^*}{K} = \frac{K_1^* k_3^* (k_2 + k_{\mathrm{app}})}{K_1 k_{\mathrm{app}} (k_2^* + k_3^*)} = \frac{\varphi}{\Phi} + \left(\tau - \frac{\varphi}{\Phi}\right) \frac{K^*}{K_1^*}. \tag{7.36}$$

However, neither $\frac{\varphi}{\Phi}$ nor $K^*$ can be determined because of the dephosphorylation and consequent loss of tracer metabolites.

The glucose phosphorylation rate must be calculated from the equation,

$$R = \frac{K^*}{\Lambda} C_a \tag{7.37}$$

but neither $K^*$ nor $\Lambda$ are known now with certainty. Thus, although the value of $R$ is independent of the model when the model is applied consistently, it is plain that there are no reasons to use the reversible model with its lack of steady-state. Instead, it is advisable simply to use the irreversible model, making sure that $\Lambda$ and $m^*$ are determined for the same durations of tracer circulation.

## 7.4 Operational Equations

Attempts have been made to determine the metabolism coefficients of glucose with labeled deoxyglucose, labeled fluorodeoxyglucose, or labeled glucose itself. The first attempt used labeled glucose itself (Gaitonde 1965). The first comprehensive examination of the kinetics of glucose analog uptake and metabolism in brain was carried out with labeled deoxyglucose by Sokoloff et al. (1977). Sokoloff and his coworkers assumed that deoxyglucose-6-phosphate would be unavailable for isomerization and dephosphorylation and hence would remain trapped in tissue for extended periods of time. The operational equations (7.14) and (7.15) were solved on this basis. Other solutions were soon developed, including solutions for the case of significant dephosphorylation.

### *7.4.1 Irreversible Metabolism of Deoxyglucose*

#### Sokoloff's Original Solution

Sokoloff et al. (1977) solved (7.14) and (7.15) for tracer deoxyglucose by assuming that $k_4^* = k_5^* = 0$, i.e., that the tissue allows no significant dephosphorylation of deoxyglucose 6-phosphate, because of the absence or sequestration of the phosphatase required for this process, and no significant direct loss of tracer deoxyglucose metabolites, because of little continued metabolism to compounds that cross cellular membranes. As discussed in Chap. 1, particularly the discussion associated with (1.53), the solution to the equations is the formula for the total tracer content, $m^*$,

$$m^* = V_o^* c_a^* + (\rho K_1^* - K^*) \int_0^T c_a^* e^{-(k_2^* + k_3^*)(T-t)} dt + K^* \int_0^T c_a^* dt, \tag{7.38}$$

where $\rho$ is a correction factor for vascular and other volumes close to unity, introduced in Chap. 1. Sokoloff et al. (1977) used this equation to estimate the individual

**Table 7.2** Deoxyglucose transfer coefficients for rat somatosensory cortex

| Author(s) | Year | $K_1^*$ ml g$^{-1}$ min$^{-1}$ | $k_2^*$ min$^{-1}$ | $k_3^*$ min$^{-1}$ | $V_e$ ml g$^{-1}$ | $K^*$ ml g$^{-1}$ min$^{-1}$ |
|---|---|---|---|---|---|---|
| Sokoloff et al. | (1977) | 0.19 | 0.21 | 0.049 | 0.93 | 0.037 |
| Cremer et al. | (1981) | 0.27 | 0.45 | 0.152 | 0.58 | 0.066 |
| Gjedde | (1982) | 0.35 | 1.09 | 0.090 | 0.32 | 0.027 |
| Fuglsang et al. | (1986) | 0.28 | 0.55 | 0.077 | 0.51 | 0.035 |
| Mori et al. | (1990) | 0.23 | 0.14 | 0.030 | 1.64 | 0.040 |

transfer coefficients by nonlinear regression analysis in rats, listed in Table 7.2. This equation was intended for use with autoradiography in which only a single observation of $m^*$ is available. The authors, therefore, solved the equation for the net clearance of the tracer, assuming negligible $V_o^*$ and unity $\rho$,

$$R = \frac{K^*}{\Lambda} C_a = \frac{C_a}{\Lambda} \frac{m^* - m_e^{calc}}{\left(\int_0^T c_a^* dt - \frac{m_e^{calc}}{K_1^*}\right)}, \tag{7.39}$$

where $K^*$ is the net clearance of the tracer and $m_e^{calc}$ is the tissue quantity of tracer. As only a single observation of $m^*$ is available, the additional terms of $m_e^{calc}$ and $m_e^{calc}/K_1^*$ were determined separately from tabulated estimates of the transfer coefficients according to (7.40),

$$m_e^{calc} = K_1^* \int_o^T c_a^* e^{-(k_2^*+k_3^*)(T-t)} \, dt. \tag{7.40}$$

Sokoloff et al. (1977) advocated circulating the tracer for 45 min to minimize the magnitude of $m_e^{calc}$ relative to $m^*$.

## Dynamic Applications

Sokoloff et al. (1977) did not attempt to assign any physiological meaning to the transfer coefficients. This approach created a problem because it was soon shown that the solution was over-parameterized; in reality there are only two independent parameters (transfer coefficients). This can also be understood by linearization of (7.39),

$$\frac{m_m^*}{m_e^*} = K^* \frac{\int_0^T c_a^* dt}{m_e^*} - \frac{K^*}{K_1^*} \tag{7.41}$$

which describes a straight line of slope $K^*$ and $y$-intercept $-K^*/K_1^*$. The presence of three parameters in the original solution (and four parameters in the solution that includes $V_o^*$) meant that the estimates of $K_1^*$, $k_2^*$, and $k_3^*$ were highly correlated and hence very uncertain. Solutions to this problem were devised by Cunningham and Cremer (1981) and Kuwabara et al. (1990).

**Table 7.3** Fluorodeoxyglucose transfer coefficients of human cerebral cortex

| Model/Authors | $K_1^a$ (ml g$^{-1}$ min$^{-1}$) | $K^a$ | $k_2^a$ | $k_3^a$ (min$^{-1}$) | $k_4^a$ | $1/\sigma^a$ (ratio) | $V_e(1+\sigma^a)$ (ml g$^{-1}$) | $\Lambda^a$ (ratio) |
|---|---|---|---|---|---|---|---|---|
| Irreversible metabolism | | | | | | | | |
| Gjedde et al. (1985) | 0.051 | 0.013 | – | – | – | – | – | 0.52 |
| Reivich et al. (1985) | 0.105 | 0.035 | 0.148 | 0.074 | – | – | – | 0.59 |
| Lammertsma et al. (1987) | 0.064 | 0.027 | 0.066 | 0.047 | – | – | – | 0.66 |
| Kuwabara and Gjedde (1991) | | | | | | | | |
| *Conventional method* | 0.068 | 0.026 | 0.092 | 0.054 | – | – | – | 0.62 |
| *Constrained method* | 0.087 | 0.032 | 0.203 | 0.127 | – | – | – | 0.61 |
| Hasselbalch et al. (2001) | 0.188 | 0.080 | – | – | – | – | – | 0.66 |
| Reversible metabolism (45 min) | | | | | | | | |
| Friedland et al. (1983a, b)[b] | 0.131 | 0.042 | 0.225 | 0.106 | 0.010 | 0.09 | 6.9 | 0.58 |
| Lammertsma et al. (1987) | 0.069 | 0.035 | 0.109 | 0.113 | 0.020 | 0.18 | 4.2 | 0.72 |
| Kuwabara and Gjedde (1991) | | | | | | | | |
| *Conventional method* | 0.077 | 0.031 | 0.146 | 0.097 | 0.014 | 0.15 | 4.2 | 0.64 |
| *Constrained method* | 0.085 | 0.035 | 0.197 | 0.149 | 0.018 | 0.12 | 4.0 | 0.65 |
| Reversible metabolism (~3h) | | | | | | | | |
| Phelps et al. (1979) | 0.102 | 0.033 | 0.130 | 0.062 | 0.0068 | 0.110 | 7.4 | 0.58 |
| Hawkins et al. (1983a, b)[b] | | | | | | | | |
| *Young adults* | 0.084 | 0.025 | 0.166 | 0.071 | 0.0051 | 0.07 | 7.6 | 0.58 |
| *Old adults* | 0.100 | 0.024 | 0.253 | 0.079 | 0.0060 | 0.08 | 5.6 | 0.52 |
| Reivich et al. (1985) | 0.095 | 0.034 | 0.125 | 0.069 | 0.0055 | 0.08 | 10.3 | 0.61 |
| Hawkins et al. (1986) | 0.041 | 0.014 | 0.119 | 0.064 | 0.0075 | 0.12 | 3.3 | 0.59 |

[a] Calculated with $\tau = 1.10$, $\varphi = 0.33$
[b] No corrections for the radioactivity within vessels in tissue

**Cunningham's Solution** Cunningham and Cremer (1981) showed that the solution to (7.14) and (7.15) has only two real parameters ($K_1^*$ and $k_3^*$ or $K_1^*$ and $K^*$) because the third transfer coefficient can deduced from the two first parameters, provided the relationship between plasma and tissue glucose is known. By introducing the ratios between $K_1^*$ and $K_1$ ($\tau$) and between $k_3^*$ and $k_3$ ($\varphi$), the authors modified (7.41) by substituting the transfer coefficients for glucose,

$$\frac{m_m^*}{m_e^*} = \frac{\tau\varphi K_1 K}{\tau K_1 - K(\tau - \varphi)} \left( \frac{\int_0^T c_a^* \mathrm{d}t}{m_e^*} \right) - \frac{\varphi K}{\tau K_1 - K(\tau - \varphi)}, \tag{7.42}$$

where $c_a^*$, $m_m^*$, and $m_e^*$ refer to deoxyglucose while the parameters $K$ and $K_1$ refer to native glucose. Equation (7.42) eliminated the problem of the lumped constant because the estimated transfer coefficients refer to glucose rather than the tracer deoxyglucose. The disadvantage is the need to know the ratios $\tau$ and $\varphi$. A simpler version of the same equation is,

$$\frac{m_m^*}{m_e^*} = R\left(\frac{\Lambda}{C_a}\right) \frac{\int_0^T c_a^* \mathrm{d}t}{m_e^*} - \left(\frac{\Lambda - \varphi}{\tau - \varphi}\right), \tag{7.43}$$

which shows that the glucose consumption can be estimated directly from this equation once $C_a$, the plasma glucose concentration, is known.

Cunningham's equation was used by Cremer et al. (1981) to obtain estimates of transfer coefficients on the basis of known concentrations of glucose in plasma and tissue. In Table 7.2, they have been compared with other previous and subsequent estimates. It is fairly apparent by this table that the estimates of $K_1^*$ and $K^*$ are considerably more robust (variation approximately twofold) than the estimates of $k_2^*$ (variation approximately 10-fold), $k_3^*$ (variation approximately fivefold), and $V_e$ (variation approximately eightfold). As the estimates of Cremer et al. (1981) and Fuglsang et al. (1986) were both constrained by known values of the relationship between plasma and tissue glucose, they tend to be more robust.

**Gjedde's Solution** Gjedde (1982) used the original derivation of the Gjedde–Patlak plot (Gjedde 1981b; Patlak et al. 1983) to rearrange (7.39) to a simpler expression that could be used to determine the glucose consumption of the rat brain,

$$K^* = \frac{m^* - \rho^* m_e^* - V_o^* c_a^*(T)}{\int_0^T c_a^* \mathrm{d}t}, \tag{7.44}$$

where $\rho^*$ is the "nonmetabolism" index $(k_2^*/(k_2^* + k_3^*)$, introduced earlier. This equation may be advantageous in cases in which the integral is known with certainty while the individual concentrations $c_a^*$ are less well known. The equation was further simplified by normalization against the final glucose tracer concentration in arterial plasma, $c_a^*$. This normalization introduced the concept of "apparent" (virtual) volumes of distribution (Gjedde 1981b) which is experimentally useful because it normalizes values arising from the use of different absolute amounts of tracers in different animals. Following such a division, (7.44) changes to,

$$K^* = \frac{\frac{m^*}{c_a^*} \quad \rho^* \left(\frac{m_e^*}{c_a^*}\right) - V_o^*}{\Theta^*} \tag{7.45}$$

where $\Theta^* = \int_0^T c_a^* \mathrm{d}t / c_a^*$. The term $\Theta^*$ has the dimension of time and represents the duration of the experiment when $c_a^*$ is constant. Thus, after bolus injection, this term "stretches" the apparent time of the experiment and estimates an effective exposure time, corresponding to the duration of exposure at constant plasma levels to which the uptake corresponds. The original example of the use of this equation is shown as Fig. 1.3 in Chap. 1, applied to labeled deoxyglucose and labeled methylglucose.

Which error is incurred by ignoring the amount of unphosphorylated deoxyglucose in the precursor pool at the time of termination of an experiment with the deoxyglucose method? The answer to this question depends on the duration of the experiment, be it 45 min, 30 min, or less. Equations (7.44) and (7.45) describe the magnitude of this error at steady-state, i.e., for $T \to \infty$,

$$R = \frac{K^*}{\Lambda} C_a \to \frac{C_a}{\Lambda} \frac{\frac{m^*}{c_a^*} - (\rho^* V_f^* + V_o^*)}{\Theta^*} = \frac{\frac{m^*}{c_a^*} - (V_g^* + V_o^*)}{\Theta^*} \tag{7.46}$$

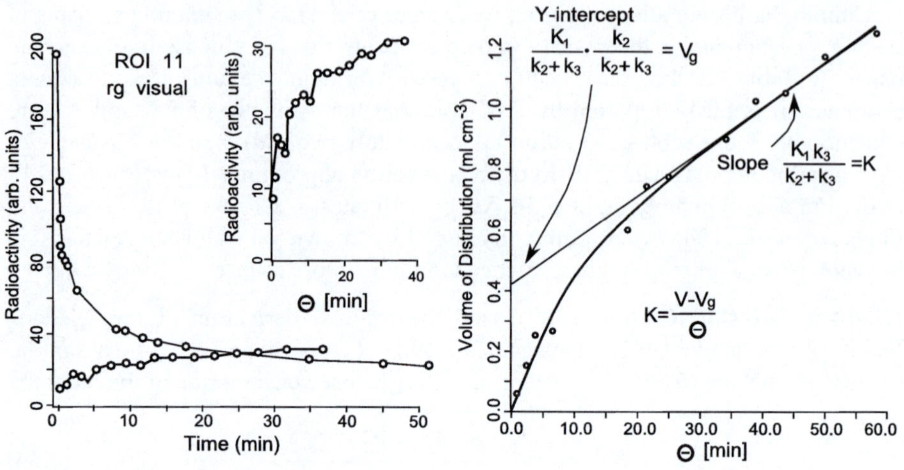


**Fig. 7.2** Gjedde–Patlak plot of normalized whole-brain tracer fluorodeoxyglucose uptake vs. normalized integrated arterial tracer fluorodeoxyglucose concentration in visual cortex of adult volunteer. *Left panel*: Abscissae: Time (min). Ordinates: Radioactivity in arbitrary units. *Right panel*: Time integral of arterial blood tracer concentration, normalized against same concentrations (unit of minutes). Ordinate: Brain net radioactivity uptake, normalized against arterial blood tracer concentration (unit of ml cm$^{-3}$). *Lower Curve* in *right graph* shows result of regression of the equation, $v^* = K^*\Theta + V_g^*(1 - e^{-(K_1^*-K^*)\Theta/V_g^*}) + V_o$ to the data reported by Gjedde et al. (1985)

in which the apparent precursor pool volume, $\rho^* V_f^*$, equals $K_1^* k_2^*/(k_2^* + k_3^*)^2$. The error incurred by ignoring these terms can be significant, particularly when the glucose phosphorylation rate is low.

The first application of the "slope-intercept" plot to the human brain was made by Gjedde et al. (1985), as shown in Fig. 7.2, who also used the relationship between the rates of uptake of methylglucose and deoxyglucose to calculate the lumped constant. The lumped constant was reevaluated by Hasselbalch et al. (2001) who confirmed that estimates had tended to increase over the years (see Table 7.3).

**Kuwabara's Solution I** From the derivation of Cunningham's solution earlier, it was clear that the solution only has two real parameters when the relationship between plasma and tissue glucose is known. Fundamentally, these two parameters are the maximal velocity ($J_{\max}^T$) of hexose transfer across the blood-tissue barrier, and the maximal velocity ($J_{\max}^{\text{Hase}}$) of hexokinase, when the half-saturation concentrations $K_m^T$ and $K_m^{\text{Hase}}$ of the Michaelis–Menten processes are known. As tissue glucose cannot be measured in the human brain in vivo, the application of Cunningham's solution to the human brain required a different approach by which the relationship was established by introducing the Michaelis half-saturation constants. On the basis of these relationships, Kuwabara et al. (1990) redefined $k_2^*$ and $k_3^*$ in (7.38) in terms of $K_1^*$ and $K^*$,

$$k_2^* = \frac{K_1^* + \mu K^*}{V_d} \quad \text{and} \quad k_3^* = \frac{K^*}{K_1^* - K^*}\left[\frac{K_1^* + \mu K^*}{V_d}\right], \tag{7.47}$$

where $\mu$ is the compound constant $\tau C_a/[\Lambda K_t]$, $K_t$ the Michaelis constant for glucose transport, $V_d$ the distribution volume, and $C_a$ is the arterial plasma glucose concentration. The lumped constant, in turn, is also a function of $K^*$ and $K_1^*$ according to (7.23). From (7.38) modified in this manner, the terms $K^*$, $K_1^*$, $V_o^*$, as well as the lumped constant, can be estimated directly by ordinary nonlinear least-squares regression. The values of $K_1^*$, $k_2^*$, and $k_3^*$, and the lumped constant, calculated from these estimates, are listed in Table 7.3.

### 7.4.2 Reversible Metabolism of Fluorodeoxyglucose

**Phelps' 'Autoradiographic' Solution** Both Phelps et al. (1979) and Reivich et al. (1979) published extensions of the deoxyglucose method to the study of brain glucose metabolism in humans by positron tomography and labeled fluorodeoxyglucose. While Reivich et al. used the method of Sokoloff et al. (1977) in its original form, including the transfer coefficients determined in rat, Phelps et al. judged the value of $k_4^*$ to be sufficiently significant in human brain to merit its inclusion in the operational equation. This is the solution for other organs, such as the liver. For $k_5^* = 0$, the solution to (7.14) and (7.15) was derived in Chaps. 1 and 2 for $\Omega$ unity,

$$m^* = V_o c_a^* + K_1^* \left[ \left( \frac{q_2 - [k_3^* + k_4^*]}{q_2 - q_1} \right) \int_0^T c_a^* \mathrm{e}^{-q_2(T-t)} \mathrm{d}t - \left( \frac{q_1 - [k_3^* + k_4^*]}{q_2 - q_1} \right) \int_0^T c_a^* \mathrm{e}^{-q_1(T-t)} \mathrm{d}t \right], \quad (7.48)$$

where $q_1$ and $q_2$ are composite relaxation constants,

$$2q_1 = k_2^* + k_3^* + k_4^* - \sqrt{(k_2^* + k_3^* + k_4^*)^2 - 4k_2^* k_4^*}$$

and

$$2q_2 = k_2^* + k_3^* + k_4^* + \sqrt{(k_2^* + k_3^* + k_4^*)^2 - 4k_2^* k_4^*}. \quad (7.49)$$

Equation (7.48) may be used to determine all four transfer coefficients, by nonlinear regression analysis of multiple time points. Such estimates are listed in Table 7.3. The net clearance of the tracer is determined as the ratio between measured values and values calculated from predetermined transfer coefficients,

$$R = \frac{K^*}{\Lambda} C_a = \bar{R}_{pop} \frac{C_a}{\bar{C}_a} \left[ \frac{m^* - m_e^{calc}}{m_m^{calc}} \right], \quad (7.50)$$

where $\bar{R}_{pop}$ is a population average of the glucose metabolic rate, chosen to be 30 $\mu$mol hg$^{-1}$ min$^{-1}$ for human whole-brain, as explained later, and $\bar{C}_a$. The equation conveniently solves the problem of the lumped constant. Equation (7.50) permits the determination of the net clearance of deoxyglucose from a single observation of $m^*$. The values of $m_{\mathrm{e}}^{\mathrm{calc}}$ and $m_{\mathrm{m}}^{\mathrm{calc}}$ were calculated from predetermined transfer coefficients, using equations derived from the solution of reversible metabolism,

$$m_{\mathrm{e}}^{\mathrm{calc}} = K_1^* \left[ \left( \frac{q_2 - k_4^*}{q_2 - q_1} \right) \int_o^T c_{\mathrm{a}} \mathrm{e}^{-q_2(T-t)} \, \mathrm{d}t - \left( \frac{q_1 - k_4^*}{q_2 - q_1} \right) \int_o^T c_{\mathrm{a}} \mathrm{e}^{-q_1(T-t)} \, \mathrm{d}t \right] \tag{7.51}$$

and

$$m_{\mathrm{m}}^{\mathrm{calc}} = \frac{K_1^* k_3^*}{q_2 - q_1} \left( \int_o^T c_{\mathrm{a}} \mathrm{e}^{-q_1(T-t)} \, \mathrm{d}t - \int_o^T c_{\mathrm{a}} \mathrm{e}^{-q_2(T-t)} \, \mathrm{d}t \right). \tag{7.52}$$

This approach does not make use of the additional information that can be gained by executing the studies dynamically, i.e., by registration of the time course of radioactivity accumulation in individual subjects.

Both assumptions, $k_4^* \neq 0$ and $m_z^* = 0$ (i.e., the entire pool of fluorodeoxyglucose metabolites is available for dephosphorylation to fluorodeoxyglucose) have been challenged by advocates of the deoxyglucose method in its original form (see Sokoloff's solution earlier), mostly on biochemical grounds while the opposite assumptions, i.e., $k_4^* = 0$ and $m_z^* \neq 0$, likewise have been disputed, mostly on mathematical or statistical grounds (see Hawkins' solution later). However, neither pair of assumptions affect the resulting estimates of glucose metabolic rate if they are treated correctly.

**Dynamic Applications** The solution presented earlier served both to calculate the glucose use rate and to determine the parameters by nonlinear least-squares regression to observations of the radioactivity as a function of time. For analysis of samples of tissue, this approach is adequate, but for regional or pixel-by-pixel analysis of radioactivity maps, the method is both time-consuming and sensitive to errors associated with samples of small size. For these reasons, more robust methods of regression exist, including multivariate linear or constrained solutions, applied to dynamic records.

**Evans' Solution** In order to obtain estimates of the transfer coefficients, Evans (1987) extended the linear, multivariate, least-squares regression method of Blomqvist (1984) to include the coefficient of dephosphorylation of FDG-6-phosphate ($k_4^*$), and a correction for the cerebral vascular volume in tissue ($V_o^*$),

$$m^* = a_1 c_{\mathrm{a}}^* + a_2 \int_0^T c_{\mathrm{a}}^* \mathrm{d}t + a_3 \int_0^T \int_0^u c_{\mathrm{a}}^* \mathrm{d}t \mathrm{d}u$$

$$+ a_4 \int_0^T m^* \mathrm{d}t + a_5 \int_0^T \int_0^u m^* \mathrm{d}t \mathrm{d}u, \tag{7.53}$$

where

$$\begin{cases} a_1 = V_o^* \\ a_2 = K_1^* + (k_2^* + k_3^* + k_4^*)V_o^* \\ a_3 = K_1^*(k_3^* + k_4^*) + k_2^* k_4^* V_o^* \\ a_4 = -(k_2^* + k_3^* + k_4^*) \\ a_5 = -k_2^* k_4^* \end{cases} . \tag{7.54}$$

In principle, as multivariate and linear, this equation converges in a single step, yielding estimates of the coefficients from which all the metabolic variables may be calculated, including the metabolic rate of glucose (Gjedde 1982),

$$R = \frac{C_a K_1^* k_3^*}{\Lambda(k_2^* + k_3^*)} . \tag{7.55}$$

**Kuwabara's Solution II** Kuwabara and Gjedde (1991) further simplified this approach. They exchanged equation (7.38) by (7.53), combining the principle of the constrained solution with the multivariate linear regression, increasing the efficiency of the optimization. The combination yielded estimates of $K_1^*$, $K^*$, and $k_4^*$. Transfer coefficients obtained with this procedure and the reversible model are listed in Table 7.3.

### *7.4.3 Metabolism of Tracer Glucose*

Glucose itself, rather than deoxyglucose, has been used as the tracer. The quantity of unmetabolized tracer in organs can be determined kinetically but the egress of labeled metabolites remains a major problem. Hawkins et al. (1974, 1979) proposed the use of tracer glucose itself rather than tracer deoxyglucose to avoid having to know the value of the lumped constant and to avoid making the assumption that $k_4 = 0$. The magnitude of $k_4$ would not affect tracer glucose because the continued metabolism of tracer glucose-6-phosphate tends to maintain $m_m^*$ at a low level and hence minimize the dephosphorylation reaction if it exists. The use of glucose meant that it was not valid to assume $k_5 = 0$. Hence the solution must include an additional term, $m_r^* = k_5 \int_0^T m_z^* dt$,

$$K = \frac{m^* - m_e^* + m_r^* - V_o c_a^*}{\int_0^T c_a^* dt - \frac{m_e^*}{K_1}} \tag{7.56}$$

in which $m_r^*$ is the unknown amount of tracer lost as tracer metabolites until the time $T$. This equation contains several unknowns that were estimated or ignored in different ways by different authors to allow the determination of glucose consumption.

**Hawkins' Solution** For brain, Hawkins et al. (1985) chose to regard $m_r^*$ and $V_o$ as negligible for short periods of time, and to calculate $m_e^*$ from known, assumed constants,

$$R = C_a K = C_a \frac{m^* - m_e^{calc}}{\int_0^T c_a^* dt - \frac{m_e^{calc}}{K_1}}. \tag{7.57}$$

This equation was designed for use with autoradiography in which only the concentration curve in the circulation and a single measurement in tissue ($m^*$) is available. The solution requires separate knowledge of $m_e^{calc}$ and $K_1$. To reduce the loss of labeled metabolites, the authors advocated shortening the circulation of the tracer to 10 min, and using glucose labeled in the C1 or, preferably, C6 position where little labeled carbon dioxide is generated during the first passage through Krebs' cycle (Hawkins et al. 1985).

The shortening of the circulation increased $m_e^{calc}$ relative to $m^*$ and necessitated a correction for unmetabolized tracer glucose as well as for delayed equilibration ($m_e^{calc}$ and $m_e^{calc}/K_1$). These corrections were accomplished either by measuring the ratio between native glucose in the tissue and plasma and using this ratio (incorrectly) to determine $m_e^{calc}$ as $V_f c_a^*$ (Duckrow and Bryan 1987) or by using predetermined values of $K_1$, $k_2$, $k_3$, and $k_4$ ($= 0$) to calculate $m_e^{calc}$ according to (7.40) (Hawkins et al. 1985).

**Blomqvist's Solution** For uniformly labeled glucose in brain for which it cannot be argued that the loss of labeled metabolites is negligible, Blomqvist et al. (1985) formulated a correction of the tracer quantity determined *in* the tissue. They devised the following modification to (7.56),

$$K = \frac{m^* \left(1 + \frac{m_r^*}{m^*}\right) - m_e^* - V_o c_a^*}{\int_0^T c_a^* dt - \frac{m_e^*}{K_1}} \tag{7.58}$$

and represented the ratio $m_r^*/m^*$ by a factor derived from the kinetics of secular equilibrium (see Chaps. 1 and 2),

$$\gamma = \frac{m_r^*}{m^*} = aT(1 - e^{-bT}). \tag{7.59}$$

For a given value of $T$, $\gamma$ was applied as constant to all values of $m^*$, assuming the egress of carbon dioxide to be proportional to the remaining radioactivity, and $V_o$ to be negligible,

$$R = C_a K = C_a \frac{m^*(1 + \gamma) - m_e^{calc} - V_o c_a^*}{\int_0^T c_a^* dt - \frac{m_e^{calc}}{K_1}}. \tag{7.60}$$

The values determined for normal human brain were 0.021 $min^{-1}$ for $a$ and 0.63 $min^{-1}$ for $b$ in the case of uniformly labeled glucose. In the same report, Blomqvist et al. (1990) introduced glucose labeled in the C1 position which delays the generation of $CO_2$. In this case, the constants were $a = 0.004$ $min^{-1}$ and $b = 0.06$ $min^{-1}$

(calculated from data presented in the paper), indicating that labeling in the C1 position delayed the loss of metabolites from brain by an order of magnitude, although not completely.

**Brøndsted's Solution** Using C6-labeled glucose, Brøndsted and Gjedde (1988) confirmed the evidence of metabolite loss from the brain in rats. To quantitate this loss, the authors solved (7.57) slightly differently, using the normalized form equivalent to (7.46),

$$K = \frac{\left(\frac{m^*}{c_a^*} - V_g\right)\left(1 + \frac{m_r^*}{m^* - c_a^* V_g}\right)}{\Theta^*}, \tag{7.61}$$

where $V_g$ is the product $\rho V_f$ for glucose. The term $m_r^*/(m^* - c_a^* V_g)$ was shown to rise at a rate consistent with the expression $\exp[K\Theta^*/V_k] - 1$. The expression was substituted into (7.61) to yield the formula,

$$R = C_a K = C_a \left[\frac{\frac{m^*}{c_a^*} - V_g}{\Theta^*}\right] e^{k_{glc}\Theta^*}, \tag{7.62}$$

where $k_{glc}$ is the apparent (and time-dependent) average relative rate of loss of glucose metabolites, equal to the ratio between $K$ and $V_k$, the latter an apparent volume of distribution of glucose metabolites. The equation was designed mainly for use with autoradiography. The constants were determined in advance by nonlinear least-squares optimization to (7.62) which yielded the estimates 0.21 ml g$^{-1}$ for $V_g$, 0.100 ml g$^{-1}$ min$^{-1}$ for $K$, and 12.9 ml g$^{-1}$ for $V_k$ in rat brain.

The value of $V_g$ was close to the value of $m_e^*/c_a^*$ determined for Hawkins' solution earlier. The value of the rate constant $k_{glc}$ (0.007 min$^{-1}$) was only 10% of the value of $b$ determined in human brain by Blomqvist et al. above for glucose labeled in the C6-position. In other words, the glucose metabolite pools are so large that $k_{glc}$ ($\cong k_5$) is small in the beginning. It is not known to which extent the glucose metabolite pools change when glucose metabolism or glucose concentration in plasma change. For this reason, $k_{glc}$ is not likely to change in exact proportion to any change of $K$. In actual practice, $k_{glc}$ was used as a constant in (7.62).

## 7.5 Glucose Metabolic Rates

Only organs with a low rate of glucose 6-phosphate dephosphorylation and a known relationship between glucose and deoxyglucose metabolism can be studied accurately with the deoxyglucose method. The use of glucose itself in organs other than brain may have fewer limitations but has not been explored in detail. The heterogeneity of the whole-body trapping of fluorodeoxyglucose is plain in tomograms such as the ones shown in Fig. 7.1. The determinant of this distribution is first the rate of transport of hexoses into the tissue ($K_1$), second the rate of metabolism relative to the rate of transport, as expressed in the true lumped constant ($\Lambda$), and third the degree of trapping of the metabolite, as expressed in the ratio between the activities of glucose-6-phosphatase and hexokinase ($k_4^*/k_3^* = 1/\sigma^*$), and in the magnitude of the apparent lumped constant ($\Lambda_{app}(t)$).

Together, these factors create the conditions responsible for an image like Fig. 7.1. Transport may be rapid, and the lumped constant high, but it would be to no avail if the phosphatase were also of high activity. Conversely, the lumped constant may be high, and phosphatase activity low, but little tracer can enter the tissue if the hexose transport across cell membranes is not very active.

In establishing the usefulness of the deoxyglucose method in different organs it was necessary to examine the presence of these factors in the reverse order, i.e., first identify organs in which the activity of glucose 6-phosphatase is low relative to the activity of hexokinase ($1/\sigma^*$). This was done either by determining the enzyme activities directly, using standard biochemical methods, or by regression of the model equations obtained from one of the solutions presented earlier to estimate the value of the parameters or rate constants.

An example of the former approach was given by Suolinna et al. (1986) who mapped the relationship between the activity of the two enzymes in rat and discovered a low ratio in brain, heart, and tumor, an intermediate ratio in kidney and skeletal muscle, and a high ratio in liver. According to these results, the deoxyglucose method could be used to study glucose metabolism in brain. heart, and certain tumors, but not in skeletal muscle, liver, and kidney. To confirm this, the authors noted that 90% of the radioactivity registered in brain, heart, and tumor after 45 min of circulation of the tracer represented [$^{18}$F]fluorodeoxyglucose 6-phosphate while in kidney and liver most of the radioactivity represented intact [$^{18}$F]fluorodeoxyglucose.

Using the same approach specifically for kidney, Paul et al. (1985) found an inverse relationship between the activity of glucose-6-phosphatase and the retention of the tracer such that retention was high in the tumor regions which displayed no enzyme activity, while retention was low in normal kidney regions which displayed a high activity of this enzyme.

Applying the regression approach to liver, Okazumi et al. (1992) determined the rate constants $k_3$ and $k_4$ for fluorodeoxyglucose shown in Fig. 7.4. The ratio between the two rate constants ranged from 10 in certain malignant tumors of the liver to one-tenth in normal liver tissue, as shown in the figure.

Once it was established that brain, heart, skeletal muscle, and certain tumors, had relatively high magnitudes of $\sigma^*$, it next became necessary to evaluate the lumped constant in these organs. As noted earlier, the time-dependent apparent lumped constant is more useful for this purpose because it approximately corrects for the dephosphorylation that may occur in the organ.

It is clear from the discussion earlier that all measurements of regional glucose metabolism depend fundamentally on the value of the lumped constant, in addition to the many methodological factors that influence tracer measurements, i.e., partial volume effects, power of resolution, and heterogeneity of tissue.

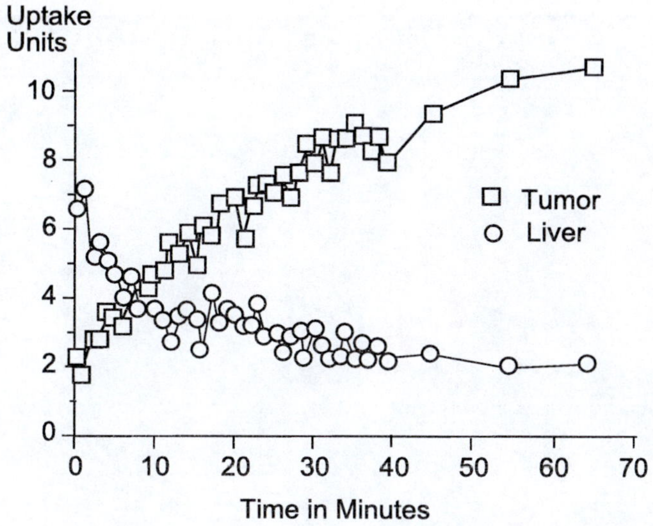


**Fig. 7.3** Graph illustrates the changes in contrast as a function of time after injection of FDG for a metastatic lesion and normal liver. Typical time activity curve graphically depicts increasing contrast between a liver metastasis (colon carcinoma) and normal liver with time. Uptake units are relative to dose and subject weight, known as differential uptake ratio (DUR) or standardized uptake value (SUV). From Goldberg et al. (1993)

### *7.5.1 Lumped Constant Variability*

The lumped constant for deoxyglucose in rat brain, determined by Sokoloff et al. (1977) as the ratio between the net extraction fractions, averaged 0.48. In human brain, the lumped constant has been measured directly only *once*, although it has been measured repeatedly in many animal species. The problem is that lumped constant measurements according to the original method of Sokoloff et al. (1977) are difficult, particularly in humans, in whom steady-state concentrations of the tracer must be maintained for extended periods of time. Nonetheless, most lumped constant *calculations* average 0.5–0.7 in the brain (see Table 7.3).

Only two direct steady-state measurements have been made in humans. In the first of these studies, the lumped constant averaged 0.56±0.09 for carbon-11-labeled deoxyglucose and 0.52±0.09 (±SD) for fluorine-18-labeled fluoro-deoxyglucose, although the clearances listed for this study in Table 7.3 yielded 0.59 and 0.61, respectively (Reivich et al. 1985). In the second study, the lumped constant averaged 0.48±0.16 for fluorine-18-labeled fluoro-deoxyglucose, at variance with the value of 0.81±0.15 calculated by the authors, and the value of 0.66 listed in Table 7.3 (Hasselbalch et al. 1998, 2001). The reason for the lower lumped constants determined from attempted steady-state extraction fractions is likely to be the absence of steady-state due to loss of labeled metabolites.

Using the values 1.10 for $\tau$ and 0.30 for $\varphi$, Kuwabara et al. (1990) obtained an average value of 0.60 for the lumped constant of fluorodeoxyglucose in human

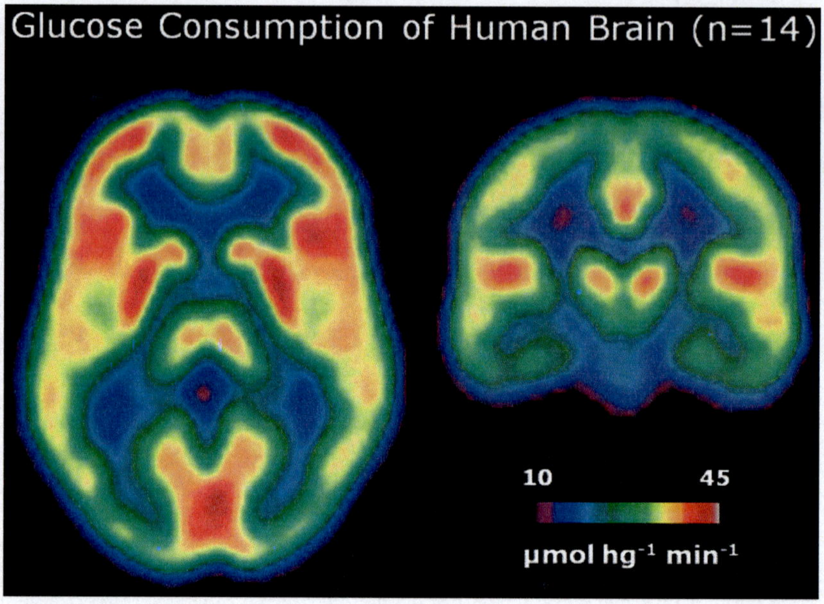


**Fig. 7.4** The case of three compartments: Map of average regional brain glucose consumption of normal adult human beings, determined with tracer fluorodeoxyglucose according to (7.38). The average map was prepared from brain glucose consumption maps of 14 normal volunteers. The glucose consumption maps were transferred into an orthogonal coordinate system corresponding to a stereotactic brain atlas. (Courtesy Christopher Bailey, PhD, Aarhus PET Center.)

brain. Kuwabara and Gjedde (1991) used a constrained approach and calculated the lumped constant for fluorodeoxyglucose in brain to be 0.65, i.e., 110% of the value determined as the ratio between the net extraction fractions according to the irreversible model. In 2002, Graham et al. determined the lumped constant of human cerebral cortex to be $0.89 \pm 0.08$. These values are examples of the observation that estimates of the lumped constant have tended to increase since the value was first directly estimated in humans in 1985.

There is no ready explanation of this phenomenon, other than the improvement in spatial resolution of positron emission tomographs which have rendered the measurements less sensitive to heterogeneity and partial volume effects, coupled with uncertainty about the true ratio between the volumes of white and gray matter. The conversion of regionally determined brain values to a whole-brain average is usually made with an assumed ratio between the volumes of white and gray matter which requires knowledge of the glucose consumption values of both, uncontaminated by heterogeneity and partial volume effects. To preserve a lumped constant of 0.5 and an average whole-brain glucose consumption of 30 $\mu$mol hg$^{-1}$ min$^{-1}$ with current tomography, the volume ratio between white and gray matter must be adjusted upward to a significant extent, perhaps as much as 12:1, as suggested by Brooks et al. (1987).

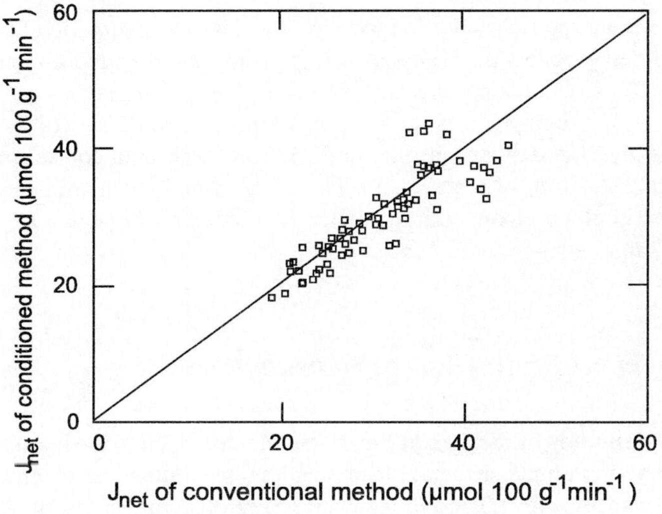


**Fig. 7.5** Regional variation of human brain glucose consumption estimated by two different deoxyglucose methods, one the conventional method, the other a constrained solution that simultaneously yields lumped constant and glucose consumption, introduced by Kuwabara et al. (1990). In the constraint, the average glucose consumption was adjusted to 30 μmol hg$^{-1}$ min$^{-1}$ through the definition of the lumped constant as a function of the constants $\tau$ and $\varphi$ expressed in (7.23). From Kuwabara et al. (1990)

## 7.5.2 *Whole-Brain Glucose Consumption*

With the single lumped constant measurement cited earlier, the whole-brain glucose consumption rate averaged 31±6 μmol hg$^{-1}$ min$^{-1}$ when measured with FDG and 28±3 (±SD) μmol hg$^{-1}$ min$^{-1}$ when measured with 2DG, using *Phelps' Solution* (Reivich et al. 1985).

Although the combination of an independently measured lumped constant in the assumed presence of dephosphorylation ($\Lambda_{app}(t)$) and a true net clearance, corrected for the assumed dephosphorylation ($K^{*}$ rather than $K^{*}_{app}(t)$), commingled methods that are not compatible in theory (*Sokoloff's Solution* vs. *Phelps' Solution*), the result is in excellent (but perhaps fortuitous) agreement with the average whole-brain glucose consumption rate measured with the Kety-Schmidt arteriovenous deficit method which averaged 31 and 30 μmol hg$^{-1}$ min$^{-1}$ in two separate series (Sokoloff 1960; Phelps et al. 1979).

All other determinations in humans have assumed an average whole-brain glucose consumption rate of 30 μmol hg$^{-1}$ min$^{-1}$ and have adjusted the lumped constant or the ratio on which it depends ($\tau$ and $\varphi$) to reflect this consensus. Therefore, the deoxyglucose method has made no substantially novel contribution to the knowledge of the whole-brain glucose consumption in normal human beings.

The constrained method based on *Kuwabara's Solution I* yields an estimate of the lumped constant as well as an estimate of the glucose metabolic rate. The estimate of the lumped constant is made from known or assumed values of $\tau$ and $\varphi$ which ultimately reflect the assumed average whole-brain glucose metabolic rate. Nonetheless, these estimates show that the lumped constant may vary slightly with the metabolic rate, being lowest at the highest metabolic rates. This slight variation suggests that the delivery of glucose increases more than consumption during physiological elevations of metabolism. This suggestion in turn points to a possible role of the brain glucose concentration in the regulation of the regulation of glucose phosphorylation.

### *7.5.3 Regional Brain Glucose Consumption*

The chief contribution of the deoxyglucose method has been to the knowledge of the regional variation of glucose metabolism in animal and human brain, as exemplified in the brain maps of Fig. 7.3 and in the average regional values of Fig. 7.5 in which the human regional values range from 20 to 40 μmol hg$^{-1}$ min$^{-1}$, using the so-called constrained method (*Kuwabara's Solution I*).

The use of labeled glucose rather than labeled deoxyglucose (*Blomqvist's Solution*) has yielded glucose consumption rates that are somewhat lower, as illustrated in Fig. 7.6, averaging 20–25 μmol hg$^{-1}$ min$^{-1}$. It is possible that the correction for lost metabolites is insufficient (Lear and Ackermann 1988). A more complex version of this explanation assigns the difference to the loss of labeled lactate and

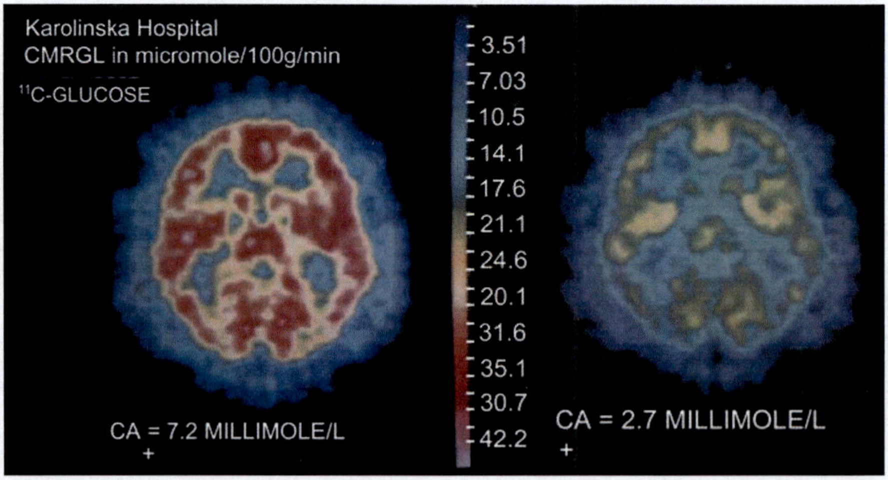


**Fig. 7.6** Regional map of human brain glucose consumption estimated by means of labeled glucose, according to *Blomqvist's Solution* in which the loss of labeled glucose metabolites is estimated by a monoexponentially increasing correction term. *Left Panel*: normoglycemia. *Right Panel*: hyperglycemia. Note uniform decline induced by hyperglycemia. From Blomqvist et al. (1991)

other nonvolatile metabolites (since the users of the method based on *Hawkins'* or *Blomqvist's Solutions* claim that the loss of volatile metabolites is either negligible or adequately corrected for). According to this view, labeled glucose traces only the glucose that fuels oxidative phosphorylation while labeled deoxyglucose traces total glucose phosphorylation (Ackermann and Lear 1990).

Modern methods of image analysis and co-registration with other image modalities such as magnetic resonance images provide the means to generate population averages of regional glucose consumption images as shown in Fig. 7.6. This stereotypic population image reveals a rather uniform distribution of glucose consumption, as expected from the averaging. It suggests that much of the variation seen in individual images is the result of methodological factors and random noise rather than true regional variation of energy metabolism under "baseline" conditions of neuronal work.

# Chapter 8
# Neuroenergetics*

## 8.1 Brain Work

The close coupling of neuronal function to rates of glucose and oxygen metabolism is well established. The processes that consume the most energy in activated cells under different conditions, the regional- and activation-dependent differences in stoichiometry of oxygen and glucose utilization, and the mechanisms that supply the energy for the work of the work of the brain are increasingly well understood. Yet, emerging evidence has required that the energy budget of different conditions of localized and general brain activity must be updated in the light of this evidence.

The brain derives most of its energy from the combustion of glucose. About 90% of the glucose is metabolized to carbon dioxide, and the oxidative metabolism of glucose in turn covers 99.5% of the brain's energy budget under normal stationary circumstances, also known as the normal (default) steady-state. Under special circumstances, the brain can cover larger fractions of its energy turnover by nonoxidative metabolism of glucose, or by oxidative metabolism of the monocarboxylic acids lactate, beta-hydroxybutyrate, acetoacetate, and acetate. The normal steady-state energy yield from these sources on the average is 10 μmol ATP for every gram of brain tissue per minute.

Although it accounts for only about 1–2% of the total body mass, the brain consumes 10% of the active human body's glucose and oxygen supplies and receives 10% of its blood supply. It is generally assumed that blood flow and energy metabolism are linked to the functions of the brain but the precise link is neither well defined nor is it known to which extent the link is important to the understanding of the organization of the brain and its functions. Therefore, to clarify the mechanisms linking the signals from brain images to the brain's electrochemical work, it is important to consider:

---

* Adapted from Gjedde (2007) Coupling of brain function to metabolism: Evaluation of energy requirements. In *Handbook of Neurochemistry*: Brain energetics from genes to metabolites to cells: Integration of molecular and cellular processes. Editor-in-chief: Lajtha, Abel. Gibson, Gary; Dienel, Gerry (Volume Eds.), 3rd ed., 2006, 400 p., 40 illus., Hardcover. ISBN: 0-387-30366-9, with permission from Springer-Verlag, Berlin Heidelberg.

A. Gjedde et al., *Neurokinetics: The Dynamics of Neurobiology In Vivo*,
DOI 10.1007/978-1-4419-7409-9_8, 

- The kind of work carried out in the brain.
- The cells which carry out the brain's work.
- The mechanism linking the work of these cells to the relative and absolute magnitudes of oxidative and nonoxidative energy metabolism in brain tissue.
- The mechanism relating the blood supply of the brain to the magnitudes of the brain's oxidative and nonoxidative metabolism.

The brain's work commonly is held to be the transfer, processing and exchange of information. The concept of information is not entirely clear, however, but it is generally held to be a function of the number of different states that a system can occupy. Thus, for $n$ different states, each state is said to contain $\log_2(n)$ bits, or $\log_2(n)/8$ bytes, of information. To the extent that a synapse can be considered a simple on–off device (switch), the central nervous system in human beings may have as many as $10^{15}$ switches, allowing the brain to occupy at least $2^{10^{15}}$ different states and thus to hold $10^{15}$ bits of information. Changes of this information occur when the on–off switches are toggled. The toggle of a switch represents a single binary operation, and the toggling incurs a cost to the system's energy supplies that qualifies as work but is probably unrelated to the information begin changed.

The energy cost of the information transfer can be calculated from the concept of entropy, according to which the information locally reduces the entropy of the system, in this case the brain. Entropy is reduced only by the supply of energy. Thus, the energy cost can be estimated from the combustion of fuels measured during functional activity of the brain. According to one such calculation, a single binary operation requires a minimum energy supply of $3 \cdot 10^{-24}$ kJ (Morowitz 1978). The hydrolysis of adenosine triphosphate (ATP) yields free energy of about $3 \cdot 10^{-5}$ kJ $\mu$mol $^{-1}$, and the human brain hydrolyzes 10 $\mu$mol ATP $g^{-1}$ $min^{-1}$. Assuming an upper limit of thermodynamic efficiency of 50%, brain tissue has the capacity to perform binary operations at the maximum rate of $10^{11}$ megabytes $g^{-1}$ $s^{-1}$, or about $10^{18}$ operations per second per brain.

## 8.2 Ion Homeostasis

Information transfer and processing in the brain occurs as propagated changes of membrane potential, brought on by changes of membrane permeability to several extra- and intracellularly concentrated cations. This section deals with the mechanisms that maintain the concentration of these ions in the face of changing membrane potentials. Neurons impose metabolic needs on the brain by being subject to excitation- and inhibition-induced changes of the membrane permeabilities to sodium, potassium, chloride, and calcium ions. Increases of the intracellular concentration of $Na^+$or the extracellular concentration of $K^+$stimulate the membrane $Na^+$, $K^+$-ATPase activity. How much of the ATPase activity subserves the transfer of information, and how much is necessary to keep the cells intact, is not known with certainty. The evidence cited later suggests that the ATPase activity associated

**Table 8.1** Ion concentrations in nerve cells

| | | Ion | | | | |
|---|---|---|---|---|---|---|
| | | Sodium | | Potassium | | Chloride |
| Variable | Unit | E&S | M | E&S | M | M |
| Equilibrium potential | mV | +41 | +40 | −84 | −100 | −75 |
| Intracellular concentration | mM | 27 | 30 | 80 | 140 | 8 |
| Extracellular concentration | mM | 133 | 150 | 3 | 3 | 130 |

From Erecinska and Silver (1989) ("E&S"), McCormick (1990) ("M")

with the functional activity of information transfer varies, from a value that is twice the absolute minimum required to maintain cellular integrity, to a value that is an order of magnitude greater.

The "sodium theory" explains the origin of the membrane potential and the graded or alternating depolarization of cells by the presence of sodium, calcium, potassium, and chloride equivalents as free ions in the intra- and extracellular spaces, and by the action on and of specific ion channels in the plasma membranes across which the ions are exchanged. The theory ascribes the electrical properties of the membrane to diffusion potentials established by the membrane conductances controlled by these channels (Hodgkin and Huxley 1952). The conductances of sodium and potassium underlying the resting membrane potential as well as the increased conductances, associated both with excitation above the nonexcited baseline and excitation above the resting, or "default," average, must be matched by appropriately active ion pumping to maintain constant ion concentrations.

The P-type (phosphorylation-type) $Na^+$,$K^+$-ATPase combines with ATP, $Mg^{2+}$, $Na^+$, and $K^+$ to form an enzyme-substrate association during which the enzyme is phosphorylated, and $Mg^{2+}$ and ADP are released (Skou 1960). As the phosphorylated enzyme splits into inorganic phosphate and the original enzyme, $Na^+$ and $K^+$ are translocated in the appropriate directions, outward for sodium and inward for potassium. The energy released by the hydrolysis of 5 $\mu$mol $g^{-1}$ $min^{-1}$ ATP (half the ATP turnover) is sufficient to transport 15 $\mu$mol $Na^+g^{-1}$ $min^{-1}$. The resulting half-life of sodium in the cells is less than a minute, and considerably less than the 20 min half-life of sodium in stimulated squid axons (Hodgkin and Keynes 1956).

The apparent average sodium permeability-surface area product in the steady-state can be calculated from the sodium flux. With the concentrations listed in Table 8.1, as well as assumed values of an average steady-state membrane potential and corresponding sodium flux calculated from the measured ATP turnover, and the 3:2 ratio between the net sodium and potassium fluxes in the steady-state, the permeability-surface area products of sodium and potassium of Table 8.2 were calculated by means of Goldman's flux equation (Goldman 1943; Hodgkin and Katz 1949). Details of this calculation were given by Gjedde (1993a). The equation yields the membrane potential difference on the basis of the permeability-weighted individual concentrations of sodium, potassium, and chloride. In the steady-state, the chloride flux matches the difference between the sodium and potassium fluxes, rendering the total ion flux electroneutral. The calculation ignores the requirements

**Table 8.2** Ion movements across nerve cell membranes

| Variable | Unit | Ion | | |
|---|---|---|---|---|
| | | Sodium | Potassium | Chloride |
| Transmembrane leakage | $\mu$mol g$^{-1}$ min$^{-1}$ | 15 | 10 | 5 |
| PS product at −65 mV | ml g$^{-1}$ min$^{-1}$ | 0.038 | 0.404 | 0.549 |
| PS product at −55 mV | ml g$^{-1}$ min$^{-1}$ | 0.044 | 0.285 | 0.246 |

From Gjedde (1993b), assuming 50% of ATP turnover dedicated to ion transport, calculated from the concentrations listed in Table 8.1 ("M"). To estimate the chloride permeability, it was necessary to use a simplified form of the equation

imposed by calcium ion fluxes but the calcium ion flux is a minor requirement in energetic terms, although it is now known to be of major functional significance (Lauritzen 2005), as discussed later.

## 8.3 Brain Energy Metabolism

The precise link between the work performed by brain tissue and the rate of metabolism is not known. Several possibilities have been raised, including the claim that the concept of a link itself is obsolete; the metabolism *is* the work. Of the putative links, the simplest but least plausible is feed-back regulation by the changing concentration of metabolites such as ATP as ADP. This subsection deals with the attempts to identify a regulatory mechanism which elicits the appropriate changes of metabolite flux in association with, rather than in response to, the changes of the work of the brain.

### *8.3.1 Definition of Brain Activity Levels*

The definition of the brain's baseline activity is in dispute and the issue of the brain's normal activity is complex. Recent theories of brain functional organization distinguish between activation, default, resting, and baseline states, but the definitions of these states are not yet universally agreed upon.

The question of the range of work between baseline (or "household") and normal (default) activities of brain tissue hinges on the definitions of average and baseline. In the work of Sibson et al. (1998), Shulman and Rothman (1998), and Shulman et al. (2004), functional activity is defined as the metabolism related to the release of the excitatory neurotransmitter glutamate. The close reading of this work suggests that the maximum functional activity of both rodent and human brain is about twice the normal (default) activity, which in turn is the twice the "resting" activity measured in apparently unconscious but otherwise intact brains. This resting activity turns out to be about twice the activity associated with the absolutely nonfunctioning baseline, which in turn is about twice the activity associated with the state of absent ion transport.

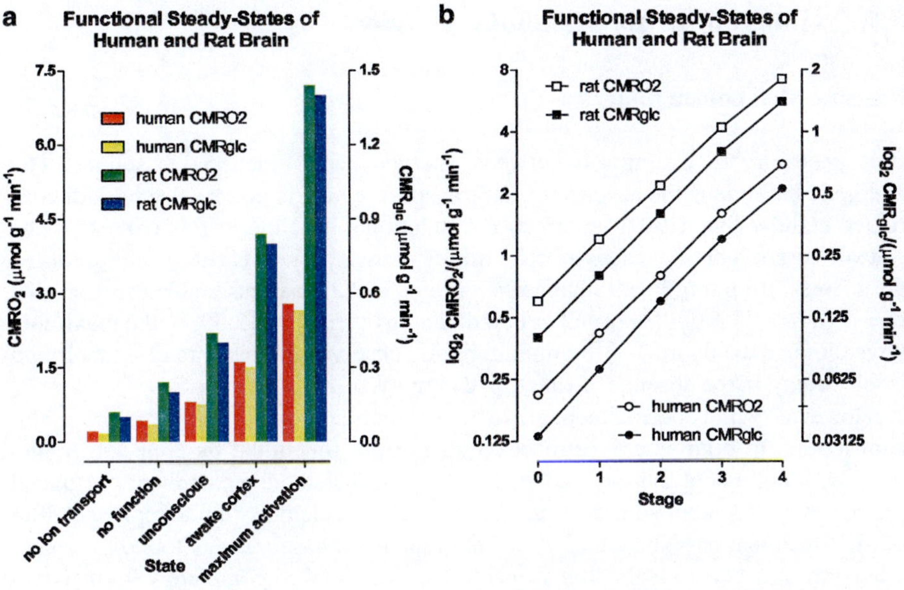


**Fig. 8.1** Relationship among functional activity of mammalian brain and estimates of rates of energy metabolism of oxygen and glucose. From studies of rat or human brain summarized by Gjedde et al. (2002), Shulman et al. (2004), and Hyder et al. (2005). *Panel A*: Ordinates show oxygen (*left*) and glucose (*right*) metabolic rates for five distinct levels of functional activity. *Panel B*: Abscissa: Functional stage on a scale from 0 to 4. Ordinates: $\text{Log}_2$ scale of oxygen (*left*) and glucose (*right*) metabolic rates in units of $\mu$mol $g^{-1}$ $min^{-1}$

The evidence suggests as a rule of thumb that brain energy metabolism doubles for every standard increment of functional activity, as shown in Fig. 8.1. The standard increment means a change from one fundamental stage of functional level to the next. These stages are illustrated in Fig. 8.1 and include the states of no ion transport (stage 0), no functional activity (baseline stage 1), low functional activity without consciousness (baseline stage 2), high functional activity of the conscious cerebral cortex (default stage 3), and elevated functional activity associated with the highest degree of physiological activation (stage 4). Activity above stage 4 would then be considered pathological. Panel B shows that the metabolic or work rates associated with these stages can be described by the simple formula,

$$\text{CMR}(S) = \text{CMR}(0)\,2^S, \tag{8.1}$$

where CMR is the cerebral metabolic rate, CMR(0) the cerebral metabolic rate of stage 0, and $S$ is one of the five functional stages (0–4) defined earlier. The equation has an interesting similarity to the formula (see earlier) for a system that holds $S$ bits of information.

### 8.3.2 *Stages of Brain Metabolic Activity*

#### Baseline Metabolism (Stages 0–1)

It is necessary to distinguish between energetic and functional baselines. The energetic baseline is the absence of ion transport, which is not consistent with continued cellular integrity. In contrast, the functional baseline in this context is the state of absent functional activity. The rule-of-thumb derived earlier from recent experiments with living intact mammalian brains makes it a reasonable estimate that one-eighth or 25% of the normal or default energy turnover (12.5% of the maximum physiological excitation) of the human brain subserves the basic ion transport and isoelectricity in the absence of functional activity.

However, while the maintenance of the membrane potential makes the major contribution to brain energy utilization also at the functional baseline, other factors also make a contribution. Whittam (1962) estimated the metabolism of isolated brain tissue associated with the transport of sodium and potassium to represent 40% of the functional baseline, and 50% of the functional baseline was found to remain when ion transport in isolated nervous tissue was blocked completely by inactivation of the $Na^+$,$K^+$-ATPase (Baker and Connelly 1966; Ritchie 1967; Hertz and Schousboe 1975; Mata et al. 1980).

#### Resting Metabolism (Stage 2)

The term "resting" state for normal brain tissue has little meaning in vivo, as the healthy brain never rests, but for the present purposes it is close to a state of lowered but not completely absent functional activity, in which no particular excitation is present. The energy metabolism of brain tissue in any one of a variety of conditions of absent higher cognitive activity (severed connections, coma, persistent vegetative state, and anesthesia) are all about 50% of the normal default average (25% of the maximum physiological excitation; McIlwain 1951; Shalit et al. 1970, 1972; Brodersen and Jørgensen 1974; Sokoloff et al. 1977; Levy et al. 1987; Alkire et al. 1995, 1997, 1999). These observations show that the work associated with lowered but not absent depolarization of neuronal membranes is about half of the work imposed by the average degree of depolarization.

#### Normal (Default) and Physiologically Elevated Metabolism (Stages 3–4)

The normal or default metabolic stage refers to the awake and normally functioning mammalian brain. For ethical reasons, this stage has been studied more closely in awake humans than in other mammals. Stage 3 metabolism refers more accurately to human cerebral cortex, while whole-brain values tend to represent a mixture of stage 2 and stage 3 metabolic conditions.

**Table 8.3** Average properties of human whole-brain and cerebral cortex

| Variable [unit] | Whole-brain | Cerebral cortex |
|---|---|---|
| $CMR_{glc}$ [μmol $g^{-1}$ $min^{-1}$] | 0.25 | 0.30 |
| $CMR_{O_2}$ [μmol $g^{-1}$ $min^{-1}$] | 1.40 | 1.60 |
| CBF [ml $g^{-1}$ $min^{-1}$] | 0.43 | 0.50 |
| OGI [ratio] | 5.6 | 5.3 |
| ATP turnover [$J_{ATP}$, μmol $g^{-1}$ $min^{-1}$] | 9.4 | 10.2 |
| Pyruvate turnover [$J_{pyr}$, μmol $g^{-1}$ $min^{-1}$] | 0.5 | 0.6 |
| Lactate efflux [$J_{lact}$, μmol $g^{-1}$ $min^{-1}$] | 0.035 | 0.07 |
| LGI [ratio] | −0.14 | −0.23 |

Modified from Kuwabara et al. (1992), Vafaee et al. (1999), and Gjedde et al. (2005a), using (8.6–8.8)

Using modern neuroimaging methods, Kety (1949) and Lassen (1959) reported the average magnitudes of normal brain energy metabolism and blood flow in living human brains. More recently determined but similar steady-state values of energy metabolism and blood flow of the human brain are listed in Table 8.3, together with the steady-state turnover rates of ATP, pyruvate, and lactate.

The whole-brain molar oxygen-glucose ratio or index (OGI) is 5.6, indicating that about 90% of the glucose is fully oxidated (see also Himwich and Himwich 1946). At the normal (default) stage, total glucose consumption of cerebral cortex is about 30 μmol $hg^{-1}$ $min^{-1}$, with an OGI of 5.3. The 10% nonoxidative metabolism of glucose leads to a lactate production of about 5–7 μmol $hg^{-1}$ $min^{-1}$. The lactate flux is about 25% of the $T_{max}$ of the blood–brain barrier monocarboxylic acid transporter (MCT-1, see later), consistent with a tissue lactate concentration of about 1.5 mM, as listed in Table 8.3. The corresponding ATP turnover calculated from (8.3) is 10 μmol $g^{-1}$ $min^{-1}$. Altogether, the brain tissue metabolite stores, including ATP itself, represent about 1 min's worth of ATP turnover (Table 8.4).

Attwell and Laughlin (2001) reevaluated the energy turnover involved in the different processes contributing to functional activity, now known as the energy "budget" of the brain, including the energy requirements of processes such as biosynthesis during functional activity in vivo, neurotransmitter vesicle formation, fusion, and release. According to this budget, 90% of the energy turnover is devoted to "synaptic" and hence functional activity in general in the brain, and of this turnover, 80% occurs in neurons and the rest in glial cells. The authors also conclude that the almost all of the energy is spent on the restoration of ion gradients with the $Na^+$-$K^+$-ATPase.

There is early and more recent evidence that the oxygen consumption of human brain may rise to as much as 3 μmol $g^{-1}$ $min^{-1}$ under some physiological circumstances, accompanied by increases of glucose consumption to as much as 0.5 μmol $g^{-1}$ $min^{-1}$ (Roland et al. 1987; Shulman et al. 2004), judged from magnetic resonance spectroscopic measurements of total oxidative metabolism of pyruvate which reach as high as 0.8–0.9 μmol $g^{-1}$ $min^{-1}$ in normal human cerebral cortex.

**Table 8.4** Average metabolites in human brain

| | Cytosol | | Glycolytic equivalents | |
|---|---|---|---|---|
| | Concentration | Content | ATP | Lactate |
| Metabolite | (mM) | ($\mu$mol g$^{-1}$) | ($\mu$mol g$^{-1}$) | |
| PCr | 5.0 | 4.0 | 4.0 | |
| Glycogen | 3.0 | 2.4 | 3.6 | 3.6 |
| Glucose | 1.2 | 1.0 | 2.0 | 2.0 |
| ATP | 2.2 | 1.7 | 1.7 | |
| ADP | $1.2 \cdot 10^{-2}$ | $1.0 \cdot 10^{-2}$ | $5.0 \cdot 10^{-3}$ | |
| AMP | $7.1 \cdot 10^{-5}$ | $5.6 \cdot 10^{-5}$ | | |
| Pyruvate | 0.16 | 0.13 | | 0.1 |
| Lactate | 2.9 (0.75[a]) | 2.3 (0.6[a]) | | 2.3 |
| Total | | | 11.3 | 8.0 |

From Olesen (1970), Roth and Weiner (1991). The "glycolytic equivalent" is the ATP reserve that each metabolite would represent in case of complete depletion
[a]MRS measurements general yield lower values of lactate in vivo (0.5–1 mM) but corresponding pyruvate values are not reported and the determination is indirect (see Prichard et al. 1991; Sappey-Marinier et al. 1992)

## 8.4 Substrate Transport in Brain

### *8.4.1 Glucose Transport*

Glucose transport is discussed in greater detail in Chap. 6. Glucose is the source of pyruvate and enters brain tissue, neurons and astrocytes by means of facilitative insulin and sodium insensitive transport by several members of the GLUT family of membrane spanning proteins (Gjedde 1992). In brain, the important members are GLUT-1 and GLUT-3 (Drewes 1999). The 55 kDalton GLUT-1 resides in the membranes of the capillary endothelium constituting the blood–brain barrier, while the slightly modified 45 kDalton GLUT-1 resides in the membranes of astrocytes and choroid plexus. The GLUT-3 protein resides in the membranes of neurons. The transport of glucose is nonlinearly proportional to the difference between blood plasma and cytoplasm concentrations, ensuring that the glucose concentration everywhere in brain tissue is the same substantial fraction of the plasma glucose (Silver and Erecinska 1994).

The transport capacities of GLUT-1 and GLUT-3 are known in some details, which indicate that glucose provision itself is not rate-limiting for brain oxidative metabolism (Gjedde 1983, 1992). For glucose, the maximum transport rate, $T_{max}$ of the endothelial GLUT-1 glucose transport across the blood–brain barrier is about 2–4 times the rate of unidirectional glucose transport, which is about twice the net transport of glucose (see Gjedde 1992). The $T_{max}$ of the neuronal and glial glucose transporters GLUT-3 and GLUT-1 is 5,000-fold higher (Diemer et al. 1985). Thus, under normal circumstances, glucose transport is not rate-limiting for glycolysis, and glucose concentrations in the different cellular compartments of brain tissue are likely to be similar. However, as blood–brain glucose transport may become rate-limiting in pronounced hypoglycemia, it is possible that it could also be rate-limiting under conditions of extreme glycolysis, unless blood flow changes are sufficient to continuously supply the glucose. During physiological activation of rats, Silver and Erecinska (1994) found slight decreases of the extracellular glucose concentration, determined by means of glucose-sensitive microelectrodes placed in the brain tissue.

### 8.4.2 Monocarboxylate Transport

The monocarboxylic acids, pyruvate, and lactate cross the membranes of brain tissue by means of facilitative proton-dependent transport catalyzed by the MCT family of membrane spanning proteins (Oldendorf 1973; Halestrap 1975; Poole and Halestrap 1993; Halestrap and Price 1999). In brain tissue, the important transporters are MCT1, MCT2, and MCT4 (Drewes 2002; Pierre and Pellerin 2005).

The MCT1 protein spans the membranes of the capillary endothelium and astrocytes. Gerhart et al. (1997, 1998) claimed that MCT1 resides on neurons and the MCT2 protein on astrocytes, particularly their foot processes. However, the original assignment by Broer et al. (1997) of MCT1 to astrocytes and MCT2 to neurons, particularly the glutamatergic synapses, has now been confirmed (Bergersen et al. 2005). The MCT4 protein appears to be specific for astrocytes.

The transport mechanism of the MCT family is near-equilibrium proton symport, and as such it is influenced by the pH of the cells and declines when pH rises. The MCT2 is of higher affinity (1 mM) toward pyruvate and lactate than the MCT1 (3–5 mM), indicating that it is saturable by lactate at normal concentrations to a greater extent than MCT1. For this reason, it is likely that preference for MCT1 over MCT2 is controlled by the lactate concentration in brain: At higher concentrations, lactate prefers MCT1. However, the lower affinity of MCT1 also means that the transporter turnover number is higher, which makes the approach to a new steady-state faster, other factors being equal. Neither isoform restricts the exchange of pyruvate at the normal low concentration, but the MCT2 isoform may restrict the exchange between the neurons and the extracellular space of lactate at higher concentrations.

The maximal transport capacity ($T_{max}$) of the MCT1 at the blood–brain barrier is 0.2 $\mu$mol g$^{-1}$ min$^{-1}$ (Pardridge 1981), with a Michaelis constant for lactate of about 5 mM (Halestrap and Price 1999), which is substantially higher than the normal lactate content of brain (Table 8.3). Under normal conditions the efflux rate across the blood–brain barrier is therefore only 25% of the $T_{max}$. As the $T_{max}$ and $K_t$ of the MCT-1 are about the same for pyruvate and lactate (Cremer et al. 1979), the export of pyruvate to the circulation is about one tenth that of lactate (Himwich and Himwich 1946). Considering the surface area of neurons and glia, it is probable that the pyruvate and lactate exchange among the compartments of the brain is at near-equilibrium, as it is for glucose.

As discussed later, it is possible that both pyruvate and lactate are transported into mitochondria by a specific mitochondrial monocarboxylic acid transporter (mMCT, Brooks et al. 1999a). The exact nature of mMCT is not known, but the bulk of the evidence suggests that it is related to MCT-2 (Halestrap and Price 1999), the high-affinity transporter, although its identity with MCT-1 has also been reported (Brooks et al. 1999a).

That the mMCT is closer to the MCT2 than the MCT1 is supported by the $K_t$ of the mMCT, which is 0.5 μmol $g^{-1}$, i.e., somewhat higher than the cytosolic pyruvate concentration of 0.1–0.2 μmol $g^{-1}$, with a $T_{max}$ of 3 μmol $g^{-1}$ $min^{-1}$ (LaNoue and Schoolwerth 1979; Nalecz et al. 1992), depending on the mitochondrial density. This appears to be much higher than the average flux of pyruvate in human cerebral cortex (0.6 μmol $g^{-1}$ $min^{-1}$), but differences of mitochondrial density exist among the different cellular constituents of brain tissue. Thus, it only on average that the rate of pyruvate entry into mitochondria can rise fivefold in the absence of a change of the $T_{max}$.

The possibility that the mitochondrial mMCT may be rate-limiting for oxidative metabolism in tissue other than brain was explored by Halestrap (1978). Halestrap and Armston (1984) concluded that this was not the case in liver. However, Shearman and Halestrap (1984) argued that pyruvate transport is rate-limiting for pyruvate oxidation by actively respiring heart mitochondria. This could be true also for the brain, although it is more usually accepted that mitochondrial pyruvate concentration is sufficient to saturate the pyruvate dehydrogenase in keeping with its flux-generating role in oxidative metabolism, rendering the pyruvate flux a function of the total PDH activity in the tissue in question.

### 8.4.3 Oxygen Transport

Oxygen transport is discussed in greater detail in Chap. 5. Oxygen transport from blood to brain tissue is significantly limited by the hemoglobin binding and possibly by other factors as well, including a specific resistance at the endothelium of brain capillaries (Gjedde et al. 1991; Kassissia et al. 1995). Simple one-dimensional models of oxygen delivery to brain tissue (Buxton and Frank 1997; Gjedde 1996b; Vafaee and Gjedde 2000) confirm that disproportionately increased blood flow delivers more oxygen during functional activation. The models assume that the mean capillary hemoglobin saturation by oxygen is a simple function of the net extraction of oxygen, assuming a reasonably even distribution of the oxygen delivery along the length of all capillaries, according to the equation,

$$\bar{S}_{O_2} = S_{aO_2}\left(1 - \frac{E_{O_2}}{2}\right), \tag{8.2}$$

where $\bar{S}_{O_2}$ is the average capillary oxygen saturation of hemoglobin, $S_{aO_2}$ the arterial oxygen saturation, and $E_{O_2}$ is the net oxygen extraction fraction, equal to the ratio $J_{O_2}/(F\ C_{O_2})$, in which $F$ is the blood flow and $C_{O_2}$ is the arterial oxygen concentration. The mean capillary oxygen tension and hemoglobin saturation are also related by the equation for the oxygen dissociation curve,

$$\bar{S}_{O_2} = \frac{1}{1 + \left[\frac{P_{50}^{hb}}{P_{O_2}^{cap}}\right]^h}, \tag{8.3}$$

where $P_{50}^{hb}$ is the hemoglobin half-saturation oxygen tension, $\bar{P}_{O_2}^{cap}$ is the average arterial oxygen tension, and $h$ is the Hill coefficient. The resulting equation establishes the inverse correlation between the net extraction fraction and the average capillary oxygen tension,

$$\bar{P}_{O_2}^{cap} = P_{50}\sqrt[h]{\frac{S_{aO_2}(2-E_{O_2})}{2-S_{aO_2}(2-E_{O_2})}} = P_{50}\sqrt[h]{\frac{2}{E_{O_2}}-1} \text{ for } S_{aO_2}\approx 1, \quad (8.4)$$

where the second equality requires an arterial oxygen saturation of 100%. The maximum delivery is proportional to the mean capillary oxygen tension for a given effective capillary density associated with a particular diffusion capacity $L$. The average mitochondrial oxygen tension is therefore,

$$\bar{P}_{O_2}^{mit} = \bar{P}_{O_2}^{cap} - \frac{J_{O_2}}{L}, \quad (8.5)$$

where $\bar{P}_{O_2}^{mit}$ is the maximum oxygen delivery capacity, $J_{O_2}$ the rate of oxygen consumption, and $L$ is the tissue oxygen diffusion capacity for the mean distance between the capillary lumen and the mitochondria. The average tension of oxygen in brain tissue depends on the distribution of oxygen delivery and consumption sites but can be approximated by the formula,

$$\bar{P}_{O_2}^{br} = \bar{P}_{O_2}^{cap} - \frac{J_{O_2}}{2\,L}, \quad (8.6)$$

where the average tissue oxygen tension is $\bar{P}_{O_2}^{br}$. The average normal tensions of oxygen in the vasculature and tissue compartments, estimated with the compartment equations earlier, are shown in the model illustrated in Fig. 8.2 (Gjedde 2007).

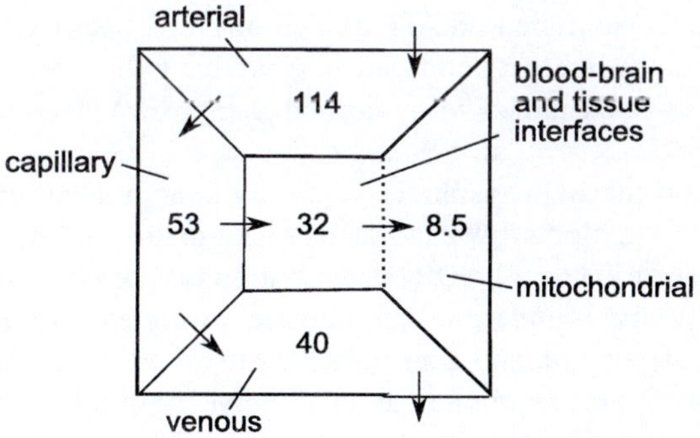


**Fig. 8.2** Compartment model of oxygen tensions in brain. The compartments include arterial, capillary, venous, and mitochondrial spaces. The interface between the capillary and mitochondrial compartments is a diffusion barrier, the exact position of which is not known with certainty, although it may be dominated by the capillary endothelium. The numbers refer to normal oxygen tensions in units of mmHg, as presented in Fig. 5.4

## 8.5 ATP Homeostasis

Adenosine triphosphate (ATP) is the energy "currency" of brain cells that links the energy utilizing and energy producing processes. This section deals with the mechanisms that maintain a constant concentration of ATP, regardless of the rate of its expenditure. Generally speaking, these processes are glycolysis, i.e., the breakdown of glucose to pyruvate, and oxidative phosphorylation, the breakdown of pyruvate to carbon dioxide and the reduction of oxygen to water.

At steady-state, the normal stoichiometric relationships between the main substrate fluxes are given by the equations,

$$J_{\mathrm{pyr}} = 2\ J_{\mathrm{glc}}, \tag{8.7}$$

$$J_{\mathrm{ATP}} = J_{\mathrm{pyr}} + 6\ J_{\mathrm{O_2}} \tag{8.8}$$

and

$$J_{\mathrm{lact}} = J_{\mathrm{pyr}} - \frac{1}{3}\ J_{\mathrm{O_2}}, \tag{8.9}$$

where $J_{\mathrm{ATP}}$ is the ATP production, $J_{\mathrm{glc}}$ the glucose consumption, $J_{\mathrm{pyr}}$ the pyruvate generation rate, and $J_{\mathrm{lact}}$ is the lactate production and efflux rate. These relationships apply only to the steady-state in which there are no changes of substrate concentrations in brain. The formulation of lactate production applies to the tissue as a whole, as pyruvate as a monocarboxylic acid is not subject to compartmentation. Thus, the lactate production is a function of the pyruvate concentration and the lactate efflux through the blood–brain barrier. It is only under nonsteady-state circumstances that changes of glucose, glycogen, pyruvate, and lactate concentrations all contribute to the turnover rates in complex ways, to be explored later.

Under normal circumstances, brain energy metabolism maintains a constant concentration of ATP. Thus, the processes which restore this metabolite are sensitive (directly or indirectly) to increased ATP utilization (feed-back), or vice versa (feed-forward). Observations in heart and brain suggest that two-to-ten-fold variations of the rate of work can be sustained with minimal change of ATP (Balaban et al. 1986; Matthews et al. 1981; Detre et al. 1990a, b; Wyss et al. 1992).

There are several mechanisms underlying the remarkable ability of brain tissue to vary chemical work, blood flow, and metabolism many-fold with little change of the ATP concentration. The real ADP concentration is much more difficult to ascertain, but it is likely that it undergoes an increase during elevations of metabolism. As the functions of enzyme and transporter proteins are kinetically similar, both enzymes and transporters are among the non- and near-equilibrium reactions which contribute to these mechanisms. However, many of the conventional explanations of cellular energetics fail if the proposed mechanisms of homeostasis depend on concentrations of nucleotides or other intermediates that do not actually change.

Near-equilibrium reactions buffer minute changes of their substrates, but flux-generating and -directing nonequilibrium reactions adjust the magnitude and direction of metabolism as dictated by extrinsic regulators. The probable targets of extrinsic regulators include the nonequilibrium hexokinase and phosphofructokinase-1 reactions of glycolysis and the nonequilibrium pyruvate dehydrogenase, citrate synthase, and oxoglutarate dehydrogenase reactions of oxidative phosphorylation in mitochondria.

### 8.5.1 *Hydrolysis of Phosphocreatine*

Creatine kinase (CK) occupies a pivotal role in the early buffering of the ATP concentration (Wyss et al. 1992). The cytosolic enzyme has tissue-specific isoforms. The brain-predominant subtype is BB-CK. There also is a form bound to the inner mitochondrial membrane, Mi-CK. The function of CK isoforms in different compartments is subject to debate, as is the near-equilibrium status of Mi-CK. However, there is agreement that the cytosolic creatine kinase reaction is in near-equilibrium in living human brain (Roth and Weiner 1991; Mora et al. 1992).

The near-equilibrium cytosolic creatine kinase reaction transfers a high-energy phosphate bond from phosphocreatine to ADP to make ATP. When the near-equilibrium of cytoplasmic creatine kinase is perturbed, the reaction buffers any increase of ADP by increased phosphorylation of ADP to ATP. Cytoplasmic phosphocreatine is replenished by mitochondrial CK, which is regenerated by hydrolysis of mitochondrially generated ATP (Fedosov 1994). Phosphocreatine diffuses an order of magnitude faster than the adenine nucleotides (Wyss et al. 1992), but under conditions of high metabolic activity ATP homeostasis may not be maintained because of rate-limitation of the CK-transphosphorylation reaction in the mitochondria (Fedosov 1994).

### 8.5.2 *Glycolysis*

Aerobic glycolysis is the breakdown of glucose to pyruvate under normal aerobic conditions. Anaerobic glycolysis is the conversion of glucose to lactate under anaerobic conditions. As glucose is the preferred substrate for brain metabolism, the understanding of the control of glycolysis is integral to the appreciation of how increased energy production and utilization are linked. The rate is regulated by the nonequilibrium reactions catalyzed by hexokinase and phosphofructokinase.

The main reactions and their time constants are listed in Table 8.5, and their interactions are shown in schematic form in Fig. 8.3. The time constants dictate the half-times of change, i.e., the times it takes the system to reach half-way to a new steady-state. It is apparent from Table 8.5 that glycolysis can respond to change with time-constants of the order of milliseconds, while oxidative metabolism can only respond with time-constants of seconds or minutes. Hence, oxidative metabolism is apt to respond to any stimulus with a considerable delay, compared to glycolysis.

**Table 8.5** Selected brain metabolic reactions and metabolite transporters

| Reaction or transporter | Equilibrium status | Activity [μmol g$^{-1}$ min$^{-1}$] | Substrate concentration [μmol g$^{-1}$] | Time constant [s] | Time constant [ms] |
|---|---|---|---|---|---|
| HK[a] | Flux-generating | 0.3 | 2 | 400 | |
| PFK -1[b] | Flux-generating | 0.3 | 0.1 | 20 | |
| PDH[c] | Flux-generating | 0.3 | 0.01 | 2 | |
| GLUT-1[d] | Flux-limiting | 2 | 5 | 150 | |
| MCT (BBB)[e] | Flux-limiting | 0.2 | 0.1 | 30 | |
| mMCT (mitochondria)[f] | Flux-limiting | 3 | 0.1 | 3 | |
| GLUT-3[g] | Near-equilibrium | 1,000 | 1 | | 60 |
| LDH (lactate)[h] | Near-equilibrium | 2,000 | 1 | | 30 |
| MCT (cell membranes)[i] | Near-equilibrium | 1,000 | 0.1 | | 6 |
| LDH (pyruvate)[h] | Near-equilibrium | 2,000 | 0.1 | | 3 |

[a]Hexokinase (Gjedde 1983),
[b]Phosphofructokinase-1 (Gjedde 1983),
[c]Pyruvate dehydrogenase complex (Katayama et al. 1998),
[d]Glucose transporter-1 (Gjedde 1992),
[e]Monocarboxylic acid transporters (Cremer et al. 1979),
[f]Mitochondrial monocarboxylic acid transport (Poole and Halestrap 1993),
[g]Glucose transporter-3 (Diemer et al. 1985),
[h]Lactate dehydrogenase (Salceda et al. 1998),
[i]Monocarboxylic acid transporters (Desagher et al. 1997)

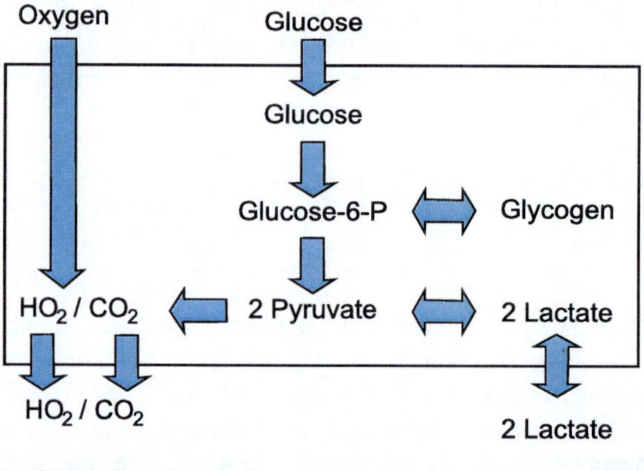


**Fig. 8.3** Simplified illustration of the main metabolic pathways of the mammalian brain

## Hexokinase and Phosphofructokinase-1

Phosphate and citrate ions, AMP, ammonium and hydrogen ions, and phosphocreatine itself, are among the classic regulators of the glycolytic enzymes hexokinase (HK) and phospho-fructokinase-1 (PFK-1), which catalyze the nonequilibrium reactions and irreversible phosphorylation of glycolytic intermediates. The classical list of extrinsic regulators of these enzymes includes $Mg^{++}$, changes of the concentration of which may accompany increased ATP turn-over and increased creatine kinase activity. Although 2 mol ATP are consumed during the first stage of glycolysis, per mol of glucose metabolized, 4 mol ATP are generated during the second stage of glycolysis, for a net return of 1 mol ATP and two hydrogen ion equivalents per mole of pyruvate synthesized.

### Pyruvate and Lactate Dehydrogenases

Pyruvate participates in at least three different reactions in brain tissue. It can be reduced by conversion to lactate, transported into mitochondria, or transported out of brain by the monocarboxylate transporter in cell membranes and the endothelium of brain capillaries. The last of these processes is minimal for pyruvate and can probably be ignored because of pyruvate's low concentration compared to lactate.

As a near-equilibrium reaction, lactate dehydrogenase (LDH) buffers changes of the concentration of pyruvate. The near-equilibrium reaction catalyzed by LDH is strongly balanced toward lactate. Synthesis of 1 mol of lactate from 1 mol of pyruvate removes one cytosolic reducing equivalent, so the net yield is only 2 mol ATP per mole of glucose converted to lactic acid.

The brain has at least two isozymes of LDH, the "heart" isozyme $H_4$ ($LD_1$), to which pyruvate has a low affinity, characteristic of oxidative tissues, and the "muscle" isozyme $M_4$ ($LD_5$), to which pyruvate has a high affinity, characteristic of less oxidative tissues such as liver and some muscles (Kaplan and Everse 1972). The measured lactate-pyruvate ratio ($\lambda$) of 10–15 suggests that the prevailing LDH isozyme pattern is closer to that of $LD_1$ than that of $LD_5$, because the relatively low ratio reflects a relatively low affinity toward pyruvate. The entities controlling the steady-state lactate-pyruvate ratio, in addition to the inherent affinities of the enzyme for the two substrates, which must have the same ratio for all the subtypes for thermodynamic reasons, include both pH and the $NAD^+$-NADH ratio,

$$\frac{[\text{lact}]}{[\text{pyr}]} = K_{eq} \frac{[\text{NADH}]\,[\text{H}^+]}{[\text{NAD}^+]},$$

where $K_{eq}$ is the equilibrium constant of the LDH reaction. The ratio between the concentrations of lactate and pyruvate is also the apparent ratio between their affinities. At steady-state, the ratio must be the same everywhere, given the near-equilibrium and facilitated diffusion nature of the proton symporters of lactate and pyruvate. The constant ratio implies that the different inherent kinetic properties of LDH exist to accomodate differences of the $NAD^+$-NADH ratio in different parts of the tissue (Newsholme and Crabtree 1979). The affinity of an enzyme for a substrate is a function of, among other factors, the turnover number of the enzyme, indicating that the velocity of the reaction is related to the affinity constant; the higher the affinity constant is (i.e., the lower the affinity), the higher the turnover number and hence the reaction velocity are, other factors being equal. The cytosolic $NAD^+$-NADH ratio is an indicator of the oxidation status of the tissue, which is low when the ratio is lower than normal (< 1,000) and high when it is higher than normal (>1,000).

It is a common claim that the relatively low affinity of pyruvate for $LD_1$ makes this enzyme particularly useful to a tissue of high oxidative capacity because it allows rapid build-up of pyruvate, while $LD_5$ is more effective at buffering the increase of pyruvate in a tissue of lower oxidative capacity, because of the relatively high affinity toward pyruvate. Thus, there is evidence that $LD_1$ and the mRNA for the subtype are found exclusively in neurons, while astrocytes appear to possess both $LD_1$ and $LD_5$ and their corresponding mRNA (Tholey et al. 1981; Bittar et al. 1996; Laughton et al. 2000).

However, it is probably incorrect to infer a net direction of the LDH reaction simply from the presence of a specific subtype in a population of cells (Van Hall 2000). For practical purposes, it is more useful to regard the properties of the LDH subtypes $LD_1$ and $LD_5$ as states of sensitivity to transient events during activation or deactivation of the tissue, as will be explored later.

### *8.5.3 Oxidative Phosphorylation*

At steady-state, brain metabolism has a respiratory quotient of unity, consistent with the oxidation of glucose (Himwich and Fazekas 1937; Gibbs et al. 1942; Himwich and Himwich 1946) and with the integration of glycolysis and oxidative metabolism. The net effect of the metabolism of pyruvate to $CO_2$ is to provide the electron chain complexes with the nicotinamide (NADH) and flavin ($FADH_2$) adenine dinucleotides necessary for electron transport and oxygen metabolism. This subsection presents the primary regulatory steps for oxidative metabolism in the mitochondria.

#### Pyruvate Dehydrogenase Complex and the Tricarboxylic Acid Cycle

Recent evidence suggests that mitochondria may import both pyruvate and lactate (Brooks et al. 1999a). Inside the mitochondria, the lactate is then reconverted to pyruvate in another LDH-catalyzed near-equilibrium reaction (Brandt et al. 1987; Brooks et al. 1999b). The pyruvate provides the mitochondrial electron carrier complexes I and II with NADH and $FADH_2$, respectively. Per mol glucose, 2 mol $NAD^+$ are reduced in glycolysis, and 2 mol $NAD^+$ are reduced by the oxidation of pyruvate in the pyruvate dehydrogenase complex. The remainder of the $NAD^+$ and all of the FAD are reduced by the flux-generating tricarboxylic acid (TCA) cycle enzymes citrate synthase and oxoglutarate dehydrogenase. Via the regeneration of guanine triphosphate (GTP), the TCA cycle leads to the nonoxidative phosphorylation of 2 mol ATP per mol glucose. In total, per mol glucose, 20 hydrogen ion equivalents are extruded from the mitochondrial matrix and join four hydrogen ion equivalents generated in the cytosol. The 24 hydrogen ion equivalents provide the driving force for the rephosphorylation of ADP. In this way, 3 mol ATP for each mol NADH oxidized to $NAD^+$, and 2 mol ATP for each mol $FADH_2$, are formed from ADP and $P_i$ by the ATP synthase.

### Cytochrome Oxidation and the Electron Transport Chain

Respiration is defined as the oxidation of cytochrome *c* by molecular oxygen which serves as the ultimate electron acceptor. Oxide ions are generated in this reaction catalyzed by the cytochrome $a, a3$ complex (complex IV), commonly known as cytochrome *c* oxidase. The transfer of protons into the matrix allows hydrogen and oxide ions to form water. While oxygen is provided by the circulation, reduced cytochrome *c* is regenerated by the NADH and $FADH_2$ adenine dinucleotides in complex near-equilibrium reactions catalyzed by proteins collectively functioning as a cytochrome reductase. In these reactions, hydrogen ions escape through complex IV to the outside of the inner membrane.

Although the mitochondrial oxygen tension is a function of the relationship between the delivery of oxygen and the rate of cytochrome *c* oxidation, its influence on oxidative metabolism in vivo is poorly understood. The "near-equilibrium" hypothesis of Erecinska et al. (1974) and Erecinska and Wilson (1982) assigned the flux-generation to the irreversible reaction between oxygen and cytochrome *c*, catalyzed by the cytochrome oxidase (Wang and Oster 1998; Springett et al. 2000). The "near-equilibrium" status refers to the entire electron transport chain with the exception of the cytochrome *c* oxidase reaction and hence also to the ATP synthesis, catalyzed by the proton-driven F-type ATPase in the inner membrane of the mitochondrial cristae, named for its discoverer Ephraim ("F") Racker (Pullman et al. 1960), and now known as ATP synthase.

Cytochrome *c* has been calculated to be 95% saturated at the oxygen tension prevailing in mitochondria in human brain in vivo, but the sensitivity to variations of oxygen tension is much greater than implied by this occupancy, because diffusion of oxygen from the microvessels is impaired at lower capillary oxygen tensions, as discussed earlier. The "near-equilibrium" hypothesis, therefore, requires that the maximum reaction rate ($V_{max}$) or apparent affinity ($P_{50}$), or both, of cytochrome oxidase toward oxygen be adjusted in response to the $[ADP][P_i]/[ATP]$, or energy charge, ratio in the cytosol, to allow the enzyme to maintain an adequate rate of ATP synthesis. The hypothesis predicts that increases of this ratio change the properties of cytochrome oxidase in such a way that cytochrome *c* continues to react with oxygen at the rate which matches the rate of cytosolic ATP utilization. However, as it is shown later, there is a theoretical limit to the efficacy of this adjustment, particularly when the cytosolic energy charge is unchanged (Gjedde et al. 2005a). In reality, therefore, the near-equilibrium hypothesis assigns the ultimate maintenance of oxygen consumption to the regulation of oxygen delivery, particularly in the situations in which the mitochondrial oxygen tension threatens to fall below a minimum threshold (Gjedde et al. 2005a, b).

Using rat heart mitochondria as a model, LaNoue et al. (1986) showed that near-equilibrium of oxidative phosphorylation exists when respiration is very slow (state 4) but that mitochondrial ATP synthesis occurs far from equilibrium when respiration is active (state 3). These observations show that the near-equilibrium of oxidative metabolism could fail in normally and rapidly respiring brain tissue. As a result of the imbalance between the delivery of oxygen and the cytochrome oxidase activity in these states, cytochrome *c* oxidase does not remain saturated when the mitochondrial $P_{O_2}$ declines relative to the average capillary $P_{O_2}$, $P_{O_2}^{cap}$. The imbalance can be expressed as a simple Michaelis–Menten relationship between the mitochondrial oxygen tension and the kinetic properties of the cytochrome oxidase,

$$J_{O_2} = V_{max}\,\sigma_e\,\sigma_{O_2}, \tag{8.10}$$

where $J_{O_2}$ is the net oxygen consumption, $V_{max}$ the maximum cytochrome *c* oxidase activity, and $\sigma_e$ and $\sigma_{O_2}$ are the cytochrome *c* oxidase saturation fractions for electrons and oxygen, respectively. The oxygen tension in mitochondria ($P_{O_2}^{mit}$) is the tension consistent with the observed oxygen consumption rate, given the magnitude of the apparent affinity or half-saturation constant ($P_{50}^{cytox}$) of the enzyme,

$$P_{O_2}^{mit} = \frac{P_{50}^{cytox}\,J_{O_2}}{J_{max} - J_{O_2}}, \tag{8.11}$$

where $J_{max}$ is the product $V_{max}\,\sigma_e$. It follows that the corresponding average capillary oxygen tension ($P_{O_2}^{cap}$) required to drive the delivery is given by:

$$\bar{P}_{O_2}^{cap} = \frac{J_{O_2}}{L}\left[1 + \frac{L\,P_{50}^{cytox}}{J_{max} - J_{O_2}}\right], \tag{8.12}$$

where $L$ is the oxygen diffusion capacity. Equation (8.12) shows that the rate of oxygen consumption depends rigidly on the average capillary oxygen tension and that it fails to rise above a certain threshold despite increases of the cytochrome oxidase activity, unless the affinity or diffusibility of oxygen are simultaneously adjusted. The threshold is dictated by the mitochondrial oxygen tension and is reached when the tension declines below the level associated with sufficient oxygen saturation of the cytochrome oxidase. Only elevations of the oxygen diffusion capacity, e.g., by capillary recruitment, or the mean capillary oxygen tension, by increased blood flow, allow the rate of oxygen consumption to rise above this threshold (see section on Substrate Delivery During Activation).

Calcium ions play a role in the oxidation of pyruvate by activating the mitochondrial enzymes mediating the non-equilibrium and flux-generating reactions (pyruvate dehydrogenase complex, citrate synthase, $NAD^+$-linked isocitrate dehydrogenase, and the 2-oxoglutarate dehydrogenase complex), which supply the NADH and $FADH_2$ for the mitochondrial complexes I and II (Denton and McCormack 1985). In the resting state, mitochondria contain little calcium, but calcium accumulates after excitatory agonist amino acid stimulation of postsynaptic neurons (Zonta et al. 2003; Lauritzen 2005). The calcium entry is facilitated by the $Ca^{++}$-uniporter in the mitochondrial membrane and driven by the mitochondrial membrane potential established by the $H^{+}$extrusion mechanism. Rises in calcium concentration often occur as repeated spikes with steep upslopes and shallower downslopes which reach baseline during sustained excitation (Clapham 1995). A steady agonist level may induce pulsatile calcium release from calcium stores, and the frequency of this calcium pulsation apparently depends on the agonist concentration.

## 8.6 Metabolic Compartmentation

Brain tissue has several populations of cells, all of which undoubtedly play important roles in brain function. This section deals with the attempts to understand the properties of neurons and astrocytes that infer important joint roles in the regulation of the metabolic responses to excitation.

### *8.6.1 Functional Properties of Neurons and Astrocytes*

Classically, the compartmentation of brain metabolism distinguishes between the "large" and "small" pools of the excitatory amino acid transmitter glutamate (Cremer 1976; Cremer et al. 1979). There is quantitative conversion of glutamine to glutamate in the large pool, catalyzed by glutaminase, while the opposite process takes place in the small pool, catalyzed by the glutamine synthetase and fueled by ATP. The two pools of glutamate are now believed to largely represent neurons and astrocytes, respectively. In mammalian cortex, the transfer of glutamate between the two pools occurs by neuronal release of glutamate during excitation and subsequent import by neurons and glia.

Astrocytes play a critical role in the import of glutamate and potassium from the interstitial space surrounding the intrasynaptic clefts. Astrocytic processes engulf the synapses and possess excitatory amino acid (glutamate) transporters (EAAT-3 in humans, Vandenberg 1998; GLAST and GLT-1 in rats, Rothstein et al. 1996) in abundance. The transporters reside also on neurons (Vandenberg 1998) but gene "knock-out" studies show that the glial EAAT are indispensable for normal brain function, unlike the neuronal EAAT (Tanaka et al. 1997). One molecule of glutamate is symported from the extracellular space with three sodium ions. The transport increases the intracellular content of sodium and leads to export of sodium ions coupled to astrocytic import and accumulation of potassium ions. The result is a net increase of potassium in the astrocytes.

### *8.6.2 Metabolic Properties of Neurons and Astrocytes*

The different transporter properties of neurons and astrocytes were summarized earlier, showing that astrocytes possessed the low affinity MCT2 transporter. In addition, there are important enzymatic differences that cause the two groups of cells to react differently to the changes of metabolism. The enzymatic peculiarities of astrocytes include the presence of the higher affinity LDH isozyme $LD_5$s and the enzyme pyruvate carboxylase, which provides an alternative pathway of oxidative metabolism. In addition, astrocytes have comparatively larger stores of glycogen, and they consume acetate in addition to glucose. The low-affinity MCT2 transporter and high-affinity LD5 enzyme together establish a preference for export of pyruvate rather than conversion to lactate. In short, it appears that astrocytes are particularly well equipped to ensure both generation and oxidative metabolism of pyruvate, in the astrocytes, or after export to other cells.

The import of glutamate stimulates ATP hydrolysis by glutamine synthetase and $Na^+$-$K^+$-ATPase in glial cells (Pellerin and Magistretti 1994; Magistretti et al. 1999). As oxidative metabolism is 19-fold more productive of ATP than nonoxidative metabolism, it has attracted attention that the fraction of total brain energy metabolism now ascribed to non-neuronal cells (15%; Attwell and Iadecola 2002; Shulman et al. 2004) is of the same order as, or greater than, the fraction ascribed to nonoxidative metabolism of glucose (10%). If neuronal metabolism were entirely nonglycolytic, metabolism of glucose to pyruvate would be confined to the non-neuronal cells, and the oxidative metabolism of pyruvate would then have to occur in neurons, and perhaps other cells with known oxidative phosphorylation of ADP. In this situation, the energy derived from nonoxidative metabolism of 10% of the glucose supply would be no more than 0.5% of the total energy turnover (Fig. 8.4).

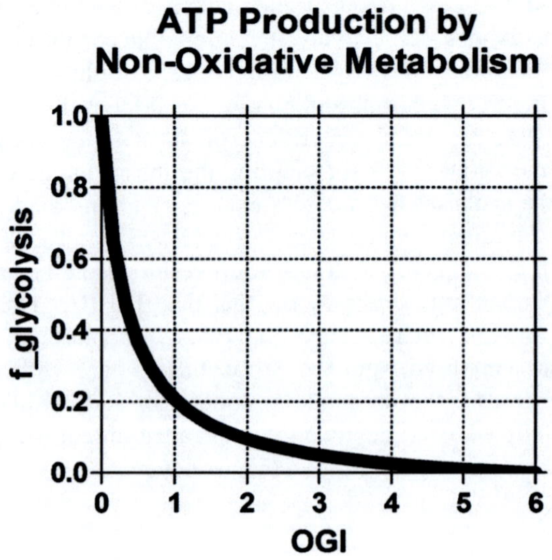


**Fig. 8.4** Illustration of behavior of (8.13) inverted to show $f_{glc}$ as function of the oxygen–glucose index (OGI). Abscissa: OGI (ratio). Ordinate: Fraction of ATP generated from that part of glycolysis that does not proceed to the TCA-cycle. At the OGI of normal human cerebral cortex of 5.3, that fraction is only 0.5%

The known differences between neurons and astrocytes nonetheless led to the hypothesis that all or most of the glucose supplied to the brain could in principle undergo glycolysis to pyruvate and lactate in astrocytes which in turn would supply all of the pyruvate to neurons which themselves used no glucose (Magistretti et al. 1999). This hypothesis was somewhat reminiscent of an early speculation that the astrocytic processes serve to siphon glucose from the microvessels to the neuron terminals (Andriezen 1893). The suggestion that glucose may enter only glial cells is not consistent with early electron microscopy by Brightman and Reese (1969) who showed that the blood–brain barrier excludes the foot-processes of astrocytes (Reese et al. 1971) and hence cannot direct glucose preferentially to glial cells. The hypothesis of phosphorylation of all of the glucose in glial cells is now replaced by a concept of a more flexible sharing of joint pyruvate and lactate pools among neurons and astrocytes, depending on the relative oxidative and glycolytic capacities of the two cell types (Gjedde et al. 2002; Hyder et al. 2005).

### Oxidative and Glycolytic Capacities

It is certain that pyruvate is oxidatively metabolized in astrocytes (Alves et al. 1995). Although increasingly, it is clear that the metabolic profile of astrocytes is not only quantitatively but also qualitatively different from that of neurons, it is also becoming evident that the most important differences are in fact quantitative. Generally, it appears that astrocytes are more glycolytic than neurons, but detailed regional measurements in mammalian brain in vivo are still lacking. Until the advent of high field strength magnetic resonance spectroscopy (Hyder et al. 2005), isolated populations of astrocytes and neurons could be studied only in vitro, where exchange of metabolites among different cell types is absent. In one such study of both cell types in isolation, Itoh et al. (2003) found that both release lactate, albeit to different extents. When both cell types release lactate, no net transfer of pyruvate or lactate from one cell type to the other is established, at the conditions present in vitro.

However, the neuronal release of lactate in vitro is probably the result of the isolation that cannot be an accurate model of the potential exchanges of pyruvate and lactate under the more active conditions prevailing in vivo. Thus, neurons and astrocytes can release pyruvate and lactate at the lower activity in vitro and still require net uptake of either at the higher activity in vivo. The significance of this point depends on the individual glycolytic and oxidative capacities of the two cell types, and on the extent to which these capacities undergo change during functional activity in vivo.

In a slightly different approach, Silver and Erecinska (1997) studied astrocytoma cells in vitro and found that 25–32% of the ATP production in these cells is glycolytic in vitro. A simple relationship relates the percentage of glycolytic ATP production to the Oxygen-Glucose Index (OGI),

$$\mathrm{OGI} = \frac{6\,(1 - f_{\mathrm{glc}})}{1 + 18\, f_{\mathrm{glc}}}, \tag{8.13}$$

where $f_{\text{glc}}$ is the fraction of glycolytic ATP production. This dependence of $f_{\text{glc}}$ on OGI is illustrated graphically in Fig. 8.4. Recent indirect studies in vivo (Nehlig et al. 2004; Hyder et al. 2005) suggest that the $f_{\text{glc}}$ fraction of astrocytes is closer to 0.20 than 0.25–0.32 in vivo. An $f_{\text{glc}}$ fraction of 0.20 indicates an OGI of 1. The glucose metabolic rate associated with this OGI can be calculated from the oxidative metabolism of the tissue,

$$J_{\text{O}_2}^{A} = f_{\text{O}_2}^{A}\, J_{\text{O}_2}, \tag{8.14}$$

where $J_{\text{O}_2}^{A}$ is the oxygen metabolic rate of the astrocytic compartment ($A$), and $f_{\text{O}_2}^{A}$ is the fraction of oxidative metabolism assigned to the astrocytic compartment, and

$$J_{\text{glc}}^{A} = J_{\text{O}_2}^{A}/\text{OGI}, \tag{8.15}$$

where $J_{\text{glc}}^{A}$ is the glucose metabolic rate. The fluxes of ATP and pyruvate and the efflux of lactate can then be computed by means of (8.2–8.4).

Combining the normal (default) metabolic rate of astrocytes of 15% of the total with this estimate of the OGI of astrocytes and the average OGI of 5.3 of cerebral cortex (Table 8.3), it is evident that neurons must have an OGI of 23. This value is of course only possible if neurons import the majority of the necessary pyruvate from external sources inside or outside the tissue. When astrocytes or extracerebral sources of lactate supply pyruvate to a joint tissue pool, neurons can draw a sizeable fraction of their needs from this pool. Estimated values of this relationship are given in Table 8.6 and illustrated in Fig. 8.5 and Fig. 8.6.

Neurons and astrocytes occupy similar volumes of cerebral cortex (50% and 33%, respectively, Norenberg et. al. (1994)). It is noteworthy therefore that the circumstances listed in Table 8.6 are consistent with a fourfold higher rate of production of pyruvate in astrocytes than in neurons, as shown by Itoh et al. (2003) in vitro.

Calculation shows that the oxidative capacity of astrocytes is low, because both of an innately higher capacity for glycolysis, and of a low rate of work. Thus, neurons remain the major sites of oxidative rephosphorylation of ATP, accounting for 80% of the oxidative metabolism but only 20% of the glycolytic metabolism of brain tissue (Silver and Erecinska 1997; Shulman and Rothman 1998; Sokoloff 1999; Hassel and Bråthe 2000; Itoh et al. 2003; Hyder et al. 2005). The estimated distribution of glycolytic and oxidative metabolism of neurons and astrocytes are shown in Fig. 8.6.

**Table 8.6** Average estimates of neuronal and astrocytic metabolism in brain

| | Flux (μmol/g/min) | | |
|---|---|---|---|
| Metabolite | Astrocytes | Neurons | Total |
| ATP | 1.92 | 8.28 | 10.2 |
| Glucose | 0.24 | 0.06 | 0.30 |
| Oxygen | 0.24 | 1.36 | 1.60 |
| OGI | 1 | 23 | 5.3 |
| Pyruvate generation | 0.48 | 0.12 | |
| Pyruvate consumption | 0.08 | 0.45 | |
| Lactate production | | | 0.07 |

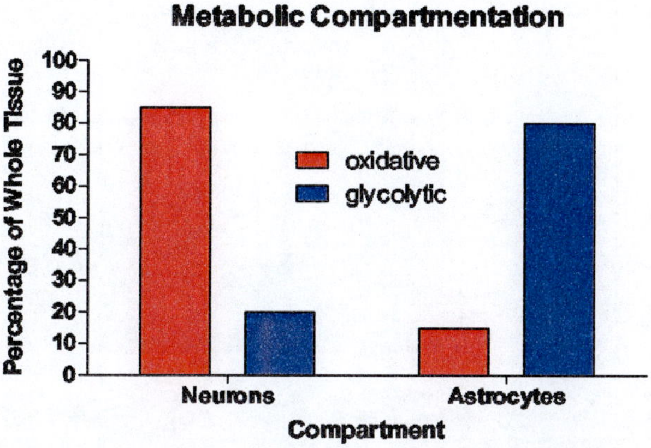


**Fig. 8.5** Metabolic compartmentation of ATP regeneration in neurons and astrocytes, modified from Silver and Erecinska (1997). The figure shows the proportions of glucose and oxygen metabolism carried out in neuronal and glial tissue compartments, respectively. The percentages were calculated from oxygen–glucose indices of glial and neuronal compartments according to (8.13) on the basis of glycolytic ATP synthesis fractions reported by or calculated from Silver and Erecinska (1997), Nehlig et al. (2004), and Hyder et al. (2005)

The issue of the rates of glycolysis and oxidative phosphorylation in neurons and glial cells hinges on the mechanisms underlying the estimates of the relative glycolytic and oxidative capacities of the respective cellular compartments. The relative capacities arise from the differential activities of hexokinase/phosphofructokinase and cytochrome oxidase in the two cell groups, respectively, which in turn appear to depend on the time constants of the metabolic fluctuations as well as on the average metabolic activity. Low time constants are likely to favor glycolysis and low average oxidative rates, while high time constants favor high oxidative rates, much as in muscle cells (Wong-Riley 1989). Although it has not been possible to obtain direct evidence that glutamate import actually enhances glycolysis in astrocytes in vivo (Hertz et al. 1998), there is increasingly firm evidence that glycolytic and oxidative rates of metabolism have different proportions in astrocytes and neurons under normal physiological conditions in vivo, such that the most of the glycolytic capacity resides in astrocytes (80% vs. 20% in neurons).

## Metabolite Cycling

While neurons and astrocytes differ with respect to metabolite enzyme and transporter distribution, there is no evidence that the pools of glucose, pyruvate, and lactate, themselves are compartmentalized in distinct populations of neurons and glia. Such a separation would require an essential impermeability of cell membranes to these substrates, or active transport, which is absent from brain tissue for these substrates. Thus, there is little direct evidence of a fundamental difference between

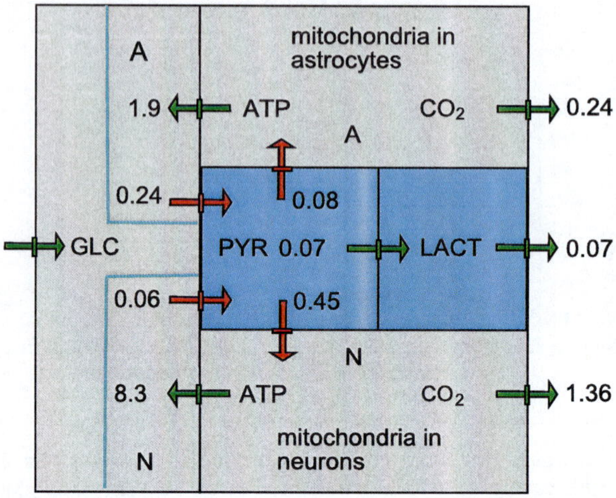


**Fig. 8.6** Estimated glycolytic and oxidative fluxes of glucose, pyruvate, lactate, $CO_2$, and ATP, among extracellular fluid and neuronal and astrocytic compartment in human cerebral cortex. Glucose, pyruvate and lactate are shown as occupying single individual pools (*light blue*). Glucose feeds into the pyruvate pool in proportion to hexokinase capacities of the two cellular compartments, and pyruvate supplies the oxidative metabolism of the two cell type (*grey*) in proportion to their oxidative capacities

the ability of neurons and glia to oxidize pyruvate, except for the greater activity of glycolytic enzymes in glial cells, and the greater density of mitochondria with the greater activity of cytochrome oxidase in neurons.

Unless pyruvate were strictly compartmentalized (which is ruled out by the abundance of MCT-1 and MCT-2), it is kinetically impossible for neurons to prefer pyruvate of non-neuronal origin, whether directly or indirectly imported, over pyruvate of neuronal origin, despite claims to the contrary (Poitry-Yamate et al. 1995). The evidence summarized earlier suggests that the two populations of cells contribute differentially to joint pyruvate and lactate pools when glycolysis is stimulated to a greater extent than oxidative metabolism, but the extent of the differential distribution would be a function of the degree of activation of the two cell types.

The simplified model of metabolite pathways gleaned from these considerations is shown in Fig. 8.7, in which the control points of the differential glycolytic and oxidative capacities are numbered 1 and 2 for the hexokinase/phosphofructokinase (HK/PFK) and pyruvate dehydrogenase (PDH) steps, respectively. The capacities of the separate PDH steps in neurons and astrocytes are most likely related to the number of calcium-stimulated mitochondria, while the capacities of the HK/PFK steps are subject to the temporal requirements of glutamate transport and metabolism in astrocytes and glycolysis in neurons. This analysis suggests that the functional ranges of the activities are very different in the two cell types, in keeping with the very different functional contingencies facing the two cell types in normally functioning brain tissue.

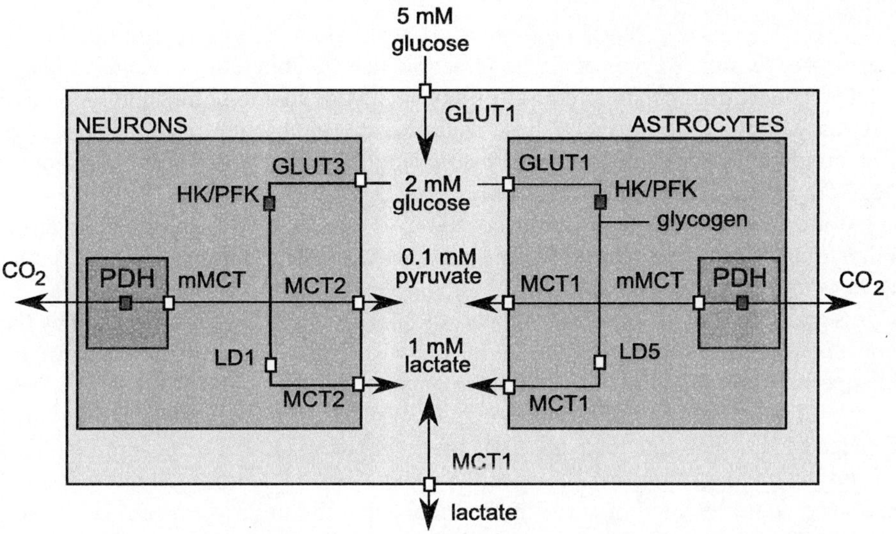


**Fig. 8.7** Metabolic compartmentation of control points of hexokinase and phosphofructokinase enzymes, and pyruvate dehydrogenase complex (*dark square points*), and near-equilibrium transporters and enzymes responsible for branching among pools (*light square points*). Metabolites linked by near-equilibrium transporter and enzyme reactions are shown with *bidirectional arrows*. Metabolic fates of pyruvate and lactate depend on relative preferences for each transporter or enzyme, which are regulated by affinities in relation to pyruvate and lactate concentrations

The evidence cited earlier shows that the glial energy metabolism in general is low compared to neuronal metabolism in cerebral cortex, where glial cells occupy 33% of the volume but generate only 20% of the ATP. Per unit cell volume, this amounts to a threefold difference in energy metabolism in normally functioning brain. Sustained stimulation of glial metabolism taxes the oxidative capacity of astrocytes beyond its normal limit. During some forms of excitation, for example, the neuronal firing frequency increases substantially and may require removal of excitatory neurotransmitters by glial cells at rates incompatible with the time constants of their oxidative capacity. By analogy, the presence of the LDH subtype $LD_5$ in astrocytes also suggests that the average oxidative metabolic rate is low, while the glycolytic rate is subject to fluctuations in the form of brief, intense activations.

## 8.7 Activation

In neuroscience, activation and deactivation are conventional labels of changing brain function. Understanding brain function is the goal of the neuroscientific explorations of brain activity, but it is sometimes not clear which specific hypothesis the explorations actually test. To be sure, activation and its opposite represent a range of observations of a change of any number of different physiological measures of brain activity, such as action potential frequency, local field potentials, membrane polarity, ATP turnover, blood flow, glucose consumption, oxygen consumption, tissue oxygen tension, oxygen extraction fraction, hemoglobin saturation, absolute and relative concentrations and amounts of oxy- and deoxyhemoglobin, and blood and vascular volumes. Collectively, these measures can contribute to a comprehensive description of activity in brain tissue but no single measure can be said to unequivocally represent activation, or even brain function, per se.

At the core of the conundrum is consciousness, of course. The emerging discipline of neuroenergetics focuses on measures of the energy turnover of brain tissue as keys to the functional integrity of brain tissue, including consciousness. The insights gleaned from the accomplishments of this discipline have been limited by the convention that changes of energy turnover are a consequence of, or response to, changes of brain function, rather than the opposite (Fodor 1983, 2000).

It is possible that this relationship in reality could be turned around, as suggested by Fig. 8.1. It is possible that conscious brain function is a phase transition induced by sufficient energy turnover in specific populations of neurons. Much empirical evidence could be cited in favor of the theory that the unitary experience of consciousness is related to the global level of energy turnover in the cerebral cortex rather than to activity in specific regions. Among this evidence is the correlation between reported magnitudes of global brain energy metabolism and levels of consciousness (Shulman et al. 2004).

### 8.7.1 Ion Homeostasis During Activation

The sodium theory explains how neuronal excitation increases the work of the brain by increasing the leakage of ions across cell membranes, and how depolarization of neuronal membranes leads to increased oxygen uptake (Erecinska et al. 1991). This empirical finding is important, because cells in theory could undergo steady-state de- or repolarization without a change in ATP utilization when an average steady-state change of sodium conductance is paired with substantial declines of potassium and chloride conductances (Gjedde 1993b). Table 8.2 illustrates this point by listing the identical ATP requirements at two different membrane potentials (−65 and −55 mV).

The Goldman-Hodgkin-Katz constant field equation predicts the changes of the membrane potential that occur when ion permeabilities change. Using the permeabilities calculated in Table 8.2, the corresponding membrane potentials were calculated as shown in Fig. 8.8. The membrane potentials reflect changes of both sodium and potassium (adjusted to preserve the 3:2 flux ratio required by the P-type $Na^+$-$K^+$-ATPase), or chloride, permeability. The resulting $Na^+$-$K^+$-ATPase activity was calculated as the flux required to preserve ion homeostasis. The glucose demand was in turn calculated as the nutrient delivery required to compensate for a steady-state ATPase activity of this magnitude by oxidative phosphorylation. Figure 8.8 reveals the metabolic consequence of sodium and potassium leakage: For an arbitrary depolarization from −70 to −60 mV, the ATP turnover must

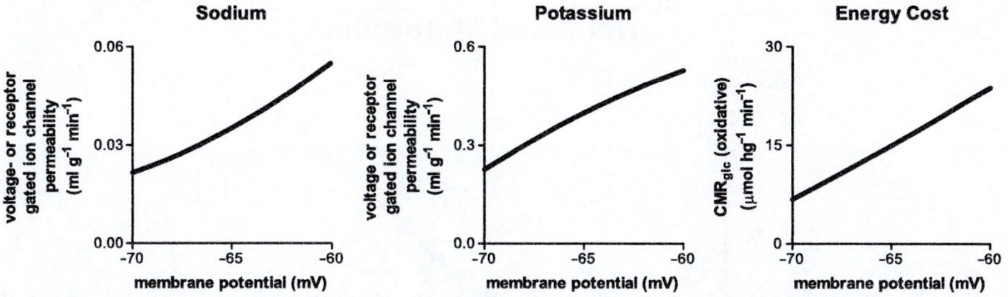


**Fig. 8.8** Energy cost of depolarization, redrawn from Gjedde (1993b). *Left and Center Panels*: Steady-state neuronal membrane potential change as function of altered sodium and potassium ion membrane permeabilities at 0.549 ml g$^{-1}$ min$^{-1}$ constant chloride ion permeability. Ordinates: Ion permeability [ml g$^{-1}$ min$^{-1}$]. Abscissae: Membrane potential [mV], calculated from the Goldman-Hodgkin-Katz constant field equation. *Right Panel*: Steady-state metabolism permitting membrane depolarization of the magnitude dictated by the increased sodium and potassium permeability. Abscissa: Membrane potential [mV], calculated from the Goldman-Hodgkin-Katz constant field equation on the basis of chosen changes of sodium and potassium ion permeabilities. Ordinate: Steady-state glucose consumption [µmol hg$^{-1}$ min$^{-1}$], calculated from steady-state ion flux, assuming constant chloride ion permeability

increase fourfold from 2.5 to 9 µmol g$^{-1}$ min$^{-1}$ to preserve ion homeostasis. With a sodium-ion-transport-stimulated glucose metabolic rate of 0.15 µmol g$^{-1}$ min$^{-1}$, the total glucose demand would be expected to increase from 0.2 to 0.4 µmol g$^{-1}$min$^{-1}$ to fuel this turnover of ATP. For further depolarization to a firing level of −55 mV, the glucose supply would have to increase to as much as 0.6 µmol g$^{-1}$ min$^{-1}$ to fuel an ATP turnover of 20 µmol g$^{-1}$ min$^{-1}$. In the absence of oxygen (or with no increase of oxygen consumption), the glucose supply would have to increase to as much as 10 µmol g$^{-1}$ min$^{-1}$, a 30-fold increase, to cover the same need for ATP.

## *8.7.2 Brain Energy Metabolism During Activation*

During excitation of brain tissue in vivo, local increases in metabolism reach 100% or more of the metabolism of the normal default state, depending on the intensity of stimulation (Bowers and Zigmond 1979; Yarowsky and Ingvar 1981; Kadekaro et al. 1985; Shulman and Rothman 1998). The evidence for the increase of brain energy metabolism in vivo comes in three different forms, i.e., as global steady-state changes brought on by general reductions or increases of brain activity; as localized responses to specific stimulation, mostly of cerebral cortex; and as changes of substrates and metabolites in the circulation and in brain tissue.

### Global Steady-State Changes of Energy Metabolism

Average global steady-state changes are given in Fig. 8.9, in which recent magnetic resonance spectroscopic and allied studies in rodent cerebral cortex show a linear

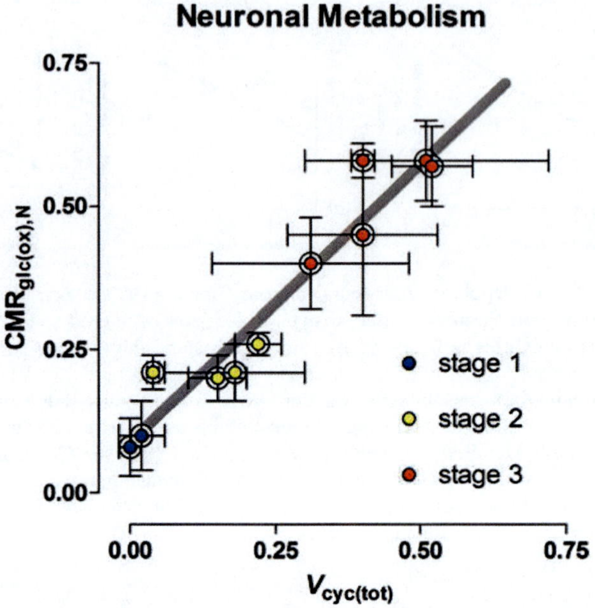


**Fig. 8.9** Stages of rat brain oxidative metabolism vs. rate of glutamate cycling measured by Sibson et al. (1998), Choi et al. (2002), Henry et al. (2002), de Graaf et al. (2004), Patel et al. (2004), Oz et al. (2004), modified from Hyder et al. (2005). Abscissa: Combined neuronal and astrocytic glutamate turnover rate ($\mu$mol $g^{-1}$ $min^{-1}$). Ordinate: Oxidative metabolism of glucose in neurons ($\mu$mol $g^{-1}$ $min^{-1}$). Functional stages were inferred from type and level of general anesthesia used in each experiment. Slope of line is 0.97$\pm$0.09 (unitless), and $y$-intercept is 0.09$\pm$0.03 $\mu$mol $g^{-1}$ $min^{-1}$

relationship between calculated cycling rates of glutamate between the larger (neuronal) and smaller (glial) metabolic compartments and the calculated turnover rates of the tricarboxylic acid cycle in the larger (neuronal) compartment. In Fig. 8.9, these findings were assigned to hypothetical functional stages on the basis of the brain states of the animals, as outlined in Fig. 8.1.

The relation between glutamate cycling and oxidative glucose metabolism shown in Fig. 8.9 is consistent with the oxidative metabolism of one mole of glucose for each mole of glutamate released and recycled. The corresponding relations for oxygen and total glucose consumption (oxidative and nonoxidative) in human cerebral cortex are predicted in Fig. 8.10, including the functional stages that they are likely to reach. The prediction is based on the two speculative but plausible hypotheses that, first, the quantity of glutamate released and recycled is proportional to the percentage of active cells or time associated with activation, and, second, that the oxygen–glucose index declines linearly from 6 at stage 1 to 5.2 at stage 4, for reasons to be discussed later.

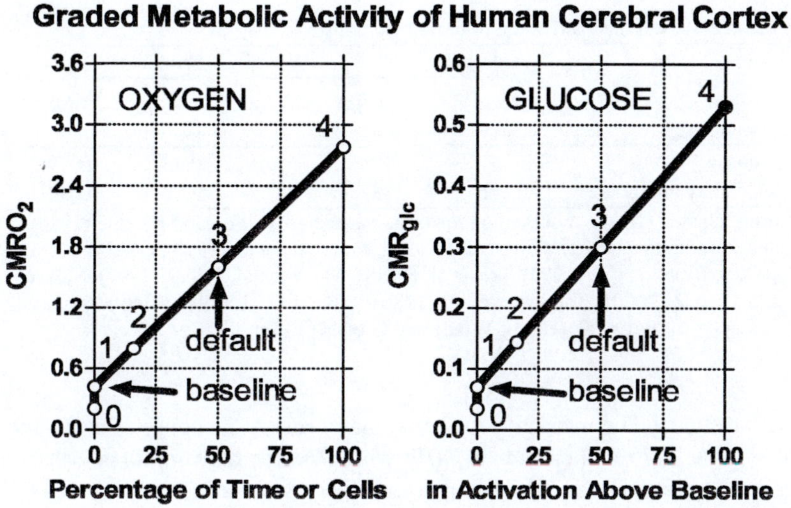


**Fig. 8.10** Model of stage transitions of oxidative and glycolytic metabolism in human cerebral cortex during activation. Abscissa: Percentage of time or cells associated with activation. Ordinates: Oxygen or glucose metabolic rates in units of μmol $g^{-1}$ $min^{-1}$. *Left panel*: Increments of oxygen consumption using the formula, $J_{O_2} = 0.42\ (1 - A) + 2.78\ A$ (μmol $g^{-1}$ $min^{-1}$), and *Right panel*: Increments of glucose consumption using the formula, $J_{glc} = 0.07\ (1 - A) + 0.53\ A$ (μmol $g^{-1}$ $min^{-1}$), where $A$ is the percentage of activation

### Localized Changes of Energy Metabolism in Response to Stimulation

While global changes can be substantial, focal changes brought on by specific stimulation of cerebral cortex generally are small, suggesting that the cortex normally functions at the same level most of the time. Localized activation responses from a number of stimulation studies of human brain are summarized in Table 8.7 (Gjedde et al. 2002), divided into two fundamentally different types of stimulation. Simple kinds of stimulation generally show little or no change of oxygen consumption (Fox and Raichle 1986; Fox et al. 1988; Seitz and Roland 1992; Ohta et al. 1999). For example, with the single-inhalation method of measuring oxygen consumption, Fujita et al. (1999) compared changes of blood flow and oxygen consumption during 30 min of vibrotactile stimulation of one hand's fingers. In primary sensory cortex, the blood flow change was 18% at the onset of stimulation and still 11% after 20 min of stimulation, but the oxygen consumption failed to increase for as long as 30 min. With more complex stimulation, significant increases of oxygen consumption are noted together with the change of blood flow (see references to Table 8.7).

The conclusion that there are more than one kind of response of oxidative metabolism to brain activation is consistent with a primary somatosensory response, in which the rise of oxygen consumption averages only 10% of the rise of blood flow, and a motor or more complex secondary somatosensory response, in which the rise of oxygen consumption averages 75% of the rise of blood flow, both relative

**Table 8.7** Neuronal activation of brain metabolism

| | Supply | | | Products | |
|---|---|---|---|---|---|
| Stimulus | ΔCBF | $\Delta J_{glc}$ [%] | $\Delta J_{O_2}$ | $\Delta J_{ATP}$ [μmol $g^{-1}$ $min^{-1}$] | $\Delta J_{lact}$ |
| Primary | 29 | 19 | 3.4 | 0.43 | 0.089 |
| Secondary and motor | 36 | 38 | 25 | 2.89 | 0.090 |

From Gjedde (2001), modified on the basis of studies by Fox and Raichle (1986), Seitz and Roland (1992), Fujita et al. (1999), Kuwabara et al. (1992), Ginsberg et al. (1988), Ribeiro et al. (1993), Fox et al. (1988), Marrett and Gjedde (1997), Vafaee and Gjedde (2000), Katayama et al. (1986), Roland et al. (1987, 1989), Raichle et al. (1976), Iida et al. (1993), Vafaee and Gjedde (2000)

to their baselines. The average primary somatosensory response is a 3% increase of $CMR_{O_2}$ for a 29% increase of CBF, while the average motor and secondary somatosensory response is a 25% increase of $CMR_{O_2}$ for a 36% increase of CBF, as illustrated in Fig. 8.11.

The discrepancy between the results of the two different types of sensorimotor stimulation listed in Table 8.7 suggests that the ultimate increase of oxygen consumption depends significantly on the biochemical peculiarities of the neuronal pathway mediating the response to the stimulus (Borowsky and Collins 1989). In brain as well as in muscle cells, the cytochrome oxidase activity of populations of cells is regulated by the habitual energy requirement, averaged over longer periods of time (Pette 1985; Hevner et al. 1995). In muscle cells, neural input regulates the categorization of muscle cells into types I, IIa, or IIb. Changing the oxidative capacity requires sustained stimulation for an extended period of time. Conversely, brief transient increases of energy metabolism above the habitual level of activity are not accompanied by commensurately increased oxygen consumption. In brain, this consideration leads to the conclusion that the two different kinds of response are related to the known differential oxidative and glycolytic responsiveness of neurons and glial cells.

### Vascular and Tissue Metabolite Changes During Activation

The changes governing the transient changes of metabolite pools during activation involve a vast number of reactants, but the most important, because of their concentrations or ease of measurement, are glucose, glycogen in glial cells, lactate, and $NAD^+$and NADH. The changes must be compared to the in- and efflux of glucose, lactate, and oxygen during the activation to give an accurate view of the dynamic relationships. The relationship between these metabolites can be reduced to the formula,

$$\Delta[\text{glc}] + \Delta[\text{glyc}] + \frac{1}{2}\Delta[\text{lact}] = \overline{J}_{\text{glc}} \left(1 + \frac{\overline{\text{LGI}}}{2} - \frac{\overline{\text{OGI}}}{2}\right) \Delta T, \qquad (8.16)$$

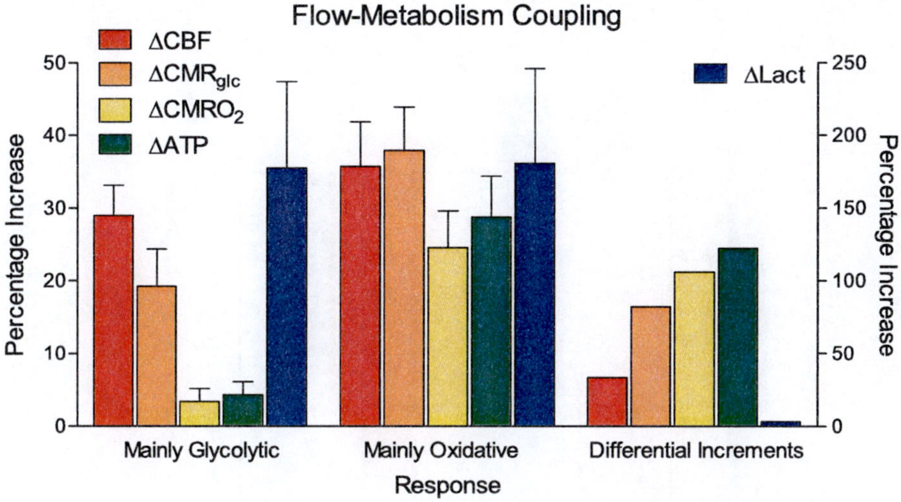


**Fig. 8.11** Flow-$CMR_{O_2}$ and flow-glycolysis coupling during stimulation of comparatively nonoxidative and oxidative cells: Relative increases of glycolysis, blood flow, and oxidative metabolism, estimated for two categories of response to stimulation (simple or primary somatosensory stimulation ("white" response); and complex or secondary somatosensory and motor ("red" response) listed in Table 8.7. Note similar flow-glycolysis coupling in the two stimulations. Also shown are rates of incremental ATP and lactate production during excitation. Stimulus and/or cells fall into two metabolic categories, one with average incremental ATP/lactate ratio of less than 10 and the other with average ratio of 300. Ordinate: Percentage increments of variables listed in graph

where the right side of the equation is an approximate Metabolite Accumulation Index (MAI), the terms [glc], [glyc], and [lact] are the tissue concentrations of glucose, glycogen, and lactate, $J_{glc}$ is the net rate of glucose transport across the blood–brain barrier, $\overline{LGI}$ is the ratio of net lactate and glucose rates of transfer across the blood–brain barrier, also known as the lactate-glucose index, and $\overline{OGI}$ similarly is the oxygen-glucose index, in this case the ratio of net rates of oxygen and glucose transfer across the blood–brain barrier, both as weighted averages over the period of observation $\Delta T$. Only by taking at least these factors into account can changes of any of them be properly evaluated during the transient changes of activation. The equation also allows metabolite changes to be estimated by measurements in the circulation, and vice versa. The following two examples of the use of this equation, one in rodents and the other in humans, are representative of this class of studies.

Dienel and Cruz (2004) reviewed metabolite and circulatory changes in rat brain during activation by manual stroking of the rat's abdomen. The measurements allowed comparison to be made between the accumulation of the metabolites and the metabolite accumulation index (MAI) included in (8.15). In Fig. 8.12, the comparison reveals that metabolites accumulate (particularly glucose and lactate) during the 5 min activation period while they decline (particularly glycogen) in the subsequent 15 min recovery period. The OGI declines during the activation and rises above normal when the rats are again at rest. These changes can be understood only by

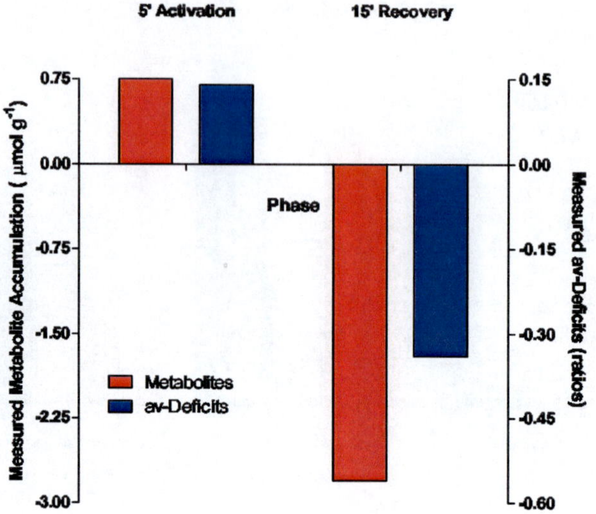


**Fig. 8.12** Changes of Metabolite Accumulation Index (MAI) and measured metabolites (metab) during 5 min stimulation and 15 min recovery of rat brain in vivo, calculated from summary published by Dienel and Cruz (2004) according to (8.16). *Right ordinate*: Metabolite-Glucose Indices according to right side of (8.16) (1 + LGI/2 - OGI/6). *Left ordinate*: Measured changes of metabolites according to left side of (8.16) ($\delta$[glc] + $\delta$[glyc] + $\delta$[lact]/2) in units of $\mu$mol $g^{-1}$. The graph shows that metabolites accumulate during stimulation and decline during recovery, and that this change is reflected in the arterio-venous deficits of glucose and lactate

invoking differential activation of cell types with different oxidative and glycolytic capacities.

Ide et al. (2000) used the measurements of arteriovenous differences across brain in humans, shown in Fig. 8.13, to similarly evaluate metabolite accumulation during and after exercise. The calculations confirm that the MAI rises substantially during the exercise and returns slowly toward zero over the next half hour. The OGI declines at the end of the exercise period, only to rise above normal after the exercise.

Kasischke et al. (2004) used fluorescence imaging to determine changes of NADH in slices of hippocampus exposed to 32 Hz electrical stimuli. The NADH signal from dendritic mitochondria declined during the first 10 s of the 20 s period of stimulation, indicating conversion of NADH to $NAD^+$, and then returned to baseline at the end of the 40 s recovery period. The NADH signal from astrocytic cytosol increased toward the end of the 20 s stimulus and continued to rise for the following 20 s of recovery, indicating greater generation than metabolism of pyruvate (oxidative as well as nonoxidative) in astrocytes in this period. Above all, these findings reveal a failure of oxidative metabolism to increase in parallel with stimulation, beyond the brief initial exhaustion of the putative oxygen reserve in mitochondria, whether because of limited oxygen delivery to the slice, or because of a constitutive inability of mitochondria to match the increased generation of pyruvate. The beginning recovery of NADH in dendritic mitochondria in the second half of the stimulation is evidence of increased pyruvate oxidation in the neurons, but it is still

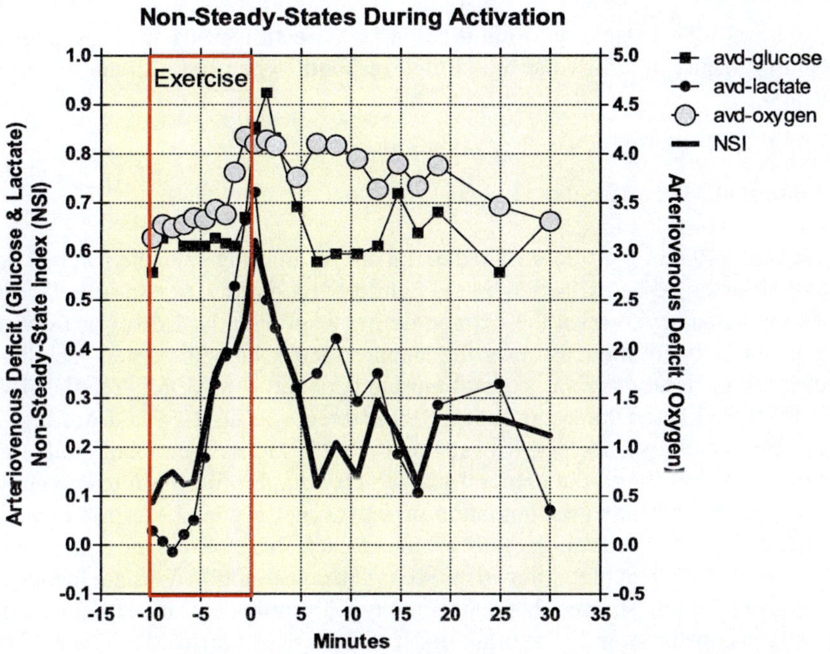


**Fig. 8.13** Changes of Metabolite Accumulation Index (MAI) and measured arterio-venous deficits of oxygen, lactate, and glucose during 10 min strenuous exercise and 35 min recovery of human whole-brain in vivo, calculated from results published by Ide et al. (2000) according to (8.16). *Right ordinate*: Arterio-venous deficits of oxygen. *Left ordinate*: Glucose and lactate arterio-venous deficits and measured changes of metabolites according to left side of (8.16) ($[\delta[glc] + \delta[glyc] + \delta[lact]/2]/\Delta T$) in units of μmol $g^{-1}$ $min^{-1}$. The graph shows that metabolites accumulate during stimulation and normalize during recovery

unclear where this pyruvate is generated. There is no indication of the fate of lactate in this study. Together, the changes of energy metabolism and metabolites show that metabolites accumulate during stimulation and decline during recovery. This paradoxical conclusion implies that oxygen availability declines during stimulation and returns to normal or above normal during recovery.

### 8.7.3 *Substrate Delivery During Activation*

The measurements underlying Table 8.7 suggest that the relation between the changes of blood flow and the generation of additional ATP is markedly variable during activation, with ratios between the relative changes ranging from unity to 20. It is not known which specific aspect of neuronal excitation is most critically dependent on the blood flow increase but surprisingly the findings suggest that the blood flow *changes* are unrelated to an immediate satisfaction of increased oxygen requirements for oxidative phosphorylation. This section deals with the attempts to identify agents to which the blood flow responds when brain tissue undergoes activation.

### Regulation of Microvascular Oxygen

Gjedde et al. (1991b) considered whether the discrepancy between the two reactions to stimulation could be explained by the failure of blood flow to rise sufficiently in some circumstances to raise the oxygen delivery sufficiently. However, restricted oxygen supply is not the reason for the inappropriately named oxygen "debt" during exercise of tissues such as skeletal muscle (Connett et al. 1985; Ye et al. 1990; McCully et al. 1991; Ohira and Tabata 1992; Reeves et al. 1992). In testing this proposal in healthy volunteers, Kuwabara et al. (1992) (also see Ohta et al. 1996; Fujita et al. 1999) found the cerebral oxygen consumption in brain to remain unchanged during vibrotactile stimulation of sensorimotor cortex, despite increased blood flow and increased capillary diffusion capacity. Because oxygen tension in the tissue must go up when blood flow rises if oxygen consumption does not increase as well, this experiment showed that deficient oxygen supply does not explain the limited oxygen consumption in the brain under these circumstances. However, Vafaee et al. (2000) tested two patients with mitochondrial encephalopathy and found that the muted oxygen consumption response to complex visual stimulation was associated with a similarly muted blood flow response.

Possible clues to the understanding of the discrepant changes of oxygen consumption and blood flow come from studies of the kinetics of oxygen delivery to brain cells (Gjedde 1996a). The diffusion-limitation imposed by the hemoglobin binding renders oxygen transport, as reflected in the extraction fraction, somewhat insensitive to blood flow increases (Honig et al. 1992), so blood flow must increase disproportionately to raise oxygen transport. The problem is that a flow-$CMR_{O_2}$ couple of unity means the same capillary oxygen tension profile and extraction fraction at all times and in all situations and hence no change of the oxygen tension gradient in the tissue. The kinetic analysis of cytochrome oxidase activity shown in Fig. 8.14 revealed that increases of blood flow above the increase of oxygen consumption are necessary for the delivery of additional oxygen during excitation.

When the decline of the mitochondrial oxygen tension threatens to reduce the oxygen saturation of cytochrome oxidase, oxygen consumption becomes dependent on the mean capillary oxygen tension for a given capillary density (Gjedde et al. 1991b). Brain activation above this threshold therefore demands disproportionately increased blood flow to increase the mean capillary oxygen tension.

The maximum oxygen delivery capacity is the upper limit of oxygen consumption, reached when the mitochondrial oxygen tension is at the minimum level sufficient to sustain the cytochrome oxidase activity. This limit at which the average oxygen tension of capillary blood drives most of the oxygen consumption previously was thought to be approached only in situations of pathologically limited blood flow but may in fact be the rule rather than the exception. The effect of the blood flow

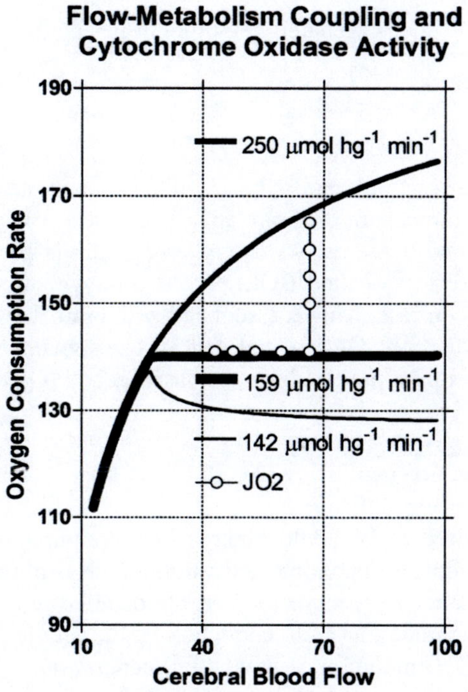


**Fig. 8.14** Oxygen consumption vs. blood flow estimated for human whole-brain according to (8.21) for three different cytochrome oxidase activities ($J_{max}$), calculated from results published by Gjedde et al. (2005a). Abscissa: Cerebral blood flow in units of ml $hg^{-1}$ $min^{-1}$. Ordinate: Oxygen consumption in units of μmol $hg^{-1}$ $min^{-1}$. Baseline oxygen consumption is 140 μmol $hg^{-1}$ $min^{-1}$, indicated by the *heavy horizontal line*. Cytochrome oxidase activities are given in the graph. Oxygen diffusibility ($L$) is 3.2 μmol $hg^{-1}$ $min^{-1}$ $mmHg^{-1}$, and the arterial oxygen concentration ($C_{O_2}^{art}$) is 8.8 mM. The cytochrome oxidase affinity for oxygen ($P_{50}^{cytox}$) as a function of cerebral blood flow is given by (8.21) when other variables are kept constant. Increases of blood flow without change of oxygen consumption (*white circles*) cause oxygen tension to rise in the tissue. At a given blood flow rate, increases of oxygen consumption require increased cytochrome oxidase activity as shown by the *white circles*, which in turn cause the oxygen tension to decline, as shown in Fig. 8.15

increase is then to raise the $J_{O_2}^{max}$. When more oxygen is needed and blood flow rises disproportionately to satisfy this need, the oxygen extraction paradoxically declines and raises the average capillary oxygen tension to the magnitude consistent with the pressure gradient required to drive the oxygen to the mitochondria.

## BOLD Signal Changes During Activation

In the absence of proportionately increased oxygen consumption, increased blood flow leads to higher capillary oxygen tensions and lower extraction fractions and reduces the fraction of deoxygenated hemoglobin in capillary and venous blood, according to (8.4),

$$\bar{P}_{O_2}^{cap} = P_{50} \sqrt[h]{\frac{2}{E_{O_2}} - 1}, \tag{8.17}$$

where the extraction fraction equals the fraction of deoxyhemoglobin in venous blood. Its decline is a measure of the relative increase of oxygenation in cerebral veins when blood flow increases. As deoxyhemoglobin is paramagnetic, its relative decline gives rise to so-called BOLD ("blood oxygenation-level-dependent") magnetic resonance contrast changes associated with brain activation (Ogawa et al. 1990a, b; Kwong et al. 1992; Ogawa et al. 1993). The signal is primarily a function of the average extraction of oxygen from the vascular bed (Gjedde et al. 1999).

## Regulation of Tissue Oxygen

If disproportionately greater blood flow increases are required to drive enough oxygen into brain tissue during functional activation then it is important to understand the mechanism that elicits the increase. Yet, the detailed relationships among the possible agents (glutamatergic and acetylcholinergic activity, synthesis of nitric oxide, extracellular accumulation of potassium, generation of lactic acid) and the functionally induced blood flow changes of brain tissue remain unknown. The elucidation of these relations is among the most pressing items on the neuroscientific agenda. Information concerning three important putative agents, nitric oxide, potassium, and prostanoids, is reviewed later.

Assuming full saturation of arterial hemoglobin and quantitatively negligible oxygen in physical solution in arterial plasma, the minimum oxygen tension in mitochondria can now be calculated as the tension commensurate with the actual delivery of oxygen, by combination of (8.4) or (8.17) and (8.5),

$$P_{O_2}^{mit} = \left( P_{50} \sqrt[h]{\frac{2}{E_{O_2}} - 1} \right) - \left( \frac{J_{O_2}}{L} \right), \tag{8.18}$$

where $P_{O_2}^{mit}$ is the average oxygen tension in mitochondria, and $L$ is the oxygen diffusibility, assumed to be constant in the absence of recruitment. Solved for the cerebral blood flow ($F$), (8.18) yields:

$$F = \frac{J_{O_2}}{2\, C_{O_2}^{art}} \left( 1 + \left[ \frac{J_{O_2} + L\, P_{O_2}^{mit}}{L\, P_{50}} \right]^h \right), \tag{8.19}$$

where $C_{O_2}^{art}$ is the arterial oxygen concentration. The equation expresses blood flow as an apparent function of the rate of oxygen consumption and the mitochondrial oxygen tension for a given arterial oxygen concentration. In this relationship, the mitochondrial oxygen tension reflects the balance between the delivery and consumption of oxygen. The tension depends on, rather than controls, the rate of oxygen consumption, according to the simple Michaelis–Menten-type kinetic expression (Gnaiger 1993; Gnaiger et al. 1998),

$$P_{O_2}^{mit} = P_{50}^{cytox} \left( \frac{J_{O_2}}{J_{max} - J_{O_2}} \right), \tag{8.20}$$

where $P_{50}^{cytox}$ is the apparent half-saturation constant of the oxygen reaction with cytochrome oxidase and $J_{max}$ is the maximum reaction rate at a given cytochrome c occupancy of the enzyme. Equation (8.20) eliminates $P_{O_2}^{mit}$ from (8.19) and yields an expanded expression of the relations among the primary factors affecting oxidative metabolism,

$$F = \frac{J_{O_2}}{2\, C_{O_2}^{art}} \left( 1 + \left( \left[ 1 + \frac{L\, P_{50}^{cytox}}{J_{max} - J_{O_2}} \right] \left[ \frac{J_{O_2}}{L\, P_{50}} \right] \right)^h \right), \tag{8.21}$$

where the independent variables are $F$, $P_{50}^{cytox}$, and $J_{max}$, and the dependent variable is $J_{O_2}$, when $C_{O_2}^{art}$ and $L$ are constants and arterial blood remains fully saturated. The independent variables may be linked, of course. Equation (8.22) combines the Hill equation of oxygen saturation of hemoglobin (Hill 1910, 1913) and the Michaelis–Menten equation of substrate occupancy of an enzyme (Michaelis and Menten 1913).

Somewhat surprisingly, (8.19) describes a compulsory flow-limitation that dictates the oxygen consumption associated with a given blood flow rate when all other parameters are constant. A change of any independent variable by necessity is reflected in the rate of oxygen consumption. No change of the oxygen consumption after a change of any of the independent variables implies a compensatory change of at least one other independent variable. Thus, invariant $J_{O_2}$ with variable $F$ most probably implies compensatory change of $P_{50}^{cytox}$ or $J_{max}$ or both.

Figure 8.14 illustrates the analysis of (8.19) published by Gjedde et al. 2005a. Previous studies of oxygen supply and delivery suggested that the affinity of cytochrome c oxidase for oxygen changes inversely with the oxidative metabolism of a tissue and thus preserves the sensitivity of the cytochrome c oxidase reaction to changes of the maximum velocity (see Brown (2001)) for a review of these findings). Gjedde et al. (2005a) hypothesized that a similar change explains the invariant cerebral oxygen consumption measured during moderate changes of blood flow to the brain, exemplified by the middle line of Fig. 8.14. Changes of the cytochrome c oxidase activity ($J_{max}$) at a given rate of blood flow would then raise or lower oxygen consumption as shown in the upper and lower curves, of which the upper curve is the completely flow-limited extreme, above which the oxygen consumption cannot rise. The effect of the respective changes of $F$ or $J_{max}$ is to raise or lower tissue and mitochondrial oxygen tensions according to (8.21). Predicted changes of oxygen consumption and $P_{O_2}$ are shown in Figs. 8.14 and 8.15 in which first blood flow and then cytochrome oxidase activity rise as shown in Fig. 8.15, causing

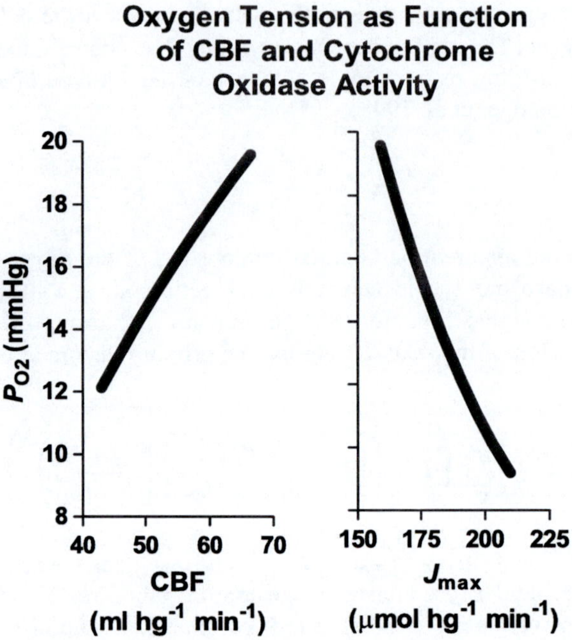


**Fig. 8.15** Biphasic change of mitochondrial oxygen tension in response to changes of blood flow and cytochrome oxidase activity ($J_{max}$) shown in Fig. 8.14, calculated from (8.20) and (8.21). Abscissas: Cerebral blood flow (ml $hg^{-1}$ $min^{-1}$) and cytochrome oxidase activity (μmol $hg^{-1}$ $min^{-1}$). Ordinate: Mitochondrial oxygen tension (mmHg). Note that oxygen tension may follow below normal when cytochrome oxidase activity rises

oxygen consumption to change as shown in Fig. 8.14. The authors also speculated that the blood flow modulator nitric oxide (NO) could be the adjustable inhibitor of cytochrome c oxidase affinity that competes with oxygen access to the enzyme. The unexpected finding that oxygen tension rises during the phase of blood flow increase and declines below normal when brain energy metabolism increases was demonstrated by Thompson et al. (2003) who determined oxygen tensions in cat visual cortex during stimulation with stripes of varying angles. Oxygen tensions increased during the stimulation, except when the angle of the stripe elicited an activation of the monitored neuron. As predicted by the model, the oxygen tension declined when that happened.

The relationship depicted in Fig. 8.14 assumes that there is no "absolute" recruitment of the capillary bed capable of increasing the diffusion capacity for oxygen by reducing the diffusion distance in the tissue, or increasing the intrinsic permeability of the capillary endothelial wall. It is the current consensus that recruitment of capillaries in brain tissue is "relative," i.e., that it occurs by reduction of capillary transit times toward greater homogeneity of transit times and a lower average transit time, rather than by an absolute increase of the number of perfused capillaries (Kuschinsky and Paulson 1992). Because the capillary surface area remains the same, relative recruitment makes very little difference to the transport of oxygen.

**Regulation of Blood Flow**

The observation that oxygen consumption sometimes fails to increase, despite adequately increased blood flow (e.g., in response to vibrotactile stimulation), suggests either that additional factors prevent neurons from using the available oxygen, or that additional oxygen is not needed. Increased blood flow elicits a number of changes. It delivers substrates such as glucose and amino acids and removes substances such as water, hydrogen ions, lactate and pyruvate. Therefore, it is possible that increased oxygen delivery is not the most important effect of increased blood flow.

A consensus is emerging that the neurotransmitters glutamate and acetylcholine are the prime but indirect agents of blood flow change when they bind to their appropriate membrane receptors, which for glutamate are placed primarily on postsynaptic neurons. Among the direct blood flow stimulators, prostaglandins, epoxy-eicosatrianoic acid (EET), carbon dioxide, adenosine, and potassium and hydrogen ions, appear to require increased synthesis of the originally endothelium-derived relaxation factor nitric oxide (NO) by means of activation of the enzyme responsible for NO synthesis, NO synthase (NOS)(Iadecola 1992; Iadecola et al. 1994; Fabricius and Lauritzen 1994; Villringer and Dirnagl 1995; Lauritzen 2005). Adenosine is a product of the hydrolysis of adenosine nucleotides that are the key elements in energy metabolism and is hence attractive as a potential signal for blood flow change, but the absence of ATP changes in physiologically functioning brain tissue rules out this role in normal brain.

**Nitric Oxide** The observations summarized above place NO in a key position as a modulator both of blood flow and of cytochrome c oxidase affinity for oxygen, and they raise the question of the link between the neurotransmitter-induced receptor stimulation and the activation of NOS. It now appears that a common link among the synthesis of NO and activation of oxidative metabolism can be the flux of calcium ions into cells through voltage- and receptor-gated channels and through the mitochondrial membrane (Lauritzen 2005).

NO causes vasodilatation of brain resistance vessels, in addition to other effects in glial cells and pre- and postsynaptic neurons. It is synthesized in endothelial cells and neurons in proportion to the cytosolic concentration of unbound calcium (Iadecola 2004). The NO is generated in reactions catalyzed by the cell-specific NO synthases that include endothelial eNOS and neuronal nNOS, of which nNOS is by far the most abundant, as well as by the inducible iNOS. Activation of eNOS is mediated by acetylcholine acting on muscarinic receptors of the $M_5$ subtype (Wang et al. 1994; Elhusseiny et al. 1999; Elhusseiny and Hamel 2000).

It is not clear to which extent NOS activation is involved in functionally induced increases of cerebral blood flow. Pharmacologic blockade of the vascular receptors involved in the synthesis of nitric oxide abolishes functionally induced blood flow increases, although the specificity of the blockade is in doubt. Focal changes of cortical blood flow induced by sensory stimulation can be eliminated by blocking endothelial acetylcholine receptors (Ogawa et al. 1994), including those involved in mediating synthesis of NO, apparently without altering the underlying neuronal activation (Ogawa et al. 1994). Other evidence suggests that the cerebral vasodilatation associated with simple somatosensory stimulation in rodents is modulated rather than mediated by NO synthesized by nNOS and not by eNOS or iNOS (Ayata et al. 1996; Ma et al. 1996; Cholet et al. 1997; Lindauer et al. 1999). Reutens et al. (1997) stimulated NO synthesis with the precursor L-arginine in humans and found cerebral blood flow to be globally increased, but the regional blood flow increase in response to vibrotactile stimulation was unaffected, either because NO is not involved in the increase, or because the increase was already maximally stimulated by NO.

Nitric oxide (NO) may also influence monoaminergic neurotransmission by inhibition of of monoamine transporters (Kiss and Vizi 2001). The synthesis of NO by nNOS is stimulated by NMDA receptors activated by glutamatergic neurotransmission. Thus, by means of NO, glutamatergic excitation is extended beyond the synapse to surrounding monoaminergic fibres and varicosities where increased accumulation of monoamines affects excitability and possibly neurovascular coupling. Geday and Gjedde (2009) and Gjedde and Geday (2009) discovered a novel link between monoaminergic neurotransmission and blood flow changes in ventromedial prefrontal cortex that appeared to reflect the emotional impact of stimuli in the form of emotive images: Depending on the degree of impact, cerebral blood flow in this part of the brain rose or fell in response to perturbations of monoamine release, in apparent inverse correlation with the impact.

**Prostanoids** Studies of circulation in skeletal muscle studies show that a correlation exists between the increase of blood flow and the increase of lactate (Connett et al. 1985; Laptook et al. 1988), suggesting that blood flow increase may reflect a glycolytic response.

Astrocytes provide a link between the extracellular space close to microvascular walls and the extracellular space close to synapses. Upon stimulation, astrocytic endfeet have been shown to mediate vasodilation by agents that appear to be products of the prostaglandin-synthesizing enzyme cyclooxygenase-2 (COX-2) in brain. Products that inhibit COX-2 are known to cause reduction of cerebral blood flow (Rasmussen et al. 2003; Gjedde et al. 2005a). Prostaglandins are released from astrocytes in response to intracellular calcium ion increases when metabotropic receptors bind glutamate. The evidence suggests that the COX product PGE2 is involved in astrocyte-mediated vasodilation (Zonta et al. 2003).

**Potassium** As a consequence of the increased potassium conductance of the cell membrane during neuronal excitation, activation raises the extracellular potassium ion concentration, which drives the potassium into astrocytes by several nonenergy-requiring routes, as well as in response to glutamate import. Paulson and Newman (1987) speculated that excess potassium in glial cells is released perivascularly where it may relax smooth muscle cells and dilate resistance vessels. They reasoned that the time constant of delivery of potassium to the perivascular space is much lower when the potassium is siphoned to the vessels inside the foot processes (66 ms), rather than when the potassium is left to diffuse through the extracellular space (2.5 s). The lower time constant would then assure a tighter regulation of the blood flow response.

The role of potassium ions in mediating functionally induced increases of blood flow in the brain was tested by Caesar et al. (1999), who found evidence for considerable heterogeneity in responses. The relative contributions of extracellularly applied potassium ions and adenosine to the blood flow regulation in cerebellum varied among the cell populations, NO and potassium playing the greatest role in parallel fiber connections, and NO and adenosine playing the greatest role in climbing fiber connections.

### 8.7.4 ATP Homeostasis During Activation

To satisfy the need for increased energy turnover during neuronal excitation, additional nutrients must be supplied from tissue stores or from the circulation. With no additional nutrients and no net loss of ATP, the PCr, glycogen, and glucose concentrations could sustain the additional energy demands of a 100% rise of ATP turnover for less than a minute (Table 8.5). However, available evidence suggests that an abundance of additional sources of energy is provided at the onset of activation and that it depends on the type of activation whether this energy is used.

#### Hydrolysis of Phosphocreatine During Activation

Hydrolysis of phosphocreatine (PCr) buffers ATP concentrations at the expense of increased energy utilization. Roth and Weiner (1991) demonstrated that a substantial reduction of the concentration of PCr (5 mM) is compatible with minimal change of ATP. This decline of PCr may be as much as threefold the normal turnover of ATP in 1 min and hence allows a threefold increase of ATP hydrolysis with little or no change of the ATP concentration during the first minute. A decrease in PCr is associated with an increase in the cytosolic free ADP concentration. Erecinska and Silver (1989) calculated that hydrolysis of 5 mM PCr may raise the pH of brain tissue by as much as 0.3 units at the prevailing buffering capacity. This finding was anticipated by Chesler and Kraig (1987, 1989) who found that astrocytes become more alkaline when depolarized during brain activation. The change is observed in several kinds of cells in which a rise in pH correlates with an increase in metabolic activity (Kraig 1990).

#### Glycolysis During Activation

Increase in ADP concentration and alkalinization of brain tissue by hydrolysis of phosphocreatine both stimulate glycolysis at the PFK step, as in skeletal muscle (Connett 1987). Thus, increases in the glycolytic rate by as much as 50% were measured during functional activation of human cerebral cortex (Fox et al. 1988;

Ginsberg et al. 1988). When maximally stimulated, the rate of pyruvate generation can rise to 3–4 μmol $g^{-1}$ $min^{-1}$ (Robin et al. 1984; Gjedde 1984a), which is close to the calculated $T_{max}$ of the mMCT symporter in mitochondria. At these rates of pyruvate generation, the pyruvate concentration rises until the rate of pyruvate removal by any route matches its generation. The increase benefits the net pyruvate conversion to lactate, as well as its transport into the mitochondrial matrix and export to the circulation. The lower the $NAD^+$-NADH ratio and hence the greater the equilibrium lactate–pyruvate ratio, the more slowly this rise occurs.

The failure of the oxygen consumption to increase during simple stimulation is frequently held to confirm that changes of oxidative phosphorylation cannot match the sevenfold increase of pyruvate production seen under the most extreme circumstances of glycolytic stimulation of the mammalian brain (van den Berg and Bruntink 1983).

On the other hand, to increase the flux of pyruvate to mitochondria, cytosolic pyruvate must increase or mitochondrial pyruvate must decline. Hence, insufficient accumulation of pyruvate due to conversion to lactate may prevent the activation of oxidative phosphorylation and instead stimulate oxidation of NADH to $NAD^+$in the cytosol. Connett et al. (1983, 1984) found that the lactate concentration (and hence also pyruvate concentration) in working dog skeletal muscle is directly rather than inversely proportional to the rate of oxygen consumption, as expected if a sufficient pyruvate level is required to sustain a given rate of oxygen consumption.

In the absence of an increase of oxidative phosphorylation, the 50% increase of glucose phosphorylation and corresponding increase of pyruvate generation listed in Table 8.7 causes the pyruvate generation and in turn the lactate efflux to rise fivefold to 0.35 μmol $g^{-1}$ $min^{-1}$ and the fraction of nonoxidative metabolism to rise from 10% at the default stage to as much as 40%, although the total ATP flux rises by a mere 5%.

The increased rate of generation of pyruvate causes pyruvate and lactate levels to rise in the tissue. It may well be asked why the concentrations must rise if pyruvate is a substrate of reactions that effectively remove the metabolite as rapidly as it is generated. However, the main transporters and enzymes responsible for the removal of pyruvate all display some saturability in the range of pyruvate concentrations, particularly the LDH isozyme $LD_5$ but also $LD_1$ and the mitochondrial pyruvate transporter mMCT. Saturability in the relevant concentration range means that concentrations must rise for the flux to increase.

When the concentration of pyruvate rises, the lactate concentration continues to rise (with the concentration of pyruvate), until the transport into mitochondria by the mMCT and oxidation by the pyruvate dehydrogenase complex, conversion to lactate, and efflux of lactate, match the rates of generation. The speed with which the pyruvate concentration rises depends on the turnover numbers of the enzymes and transporters removing pyruvate and hence on the affinities of the processes toward pyruvate: Paradoxically it is the case that the lower the affinities are, the faster the rise is, other factors being equal.

The change of the pool size of pyruvate induced by an increase of the rate of glycolysis can be approximated by equations for the tissue contents of pyruvate

($M_{pyr}$) and lactate ($M_{lact}$). The equations account for the three pathways available to pyruvate (export from cells, conversion to lactate, and transport into mitochondria) and the two pathways available to lactate (conversion to pyruvate and export from cells). For pyruvate, Gjedde and Marrett (2001) described the increase of pyruvate as:

$$\Delta M_{pyr}(t) = \frac{1}{(1+\lambda)k}\,\Delta J_{pyr}\left(1 - e^{-kt}\right), \tag{8.22}$$

where $\lambda$ is the steady-state ratio of the pyruvate and lactate concentrations, $\Delta J_{pyr}$ the incremental generation of pyruvate, and $k$ is the rate constant of approach to a new steady-state, which is a function of the kinetic properties of LDH in the presence of its substrates.

The magnitude of the rate constant of approach to a new steady-state, $k$, varies with the isozyme subtype, such that the oxidative heart form $LD_1$ ($H_4$) has the lower affinity (and hence the higher $k$) and the nonoxidative muscle and liver form $LD_5$ ($M_4$) the higher affinity (and hence the lower $k$). The magnitude of $k$ differs between the isozymes (0.5 $min^{-1}$ for $LD_1$, 0.07 $min^{-1}$ for $LD_5$, Gjedde and Marrett 2001). Thus, the isozyme $LD_1$ causes pyruvate to rise quickly for a given increase of glycolysis, while $LD_5$ causes pyruvate to rise more slowly. Essentially, $k$ is proportional to the blood–brain barrier permeability to lactate and the oxidative capacity of the tissue. A value of 0.4 $min^{-1}$ for average cortical tissue indicates that cells with $LD_1$ (neurons) predominate the energy metabolism.

The magnitude of $\lambda$, the ratio between the lactate and pyruvate concentrations, depends primarily on the NADH-$NAD^+$ ratio and the pH but does not affect the velocity of the approach to steady-state. Together $\lambda$ and $k$ determine the magnitude of the coefficient $1/([1+\lambda]k)$, which is about 0.1 min for the LDII isozymes in brain tissue at steady-state (Gjedde and Marrett 2001).

For lactate, the change of tissue content is a function of the effective ratio of the affinities of LDH toward the two substrates (Gjedde and Wong 2001),

$$\Delta M_{lact}(t) = \frac{\lambda}{(1+\lambda)k}\,\Delta J_{pyr}\left(1 - e^{-kt}\right), \tag{8.23}$$

where $\Delta M_{lact}(t)$ is the accumulated lactate in the tissue. The higher LDH's apparent affinities (i.e., the lower the ratio between the Michaelis constants) are for lactate and pyruvate, the more rapid the approach to a new steady-state, with a time constant of $1/k$.

Substantial lactate generation reflects the situation in which glutamate is released presynaptically but excitation is subliminal and postsynaptic depolarization is prevented by parallel inhibitory input. Mathiesen et al. (1998) showed that stimulation of cerebellar neurons in some cases led to increased blood flow despite inhibition of postsynaptic spiking activity by prevention of postsynaptic depolarization. Minimal postsynaptic depolarization would have two important consequences: Post-synaptic mitochondrial dehydrogenases would not be activated by calcium, but astrocytes nevertheless would be stimulated to remove glutamate at a rate exceeding the oxidative capacity of the tissue and generating lactate.

### Reaction Potential Changes During Activation

Transporters and enzymes with different affinities are affected differently when substrates rise in the tissue, because the individual saturabilities vary according to the affinity. The variable saturability of transporters and enzymes in turn influences the pathway preferences of substrates. The sensitivity of lactate's conversion to pyruvate and subsequent oxidation at increasing substrate concentrations was tested in vitro by Itoh et al. (2003) who fed different concentrations of ordinary glucose to cultures of astrocytes and neurons and observed the effect on the oxidative metabolism of carbon-14-labeled lactate. The results on the ability of glucose to inhibit the metabolism of lactate is shown in Fig. 8.16. The authors found that lactate generated from glucose inhibits the LDH in astrocytes ($LD_5$) much less than the LDH in neurons ($LD_1$). This inhibition can be expressed as the relative change of the

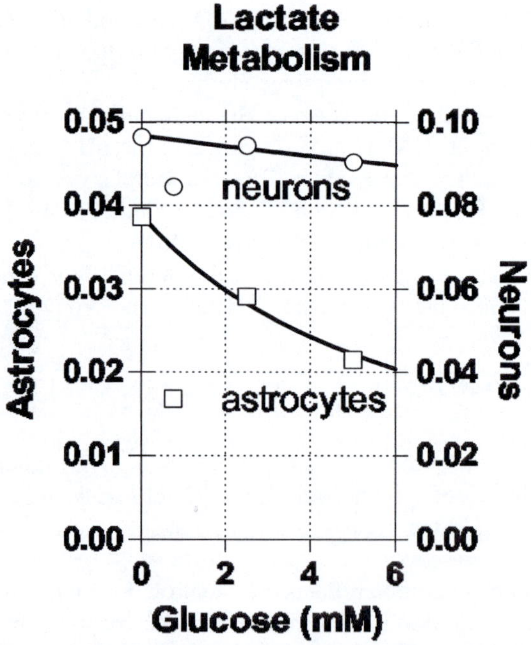


**Fig. 8.16** Sensitivity of lactate metabolism by LDH isozymes in neurons and astrocytes to substrate concentration changes in vitro, published by Itoh et al. (2003). Abscissa: Lactate concentrations approximated as shown from concentration of glucose added to medium in vitro at the concentrations shown ($\mu$mol g$^{-1}$). Ordinate: Carbon-14-labeled lactate oxidation rate in vitro, approximated from metabolism measured per mg protein. The graph shows that astrocytic LDH has a higher affinity toward the substrate than neuronal LDH

**Table 8.8** Estimated saturability of transporters and enzymes of pyruvate and lactate

| | Pyruvate | | | Lactate | | |
|---|---|---|---|---|---|---|
| | $K_{\mathrm{M_{app}}}^{\mathrm{pyr}}$ | $\Delta p_{\mathrm{R}}/p_{\mathrm{R_o}}$ | | $K_{\mathrm{M_{app}}}^{\mathrm{lact}}$ | $\Delta p_{\mathrm{R}}/p_{\mathrm{R_o}}$ | |
| | (mM) | (0.1 mM) | (0.5 mM) | (mM) | (1 mM) | (5 mM) |
| MCT1 | 5 | 0.02 | 0.09 | 5 | 0.17 | 0.50 |
| MCT2 | 1 | 0.09 | 0.33 | 1 | 0.50 | 0.83 |
| LD1 | 0.5 | 0.17 | 0.50 | 5.0 | 0.17 | 0.50 |
| mMCT | 0.5 | 0.17 | 0.50 | 0.5 | 0.67 | 0.91 |
| LD5 | 0.05 | 0.67 | 0.91 | 0.5 | 0.67 | 0.91 |

reaction potential ($p_{\mathrm{R}}$), which is defined as the velocity-substrate ratio (v/s) of the enzyme or transporter reaction,

$$p_{\mathrm{R}}(C) = \frac{V_{\mathrm{max}}}{K_M + C} = \frac{V_{\mathrm{max}}/K_M}{1 + \frac{C}{K_M}} = \frac{p_{\mathrm{R}}(0)}{1 + \frac{C}{K_M}}, \tag{8.24}$$

where $C$ is a substrate or inhibitor and $K_M$ is half-saturation or half-inhibition constant. The relative decline of the reaction potential is then a measure of the occupancy of the substrate and the saturability of the transporter or enzyme,

$$\frac{p_{\mathrm{R}}(0) - p_{\mathrm{R}}(C)}{p_{\mathrm{R}}(0)} = \frac{\Delta p_{\mathrm{R}}}{p_{\mathrm{R}}(0)} = \frac{C}{K_M + C} = \sigma_C, \tag{8.25}$$

where $\sigma_C$ is the occupancy. Estimates of saturability of the main transporters and enzymes of interest to pyruvate and lactate pathways are listed in Table 8.8. The table shows that the processes most likely to be inhibited at higher substrate concentrations are MCT2 and mMCT for lactate and LD5 for both pyruvate and lactate. This inhibition is likely to direct elevated lactate toward MCT1 and LD1, a path that keeps lactate in neurons, and to direct pyruvate toward MCT1, MCT2, LD1, and mMCT, a path that leads pyruvate generated in astrocytes toward metabolism in neurons where the density of mMCT is greatest.

### Oxidative Phosphorylation During Activation

In contrast to the results of no change of oxygen consumption upon simple primary somatosensory stimulation, both motor stimulation and more complex stimulation of visual cortex with a reversing checkerboard pattern for 5 or 10 min caused significant increases of oxygen consumption (Raichle et al. 1976; Marrett and Gjedde 1997; Vafaee et al. 1998, 1999; Vafaee and Gjedde 2000).

If pyruvate transport is rate-limiting in stages 3 and 4 activation (Shearman and Halestrap 1984), the consumption of oxygen must depend on the cytosolic pyruvate concentration and hence on the rate of glycolysis as described by equation (8.20):

$$\Delta J_{O_2}(t) = \frac{[\text{OGI}]}{2} \, \Delta J_{\text{pyr}} \left(1 - e^{-kt}\right), \tag{8.26}$$

where $k$ is the rate constant defined in (8.20) and (8.21).

The relation between the rise of lactate and the rise of oxygen consumption similarly can be derived from the ratio between (8.21) and (8.22), which eliminates the function of time,

$$\Delta J_{O_2}(t) \approx k \, \frac{[\text{OGI}]}{2} \, \Delta M_{\text{lact}}(t) \tag{8.27}$$

in which the relationship is seen to depend both on the ratio between the apparent LDH affinities ($\lambda$) and on the oxidative capacity (OGI) of the cells in which the change occurs. When other factors are constant, the change of oxygen consumption and lactate accumulation are simply proportional.

The relative ATP and lactate changes ($\Delta J_{\text{ATP}}$-$\Delta J_{\text{lact}}$ ratio) listed in Table 8.7 divides the metabolic responses to excitation into glycolytic ("white") and oxidative ("red") responses, illustrated in Fig. 8.11, of which the "white" are the primary somatosensory responses and the "red" are the motor and secondary somatosensory responses. The distribution of glycolytic and oxidative capacities among the two cell types suggests that neuronal metabolism is in part responsible for the "red" response (OGI $\approx$ 4), while astrocytes exclusively generate the "white" response (OGI $\approx$ 1).

### *8.7.5 Metabolic Compartmentation During Activation*

Taken at face value, the findings discussed above indicate activation of at least two different populations of cells, one with a lower oxidative capacity and one with a higher. The slow or absent rise during primary somatosensory stimulation is then the consequence of the slow rise of the pyruvate concentration in the presence of a substantial lactate sink. This behavior is to be expected when the bulk of the acceleration of metabolism occurs in cells of low oxidative capacity. The more substantial rise of oxygen consumption during more complex stimulation is the consequence of activation of cells with a higher oxidative capacity. These two types of response conveniently fit the properties of the astrocytic and neuronal compartments.

#### Sites of Cellular Activity During Activation

The cellular and subcellular sites of the increases of glucose and oxygen metabolism accompanying brain activation have been the subjects of much investigation (Rose 1975; Muir et al. 1986; Bachelard et al. 1991; Poitry-Yamate and Tsacopoulos 1992;

Magistretti et al. 1999). There is evidence that the site of increased glycolysis is the synaptic structures of the neuropil (Eisenberg et al. 1993; Sokoloff et al. 1996; Sokoloff 1999), including astrocytic endfeet (Kasischke et al. 2004), which is also the predominant location of lactate dehydrogenase (Borowsky and Collins 1989).

Mitochondria, on the other hand, have been observed to be particularly concentrated in the dendritic structures of the neuropil, particularly the proximal dendrites (Ribak 1981; Gonzalez-Lima and Jones 1994; Wong-Riley 1989), which stain weakly for hexokinase (Snyder and Wilson 1983). These observations suggest that the glycolytic and oxidative reactions occur in separate cellular compartments, although the compartmentation is not complete (Aoki et al. 1987).

The removal of glutamate and the consequent hydrolysis and rephosphorylation of ATP in astrocytes must be sufficiently rapid to allow the frequent firing of excitatory neurons believed to underlie the functional integration of neurons in cerebral cortex (Joliot et al. 1995; Pedroarena and Llinas 1997). During nonsteady or otherwise transiently excited states, it is possible that rapid removal of glutamate (Kojima et al. 1999) may generate pyruvate in excess of the average oxidative capacity and explain why astrocytes react more glycolytically to stimulation than neurons.

Thus, although recent measurements of the relative contributions of oxidative phosphorylation to the energy turnover in neurons and astrocytes show that 15% of the total energy turnover in brain tissue takes place in astrocytes at normal steady-state (Shulman et al. 2004), the astrocytes may contribute significantly more to the increase of nonoxidative metabolism when excitation of dendritic structures is prevented by parallel inhibition.

As brain energy metabolism is predominantly oxidative (Hevner et al. 1995), the compartmentation suggests that enhanced oxidative metabolism associated with brain activation occurs predominantly in the proximal dendritic structures of postsynaptic neurons. When there is little increase of the oxidative capacity but increased glutamate uptake, probably because of insufficient calcium ion influx (Lauritzen 2005), the generated lactate is left to be exported across the blood–brain barrier and removed by the increased blood flow. This condition is associated with the low ATP-lactate flux ratio seen with the "white" (astrocytic) response of lower oxidative capacity but higher glycolytic capacity. Conversely, in the visual cortex stimulated by a complex checkerboard pattern, the energy demand rises greatly but lactate accumulation is kept at a minimum by the rise of oxygen consumption. In this case, the ATP-lactate flux ratio is in the high range. The increased oxygen supply establishes the increased ATP turnover characteristic of the "red" (dendritic) response of higher oxidative capacity but lower glycolytic capacity.

### Model of Compartmentation During Activation

One plausible model of the interaction between neurons and glial cells during activation of the human cerebral cortex is illustrated in Fig. 8.17 in which the ranges of oxidative and glycolytic capacities of the two cellular compartments are given for the two extremes of complete rest (stage 1) and complete activation (stage 4).

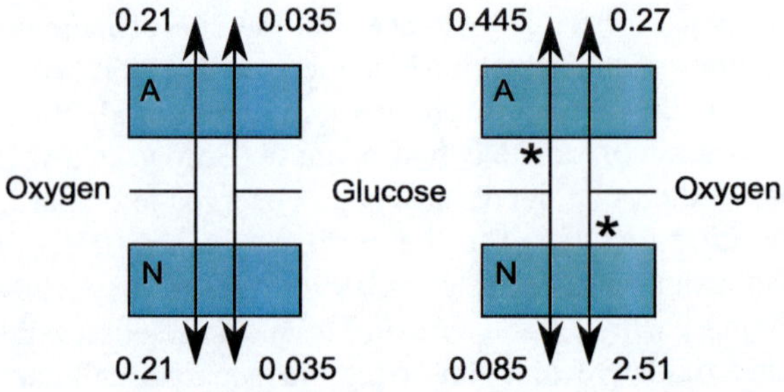


**Fig. 8.17** Model of ranges of oxidative and glycolytic metabolism in astrocytes (*top boxes*) and neurons (*bottom boxes*) in human cerebral cortex. The illustration shows two sets of cells, one at complete rest (stage 1), the other fully activated (stage 4). Numbers have units of μmol $g^{-1}$ $min^{-1}$*Asterisks* indicate main control points (HK/PFK in astrocytes, PDH in neurons), conferring specific oxidative and glycolytic capacities on the cell populations when stimulated. In the model, astrocytes and neurons can be differentially stimulated, depending on the degree of dendritic inhibition and activation. Stimulation either increases the number of cells that are activated, or the time in which all cells are activated

Differential activation of the two cell types between these two extremes result in the metabolic changes summarized earlier. If it is assumed that the normal (default) condition of stage 3 is characterized either by half of the cells in each group being active at all times, or all of the cells being active half of the time, the model predicts the ranges of pyruvate and ATP production rates associated with variable activation from 0% to 100%, as shown in Fig. 8.18. Depending on the relative activation of the two cell types, the model predicts the changes of lactate production or uptake and the changes of OGI seen during activation of cortical brain tissue.

Figure 8.19 illustrates the predicted range of relative differential activation of glial cells and neurons, relative to the normal (default) activation of neurons: When astrocytes are more active than neurons, the OGI declines and the lactate efflux increases. On the other hand, when neurons are more active than astrocytes, the OGI increases and the lactate efflux decreases or switches to influx of lactate or depletion of glycolytic stores such as glucose, glycogen, or lactate. The evidence from Ide et al. (2000), Dienel and Cruz (2004) and other recent studies of experimental or physiological activation shows that the activity of astrocytes increases more than the activity of neurons during the activation itself, while the activity of neurons is higher than the activity of astrocytes during the subsequent recovery period, possibly because the effects of the excitation persist longer in the neurons than in the astrocytes after the termination of the direct activation.

## 8.8 Conclusions

Current evidence suggests that there is no rigid association in vivo between *changes* of oxygen consumption, glucose combustion, and blood flow in the human brain. The claim that increased blood flow must occur simply to satisfy the demands for oxygen and glucose during neuronal excitation therefore is simplistic.

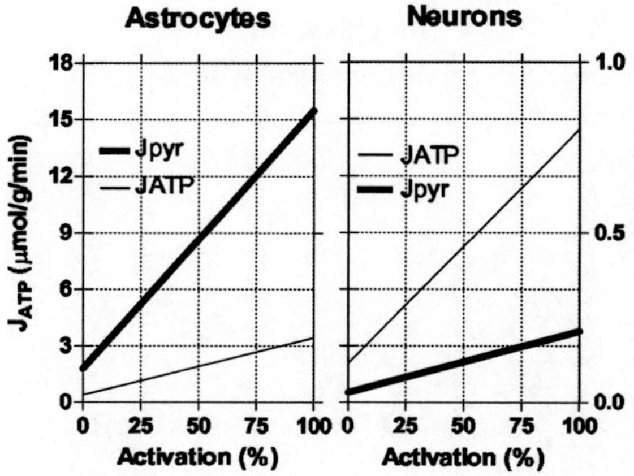


**Fig. 8.18** Rates of pyruvate and ATP generation in astrocytes and neurons, estimated from model shown in Fig. 8.17. Abscissae: Percentage time or cells associated with activation. Ordinates: ATP and pyruvate fluxes in units of μmol $g^{-1}$ $min^{-1}$, as functions of the degree of activation. For astrocytes, the lines are given by the equations $J_{pyr} = 0.07\ (1 - A) + 0.89\ A$ and $J_{ATP} = 1.96\ (1 - A) + 2.51\ A$, in units of μmol $g^{-1}$ $min^{-1}$. For the neurons, the lines are given by $J_{pyr} = 0.07\ (1 - A) + 0.17\ A$ and $J_{ATP} = 1.96\ (1 - A) + 15.23\ A$, also in units of μmol $g^{-1}$ $min^{-1}$ where $A$ is the degree of activation. The graph shows how astrocytes generate much pyruvate but little ATP, while neurons do the opposite upon stimulation

Energy budget estimates suggest that most of the cerebral energy demand reflects the steady-state level of graded postsynaptic membrane depolarization, and action potential generation and propagation. Increased energy supply is required to maintain the graded dendritic and somatic depolarization of neuronal membranes with increased sodium and potassium conductances. Increased energy turnover is not required to sustain hyperpolarization caused by decreased conductance of sodium or increased conductance of potassium or chloride.

Glucose, pyruvate, and lactate occupy single tissue compartments. For this reason, it is improbable that the effective properties of the LDH subtypes in vivo differ spatially among different populations of cells. Instead, the properties may vary temporally, as dictated by the cytosolic redox potential and pH. Substantial pyruvate and lactate generation occurs when glutamate release is not followed by sufficient dendritic depolarization, i.e., when astrocytes are activated more than postsynaptic neurons. In these cases, there is demand for glutamate removal of a magnitude exceeding the lower oxidative capacity of astroglial cells. Although the resulting pyruvate and lactate accumulation is influenced by lactate export or import across the blood–brain barrier, the pyruvate and lactate thus accumulated is available for joint use by neurons and astrocytes. To this joint pool, astrocytes produce much more pyruvate than neurons. The evidence suggests that the rapid astrocytic (“white’) response reflects glutamate release, while the bulk of the neuronal (“red’) response reflects substantial mitochondrial activation of primarily dendritic origin, induced by sufficiently influx of calcium ions.

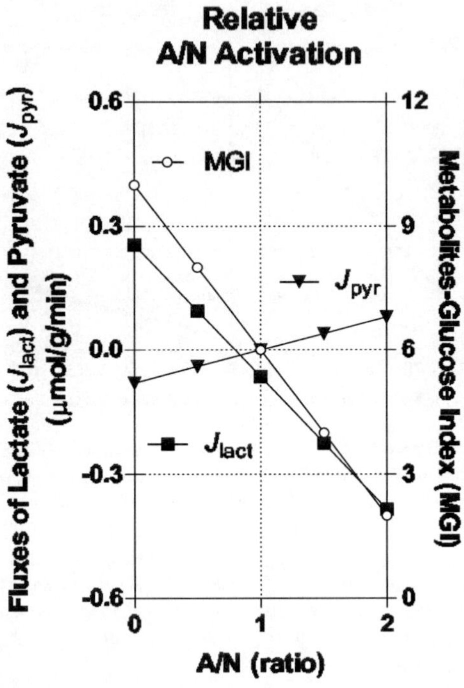


**Fig. 8.19** Lactate flux into or out of brain, and arteriovenous deficits of oxygen, glucose, and lactate, expressed as the arterio-venous oxygen-glucose and lactate-glucose indices, as function of relative degrees of activation of astrocytes and neurons. Abscissa: Degree of astrocytic activation relative to normal (default) neuronal activation of 50%. *Left ordinate*: Lactate flux (μmol g$^{-1}$ min$^{-1}$) into (positive) or out of (negative) brain. *Right ordinate*: Summed oxygen-glucose and lactate-glucose indices (ratio). Lines are given by $J_{lact} = 0.335 - 0.4\ (A/N)$ and OGI + 3 LGI = $10 - 4\ (A/N)$ relative to normal (default) neuronal activity (50%)

The blood flow increase appears to be coupled to the rate of glycolysis. There is increasing evidence that the putative mechanism underlying such a flow-glycolysis couple is an astrocytic function. The physiological reason for an astrocytically mediated blood flow increase is not very clear, as flow has only a moderate effect on glucose delivery, when not accompanied by capillary recruitment, suggesting that increased glycolysis is a signal rather than a direct demand.

The evidence suggests that the increase of oxidative metabolism in neurons is coupled to the rise of pyruvate, as dictated by the prevailing kinetic profile of LDH ($\lambda$) and the degree of mitochondrial activation. The tissue pyruvate level at which this occurs is under the control of the LDH isozyme profile and therefore differs for different populations of cells. Under some circumstances, regional "peaks" of increased blood flow and increased oxygen consumption could be dissociated by the differential activation of primary and secondary neuronal networks. The activations accompanying the most complex processing of information could be those with the tightest coupling between oxygen consumption and blood flow and hence with the least generation of lactate.

# Erratum to:

# Introduction to Compartmental Analysis

A. Gjedde et al., *Neurokinetics: The Dynamics of Neurobiology In Vivo*,
DOI: 10.1007/978-1-4419-7409-9, pp. 1–22,


**DOI 10.1007/978-1-4419-7409-9_9**

The publisher regrets that in the print and online version of this book, the equation for the chapter entitled "Introduction to Compartmental Analysis" by Albert Gjedde et al., is incorrect. Equation 1.6 should not have a minus sign.

$$\frac{\partial c}{\partial t} = D\frac{\partial^2 c}{\partial x^2} \tag{1.6}$$

The online version of the original chapter can be found at
http://dx.doi.org/10.1007/978-1-4419-7409-9_1

# Erratum to:

# Fundamentals of Compartmental Kinetics

A. Gjedde et al., *Neurokinetics: The Dynamics of Neurobiology In Vivo*,
DOI: 10.1007/978-1-4419-7409-9, pp. 23–101,


**DOI 10.1007/978-1-4419-7409-9_10**

The publisher regrets that in the print and online version of this book, the equation for the chapter entitled "Fundamentals of Compartmental Kinetics" by A. Gjedde et al., is incorrect. Equation 2.96 should not have a minus sign.

$$\frac{\partial c}{\partial t} = D\frac{\partial^2 c}{\partial x^2} \tag{2.96}$$

The online version of the original chapter can be found at
http://dx.doi.org/10.1007/978-1-4419-7409-9_2

# Glossary

In this book, we focus on compartmental models as typically applied in physiological and chemical sciences. Not uncommonly, compartment contents were given symbols of concentration such as $C_i$, where $i$ ranges from 1 to $n$ compartments that describe the system. In brain imaging, compartment contents are often described with symbols of concentration such as $C_1$ where 1 indicates the plasma input function (corrected for metabolites) and $C_2$ where 2 indicates the first compartment behind the blood–brain barrier (BBB). Examples of this are seen in early models of brain radiotracers (Mintun et al. 1984; Frey et al. 1985).

In the following, we employ a nomenclature first developed by Gjedde and Crone (1975), Gjedde (1980, 1981b, 1982), Gjedde et al. (1985, 1986, 2005b), Gjedde and Wong (2001), and Wong et al. (1984, 1986a–c 1997a, b). In some minor respects, this nomenclature is not identical to the consensus nomenclature recently presented by Innis et al. (2007), to which two of the present authors contributed. In this book, we use the original terms for consistency as well as adherence to common mathematical practice. The precise definitions of the main parameters and variables used throughout the book are given in the Glossary section, with comparison to other nomenclature conventions, including Innis et al. (2007).

As examples of conventions adopted in this book, we use the terms $m$ and $M$ for masses of tracer and native substances. Rather than the terms $F$ and $B$ (traditionally used in pharmacology), we use $M_f$ and $M_b$ for free and bound ligand quantities, and $p_B$ for binding potential, which is consistent with the nomenclature of pH for ion potential, instead of $BP_{ND}$, one of several binding potential formulations presented by Innis et al. (2007).

The term for volumes of distribution in tissue is restricted to $V$ with appropriate subscripts (e.g., $V_d$ for physical volume of distribution, however composed). We also employ the term $V_w$ for the volume of distribution of a compound in water. We retain the use of the terms $f_1$ and $f_2$ for free fractions rather than the terms $f_P$ and $f_{ND}$, and we retain the use of the term $V_e$ for the volume of passive exchange partition of tracers between brain and circulation ("partition volume") rather than the term $V_{NS}$, which commingles passive and nonspecific distributions.

A. Gjedde et al., *Neurokinetics: The Dynamics of Neurobiology In Vivo*,
DOI 10.1007/978-1-4419-7409-9, © Springer Science+Business Media, LLC 2011

## A

| | |
|---|---|
| * | Asterisk, denotes labeled tracer |
| $a_i, 1 \le i \le N$ | Parameter of multilinear regression |
| $\alpha, \alpha_1, \alpha_2$ | Solubility in sample of blood or plasma (source fluid) relative to water (ratio) |
| $A$ | (1) Area (unit of distance$^2$), (2) diffusional surface area of blood-tissue barrier (cm$^2$ g$^{-1}$ or ml$^{-1}$) |

## B

| | |
|---|---|
| $\overline{b}$, bar | denotes calculated or average rather than measured variable |
| $B$ | Quantity of specifically bound ligand in brain (pmol g$^{-1}$ or ml$^{-1}$) |
| $B_I$ | Quantity of specifically bound inhibitor(s) in brain (pmol g$^{-1}$ or ml$^{-1}$) |
| $B_{\text{max}}$ | Maximum binding capacity of tissue receptor sites (pmol g$^{-1}$ or ml$^{-1}$); (1) maximum binding capacity of protein in ligand–protein interaction (unit is mass), (2) maximum binding capacity of tissue binding sites for tracer in absence of inhibitors (pmol g$^{-1}$ or ml$^{-1}$) |
| $B'_{\text{max}}$ | Available receptors, maximum binding capacity of tissue receptor sites for ligand after blockade by inhibitors (pmol g$^{-1}$ or ml$^{-1}$), $B_{\text{avail}}$ in Innis et al. (2007). |

## C

| | |
|---|---|
| $c$ | (1) Concentration, time-variable concentration of delivery compartment (unit of mass volume$^{-1}$), (2) tracer concentration in water (nM) |
| $c_{\text{a}}$ | Time-variable concentration in arterial sample of source fluid (plasma or blood) (nM), implicitly $c_{\text{a}}(t)$, but explicit as $c_{\text{a}}(t, x)$ where there is an additional spatial variable, $x$. |
| $c_{\text{a}}^*$ | Time-variable tracer concentration in arterial sample of source fluid (plasma or blood) (MBq or μCi ml$^{-1}$) |
| $c_{\text{v}}$ | Time-variable concentration in venous sample (nM) |
| $C$ | Concentration, time-invariant (steady-state) aqueous concentration in delivery compartment (unit of mass volume$^{-1}$); steady-state concentration in water (nanomolar) |
| $C_{(i)}$ | Steady-state concentration in compartment $i$ (nanomolar) |
| $C_{\text{a}}$ | Concentration in arterial sample of source fluid (plasma or blood); in the case of glucose the steady-state native glucose concentration in arterial sample of source fluid (plasma or blood) (mM) |
| $C_e$ | Steady-state substrate or ligand concentration in tissue (mM) |
| $C_{\text{I}}$ | Inhibitor concentration in water compartment (nanomolar) |

$C_{I_a}$ Steady-state inhibitor concentration in arterial sample (nanomolar)
$C_v$ Steady-state concentration in venous sample (nanomolar); in the case of glucose the steady-state native glucose concentration in venous sample of source fluid (plasma or blood) (mM)

## D

$D$ Coefficient of diffusion (unit of distance$^2$ time$^{-1}$; cm$^2$ s$^{-1}$)

## E

$E$ Steady-state extraction fraction (ratio)
$E^*$ Fraction of tracer extracted at steady-state (ratio)
$E^*_{\text{app}}(t)$ apparent fraction of tracer extracted at time $t$ (ratio)
$E_o$ Fraction of substance extracted during "first pass," i.e., in absence of back-flux

## F

$f_1$ Fraction of "free" substance in arterial sample; reciprocal of $\alpha$ (ml ml$^{-1}$), equal to $f_P$ in Innis et al. (2007), when arterial sample is plasma.
$f_2$ "Fraction" of "free" substance in sample of brain tissue; reciprocal of $V_\text{d}$ (g or ml ml$^{-1}$), equal to $f_{ND}$ in Innis et al. (2007).
$f_b$ Bidirectionality factor (ratio)
$f_b{}'$ Bidirectionality factor for saturable ligand–protein interactions (ratio)
$\varphi$ Phosphorylation ratio between tracer and native glucose(ratio)
$\zeta$ Dummy differentiation variable. Equal to $\overline{m}_\text{e}/K_1$
$F$ source fluid (plasma or blood) flow to tissue sample (ml g$^{-1}$ min$^{-1}$)

## G

$g$ Molecular mobility
$\gamma$ Factor describing egress of radioactive carbon dioxide from brain during metabolism of labeled glucose
$G$ Spatial concentration gradient, $\text{d}C/\text{d}x$ (unit is mass volume$^{-1}$ distance$^{-1}$)

## HIJ

| | |
|---|---|
| $j$ | Time-variable flux (pmol $g^{-1}$ $min^{-1}$ or pmol $ml^{-1}$ $min^{-1}$) |
| $j_n$ | Time-variable flux (unit is mass $time^{-1}$, $n$ is any number) |
| $j^{Hase*}$ | Rate of tracer phosphorylation by hexokinase (μmol $g^{-1}$ $min^{-1}$) |
| $j^{L*}$ | Rate of tracer metabolite loss from tissue (μmol $g^{-1}$ $min^{-1}$) |
| $j^{Pase*}$ | Rate of tracer 6-phosphate dephosphorylation by glucose 6-phosphatase (μmol $g^{-1}$ $min^{-1}$) |
| $\Delta j^*$ | Net time-variable tracer flux through tissue (μmol $g^{-1}$ $min^{-1}$) |
| $J$ | Steady-state flux (pmol $g^{-1}$ $min^{-1}$ or pmol $ml^{-1}$ $min^{-1}$) |
| $J_n$ | Flux, time-invariant (steady-state) flux (unit is $mass^{-1}$ time, $n$ is any number) |
| $J^{Hase}$ | Steady-state rate of native glucose phosphorylation by hexokinase (μmol $g^{-1}$ $min^{-1}$) |
| $J^{Pase}$ | Steady-state rate of native glucose 6-phosphate dephosphorylation by glucose 6-phosphatase (μmol $g^{-1}$ $min^{-1}$) |
| $J_{max}$ | Maximum velocity (flux) of enzymatic reaction in forward direction (pmol $g^{-1}$ $min^{-1}$) |
| $J_{max}^{Hase}$ | Maximum rate of glucose phosphorylation by hexokinase (μmol $g^{-1}$ $min^{-1}$) |
| $J_{max}^{Hase*}$ | Maximum rate of tracer phosphorylation by hexokinase (μmol $g^{-1}$ $min^{-1}$) |
| $J_{max}^{Pase}$ | Maximum rate of glucose 6-phosphate dephosphorylation by glucose 6-phosphatase (μmol $g^{-1}$ $min^{-1}$) |
| $J_{max}^{Pase*}$ | Maximum rate of tracer 6-phosphate dephosphorylation by glucose 6-phosphatase (μmol $g^{-1}$ $min^{-1}$) |

## K

| | |
|---|---|
| $k, k_n$ | Relaxation or decay constant (unit is $time^{-1}$, $n$ is not any number) |
| $k_{(i)}$ | Coefficient (frequency) of transfer between two compartments, numbered consecutively (per minute) |
| $k', k'_n$ | Steady-state relaxation constant representative of multiple compartments (unit is $time^{-1}$, $n$ is any number) |
| $k_2$ | Rate constant of efflux from second compartment to circulation in three-compartment system ($min^{-1}$); in the case of glucose the relative clearance of native glucose from tissue ($min^{-1}$) |
| $k_2^*$ | Relative clearance of glucose tracer from tissue ($min^{-1}$) |
| $k_3$ | Rate constant of efflux from second compartment to third compartment in three-compartment system ($min^{-1}$); in case of glucose the true rate constant of phosphorylation of native glucose in tissue ($min^{-1}$) |
| $k_3^*$ | True rate constant of phosphorylation of glucose tracer in tissue ($min^{-1}$) |
| $k_4$ | Relaxation or decay constant of third compartment ($min^{-1}$); in case of glucose the rate constant of dephosphorylation of native glucose |

| | |
|---|---|
| $k_4^*$ | Rate constant of dephosphorylation of glucose tracer-6-phosphate in tissue ($\text{min}^{-1}$) |
| $k_5^*$ | Assumed first-order rate constant of loss of glucose racer metabolites from tissue ($\text{min}^{-1}$) |
| $k_{\text{app}}$ | Apparent steady-state rate constant of net phosphorylation of native glucose in tissue in presence of dephosphorylation ($\text{min}^{-1}$) |
| $k_{\text{app}}^*(t)$ | Apparent variable rate of net phosphorylation of glucose tracer in tissue in presence of significant dephosphorylation ($\text{min}^{-1}$) |
| $k_{\text{cat}}$ | Rate constant of catalysis, turnover constant of enzyme ($\text{min}^{-1}$) |
| $k_{\text{in}}$ | Dissociation rate constant (unit of $\text{time}^{-1}$); frequency of internalization, transport, or inactivation of ligand–receptor complex (per minute) |
| $k_{\text{off}}$ | Dissociation rate constant (unit of $\text{time}^{-1}$), frequency of dissociation of tracer from binding site ($\text{min}^{-1}$) |
| $k_{\text{on}}$ | Association coefficient (unit of mass $\text{volume}^{-1}$ $\text{time}^{-1}$), transfer coefficient for association of substance to binding sites ($\text{min}^{-1}$ $\text{nM}^{-1}$); transfer coefficient for association of ligand to receptor sites ($\text{min}^{-1}$ $\text{nM}^{-1}$); apparent transfer coefficient for association of ligand to receptor sites (g $\text{min}^{-1}$ $\text{pmol}^{-1}$) |
| $k_{\text{out}}$ | Association coefficient (unit of mass $\text{volume}^{-1}$ $\text{time}^{-1}$) |
| $\chi$ | Concentration ratio (example $C\ K_{\text{M}}{}^{-1}$) |
| $\kappa_i^*$ | Apparent clearance of tracer to compartment $i$ as function of time of circulation defined as $m_i^*(T)/\int_o^T c_{\text{a}}^*\,\text{d}t$ |
| $K$ | 1) Net clearance, spanning several compartments (unit is volume $\text{min}^{-1}$), 2) coefficient of net volume transfer (clearance) to tissue having three compartments (ml $\text{g}^{-1}$ $\text{min}^{-1}$); in case of glucose, the true steady-state clearance of native glucose (ml $\text{g}^{-1}$ $\text{min}^{-1}$) |
| $K^*$ | Steady-state clearance of tracer (ml $\text{g}^{-1}$ $\text{min}^{-1}$) |
| $K_{\text{app}}^*(t)$ | Apparent steady-state clearance of glucose tracer in presence of significant dephosphorylation of tracer-6-phosphate to exchangeable tracer (ml $\text{g}^{-1}$ $\text{min}^{-1}$) |
| $K_1$ | Unidirectional clearance of native glucose (ml $\text{g}^{-1}$ $\text{min}^{-1}$); (1) coefficient of volume transfer (clearance) from compartment 1 to compartment 2 (ml $\text{g}^{-1}$ $\text{min}^{-1}$ or ml $\text{ml}^{-1}$ $\text{min}^{-1}$); "unidirectional" clearance (unit is volume $\text{min}^{-1}$), (2) coefficient of unidrectional volume transfer (clearance) from compartment 1 to compartment 2 (ml $\text{g}^{-1}$ $\text{min}^{-1}$ or ml $\text{ml}^{-1}$ $\text{min}^{-1}$) |
| $K_1'$ | Net clearance, spanning several compartments (unit is volume $\text{min}^{-1}$) |
| $K_1^*$ | Unidirectional clearance of glucose tracer (ml $\text{g}^{-1}$ $\text{min}^{-1}$) |
| $K_d$ | Dissociation constant (unit of mass $\text{volume}^{-1}$) |
| $K_D$ | Michaelis half-saturation concentration of ligand (nanomolar) |
| $K_d^{\text{w}}$ | Michaelis half-saturation concentration of substance (nM) |
| $K_I$ | Inhibitor constant (nanomolar) |
| $K_I'$ | Concentration-dependent inhibitor constant; $\text{IC}_{50}$ of inhibitor (nanomolar) |
| $K_{\text{M}}$ | Michaelis half-saturation constant (unit of mass $\text{volume}^{-1}$) |
| $K_m^{\text{Hase}}$ | Michaelis half-saturation concentration for phosphorylation of native glucose by tissue hexokinase (mM) |

$K_m^{\mathrm{Hase}*}$ Michaelis half-saturation concentration for phosphorylation of glucose tracer by tissue hexokinase (mM)

$K_m^{\mathrm{Pase}}$ Michaelis half-saturation concentration for dephosphorylation of native glucose-6-phosphate by tissue phosphatase (mM)

$K_m^{\mathrm{Pase}*}$ Michaelis half-saturation concentration for dephosphorylation of glucose tracer-6-phosphate by tissue phosphatase (mM)

$K_m^T$ Michaelis half-saturation concentration for blood-tissue transfer of native glucose (mM)

$K_m^{T*}$ Michaelis half-saturation concentration for blood-tissue transfer of glucose tracer (mM)

$K_o$ Apparent clearance associated with diffusion (unit volume min$^{-1}$)

$K_o'$ Apparent clearance associated with diffusion and bidirectional exchange across membrane (unit volume min$^{-1}$)

## L

$l_\mathrm{D}$ Length constant of diffusion, virtual length of pathway

$\lambda$ (1) Ratio $L/l_\mathrm{D}$, (2) partition coefficient, ratio $\alpha_2/\alpha_1$

$L$ (1) Diffusion distance, physical length of pathway, (2) length of exchange vessels (mm)

$\Lambda$ Lumped constant (ratio)

$\Lambda_{\mathrm{app}}(t)$ Apparent time-dependent lumped constant measured in the presence of significant dephosphorylation of glucose tracer-6-phosphate(ratio)

## MNO

$m, m_n$ Mass, time-variable content of any compartment or compartment numbered "n" (pmol g$^{-1}$ or ml$^{-1}$)

$m^*$ Total quantity of tracer and all labeled materials derived from tracer, in sample of tissue (μCi g$^{-1}$)

$m_a$ Time-variable mass of substance in arterial compartment of a sample of tissue (pmol g$^{-1}$ or ml$^{-1}$)

$m_b$ Time-variable mass of specifically bound substance in tissue (pmol g$^{-1}$ or ml$^{-1}$)

$m_c$ Time-variable mass of substance in exchange vessels (capillaries) (pmol g$^{-1}$ or ml$^{-1}$)

$m_\mathrm{e}$ Time-variable mass of exchangeable (i.e., nonspecifically bound) substance in tissue (pmol g$^{-1}$ or ml$^{-1}$)

$m_\mathrm{e}^*$ Time-variable mass of tracer in exchange compartment, quantity of exchangeable intact tracer in tissue (MBq or μCi g$^{-1}$)

$m_e^{calc}$ Time-variable mass of exchangeable substance in tissue, calculated from tabulated rate constants (MBq or $\mu$Ci $g^{-1}$)

$m_m$ Time-variable mass of substance specifically bound in tissue, equivalent to $m_b$ (pmol $g^{-1}$ or $ml^{-1}$)

$m_m^*$ Time-variable mass of glucose or other tracer metabolites subject to dephosphorylation to exchangeable glucose tracer in tissue (MBq or $\mu$Ci $g^{-1}$)

$m_m^{calc}$ Time-variable mass of glucose or other tracer metabolites subject to dephosphorylation to exchangeable glucose tracer in tissue, calculated from tabulated rate constants (MBq or $\mu$Ci $g^{-1}$)

$m_p^*$ Time-variable mass of tracer in compartment(s) (product compartments) other than delivery and exchange compartments

$m_r^*$ Time-variable mass of tracer metabolites lost from tissue (MBq or $\mu$Ci $g^{-1}$)

$m_v$ Time-variable mass of substance in veins (pmol $g^{-1}$ or $ml^{-1}$)

$m_z^*$ Time-variable mass of tracer metabolites subject to direct loss from tissue ($\mu$Ci $g^{-1}$)

$M$ Steady-state mass of substance in a sample (pmol $g^{-1}$ or $ml^{-1}$)

$M$, $M_n$ Mass, time-invariant (steady-state) content of compartment or compartment $n$

$\mathrm{M}_{(i)}$ Steady-state mass of substance held in compartment $i$ (pmol $g^{-1}$ or $ml^{-1}$)

$M_a$ Steady-state mass of substance in the vascular compartment of a sample of brain tissue (pmol $g^{-1}$ or $ml^{-1}$)

$M_b$ Steady-state mass of substance specifically bound in the brain; equivalent to $B$ (pmol $g^{-1}$or $ml^{-1}$)

$M_e$ Steady-state mass of exchangeable (i.e., nonspecifically bound) substance in brain (pmol $g^{-1}$ or $ml^{-1}$); in the case of glucose the steady-state quantity of native glucose in tissue (pmol $g^{-1}$)

$M_m$ Steady-state quantity of native glucose-6-phosphate in tissue (pmol $g^{-1}$)

$\mathbf{M}$, $\mathbf{M}_n$ Compartment, compartment $n$

$\mu_h$ Penalty function of compartment system of $h$ compartments

# P

$p$ Apparent blood-tissue barrier permeability-surface area product, relative to tissue distribution volume, equal to $P_d A/V_d$ ($min^{-1}$)

$p_{(i)}$ Coefficient of independent variable $x_{(i)}$

$p_B$ Binding potential of ligand in brain, equivalent to $k_3/k_4$ (ratio)

$p_{NSB}$ Nonspecific and low affinity binding potential of ligand, equal to $[V_d/V_w] - 1$ (ratio)

$P$ (1) permeability (unit of volume distance$^2$ time$^{-1}$), (2) apparent blood-tissue barrier permeability-surface area product of tracer, equal to $P_d A/\alpha_1$ (ml $g^{-1}$ $min^{-1}$)

$P$ Blood–brain barrier permeability of ligand (cm $s^{-1}$)

$P_d$ Blood-tissue barrier permeability of tracer (cm $s^{-1}$)

# Q

$q$ Parameter of three-compartment model (unit of time$^{-1}$)

# R

$\rho$ (1) Trapping index of tracer subject to sequestrationin tissue; the closer to unity $\rho$ is, the lower the fraction of tracer that is trapped; equal to $k_2/(k_2 + k_3)$ (ratio). (2) Metabolism index of substrate not subject to metabolismin tissue; the closer to unity $\rho$, the higher the fraction of substrate that escapes metabolism; equal to $k_2/(k_2 + k_3)$ (ratio). (3) Volume ratio; $[V - V_o]/V_e$ or $(V_c + V_e + V_b - V_o)/V_e$ (ratio)

$\rho^*$ Time-varying ratio between integrated masses of tracer of region of interest and reference region

$\rho^*_{app}(t)$ Time-dependent apparent index of tracer not subject to metabolism in tissue; equal to $k_2^*/(k_2^* + k^*_{app}(t))$ (ratio)

$R$ Steady-state glucose consumption (μmol g$^{-1}$ min$^{-1}$)

$\bar{R}_{pop}$ Population averaged steady-state glucose consumption (μmol g$^{-1}$ min$^{-1}$)

$R_1$ ratio between clearances into region of interest and reference region

# S

$\sigma$ (1) Saturation, occupancy, fraction of binding sites occupied by ligand (example $\chi\,(1+\chi)^{-1}$). (2) Metabolite pool index of tracer in tissue, equivalent to $k_3/k_4$ (ratio)

$\sigma^*$ Apparent metabolite pool index of tracer in tissue, equivalent to $k_3^*/k_4^*$ ratio

$\varsigma_n$ Occupancy of noncompetitively interacting ligand $n$

$S$ Surface area (unit of distance$^2$); diffusional surface area of blood–brain barrier (cm g$^{-1}$ or ml$^{-1}$)

$S_{sq}$ Sum of squares to be minimized by regression analysis

# T

$t$ Time (minutes)

$\tau$ (1) Time (min). (2) Transport ratio between tracer and native glucose (ratio)

$\tau_D$ Characteristic time constant of diffusion

$\theta^*$ Transformed time, equal to ratio between integral of reference region mass and reference region mass of tracer

$\theta'^*$ Transformed time, equal to ratio between twice integrated reference region mass and integrated reference region mass of tracer

$T$ Time of termination of experiment; time point; integration limit (min)

$\Theta$ Normalized time-integral of $C_a$ (minutes); normalized time-integral of $c_a$, virtual time (min)

$\Theta^*$ Transformed time, equal to ratio between integral of arterial concentrations and arterial concentrations of tracer

$\Theta'^*$ Transformed time, equal to ratio between twice integrated arterial concentrations and integrated arterial concentrations of tracer

$T_{max}$ Maximum flux mediated by transporter protein (pmol $g^{-1}$ $min^{-1}$)

## U

$u$ Dummy time variable of integration (minutes)

## V

$v^*$, $v'^*$ Apparent volume of distribution of tracer, defined as time-variable mass-concentration ratio

$V$ Volume of distribution, (1) time-invariant (steady-state) mass-concentration ratio, (2) apparent volume of distribution, equal to $V_a + V_c + V_e + V_v$; ratio between $m$ and $c_a$ at steady-state (ml $g^{-1}$ or $ml^{-1}$), (3) measured volume of distribution; ratio between $M$ and $C_a$ (milliliters per gram or per milliliter)

$V_a$ Actual vascular volume of distribution of ligand in sample of brain; ratio between $M_a$ and $C$ (milliliters per gram or per milliliter), arterial volume of distribution of tracer in sample of tissue; ratio between $m_a$ and $c_a$ (ml $g^{-1}$ or $ml^{-1}$)

$V_b$ Measured volume of distribution of specifically bound ligand in brain; equivalent to $p_B V_e$ and $B/C_a$ (milliliters per gram or per milliliter), apparent volume of distribution established by binding, ratio between $B_{max}$ and $\alpha_1 K_d^w$ (ml $g^{-1}$ or $ml^{-1}$)

$V_c$ Capillary volume of distribution of tracer in sample of tissue; ratio between $m_c$ and $\bar{c}_c$ (ml $g^{-1}$ or $ml^{-1}$)

$V_c^w$ Actual volume of capillary plasma water in sample of tissue (ml $g^{-1}$ or $ml^{-1}$)

$V_d$ Actual volume of distribution of ligand in sample of brain relative to water; ratio between $M_e$ and $C$ (milliliters per gram or per milliliter), (1) physical (aqueous) volume of distribution, time-invariant (steady-state) mass–concentration ratio of mass of substance in compartment(s) and concentration of substance in reference fluid (water) in same compartment(s), (2) volume of extravascular distribution of tracer in sample of tissue; ratio between $m_e$ and $c_e$ (ml $g^{-1}$ or $ml^{-1}$), volume of distribution of tracer in sample of tissue relative to water; ratio between $M_e$ and $C_e$ (ml $g^{-1}$)

$V_d^w$ Actual volume of extravascular water in sample of tissue (ml $g^{-1}$ or $ml^{-1}$)

$V_e$ Equilibrium volume of distribution of exchangeable ligand in sample of brain relative to source fluid (plasma or blood); ratio between $M_e(\infty)$ and $C_a(\infty)$ (milliliters per gram or per milliliter); partition volume of exchangeable tracer in sample of tissue relative to source fluid (plasma or blood); ratio between $K_1$ and $k_2$ (ml g$^{-1}$). (1) Partition volume, time-invariant (steady-state) mass-concentration ratio of mass of substance in compartments supplied by, and concentration of substance in, delivery compartment, equal to ratio $K_1/k_2$. (2) Partition volume of exchangeable tracer in sample of tissue relative to source fluid (blood), equal to $V_d/\alpha_1$; ratio between $m_e$ and $c_a$ at steady-state (ml g$^{-1}$ or ml$^{-1}$)

$V_f$ Steady-state nonequilibrium volume of distribution of precursor of binding in sample of brain (milliliters per gram or per milliliter); steady-state volume of distribution of substrate of metabolism in sample of tissue; equal to $\rho V_e$ (ml g$^{-1}$). (1) Special partition volume, time-invariant (steady-state) mass-concentration ratio of mass of substance in exchange compartment $\mathbf{M}_2$ shown in Fig. 2.3 and concentration of substance in delivery compartment $\mathbf{M}_1$ ($\mathbf{M}_o$ in case of capillary delivery compartment), equal to ratio $K_1/(k_2+k_3)$. (2) Steady-state volume of distribution of tracer in sample of tissue; equal to $\rho V_e$, and to $K_1/(k_2+k_3)$ (ml g$^{-1}$ or ml$^{-1}$)

$V_g$ Steady-state volume of distribution of substrate of metabolism in sample of tissue; equal to $\rho^2 V_e$ (ml g$^{-1}$); special partition volume, time-invariant (steady-state) mass-concentration ratio of mass of substance in exchange compartment $\mathbf{M}_2$ shown in Fig. 2.3 and concentration of substance in delivery compartment $\mathbf{M}_1$ ($\mathbf{M}_o$ in case of capillary delivery compartment), equal to ratio $K_1 k_2/(k_2+k_3)^2$, (ml g$^{-1}$ or ml$^{-1}$)

$V_I$ Equilibrium volume of distribution of exchangeable inhibitor in sample of brain relative to source fluid (plasma or blood) (ml g$^{-1}$ or ml$^{-1}$)

$V_m$ Volume of distribution of trapped tracer in tissue; equivalent to $\sigma V_e$ and ratio between $m_m$ and $c_a$ at steady-state (ml g$^{-1}$ or ml$^{-1}$)

$V_{max}$ Maximum reaction rate of enzyme catalysis (pmol g$^{-1}$ min$^{-1}$)

$V_o$ Apparent vascular volume of distribution of tracer in tissue, equal to $V_a + V_c(E_o F)/(P_d A) + V_v(1-E_o)$ (ml g$^{-1}$ or ml$^{-1}$)

$V_o^*$ Aapparent vascular volume of distribution of tracer in sample of tissue (ml g$^{-1}$)

$V_v$ Venous volume of distribution of tracer in sample of tissue, ratio between $m_v$ and $c_v$ at steady-state (ml g$^{-1}$ or ml$^{-1}$)

$V_w$ Actual volume of water in sample of brain (ml g$^{-1}$ or ml$^{-1}$)

# XYZ

$x$ Distance

$x_{(i)}$ Independent variable $i$

$y$ (1) Ordinate, axis of orthogonal coordinate system, (2) dependent variable

$z$ Third axis of three-dimensional orthogonal coordinate system.

# References


Abi-Dargham A, Rodenhiser J, Printz D, Zea-Ponce Y, Gil R, Kegeles LS, Weiss R, Cooper TB, Mann JJ, Van Heertum RL, Gorman JM, Laruelle M (2000) Increased baseline occupancy of D2 receptors by dopamine in schizophrenia. Proc Natl Acad Sci USA 97: 8104–8109.

Ackermann RF, Lear JL (1990) Glycolysis-induced discordance between glucose metabolic rates measured with radiolabeled fluorodeoxyglucose and glucose. J Cereb Blood Flow Metab 9: 774–785.

Agnew WF, Crone C (1967) Permeability of brain capillaries to hexoses and pentoses in the rabbit. Acta Physiol Scand 70: 168–175.

Akaike H (1974) A new look at the statistical model identification. IEEE Trans Automatic Contrl 19: 716–723.

Alkire MT, Haier RJ, Barker SJ, Shah NK, Wu JC, Kao YJ (1995) Cerebral metabolism during propofol anesthesia in humans studied with positron emission tomography. Anesthesiology 82: 393–403.

Alkire MT, Haier RJ, Shah NK, Anderson CT (1997) Positron emission tomography study of regional cerebral metabolism in humans during isoflurane anesthesia. Anesthesiology 86: 549–557.

Alkire MT, Pomfrett CJ, Haier RJ, Gianzero MV, Chan CM, Jacobsen BP, Fallon JH (1999) Functional brain imaging during anesthesia in humans: effects of halothane on global and regional cerebral glucose metabolism. Anesthesiology 90: 701–709.

Alves PM, McKenna MC, Sonnewald U (1995) Lactate metabolism in mouse brain astrocytes studied by [13C]NMR spectroscopy. Neuroreport 6: 2201–2204.

Andriezen WL (1893) The neuroglia elements in the human brain. Br Med J ii: 227–230.

Aoki C, Milner TA, Berger SB, Sheu KF, Blass JP, Pickel VM (1987) Glial glutamate dehydrogenase: ultrastructural localization and regional distribution in relation to the mitochondrial enzyme, cytochrome oxidase. J Neurosci Res 18(2): 305–318.

Apostel L (1960) Towards the formal study of models in the non-formal sciences. Synthase 12: 125–161.

Armstrong BD, Noguchi KK (2004) The neurotoxic effects of 3,4-methylenedioxymetham phetamine (MDMA) and methamphetamine on serotonin, dopamine, and GABA-ergic terminals: an in-vitro autoradiographic study in rats. Neurotoxicology 25: 905–914.

Arnett CD, Wolf AP, Shiue CY, Fowler IS, MacGregor RR, Christman DR, Smith MR (1986) Improved delineation of human dopamine receptors using [$^{1}$8F]-N-methylspiroperidol and PET. J Nucl Med 27: 1878–1882.

Asano T, Shibasaki Y, Kasuga M, Kanazawa Y, Takaku F, Akanuma Y, Oka Y (1988) Cloning of a rabbit brain glucose transporter cDNA and alteration of glucose transporter mRNA during tissue development. Biochem Biophys Res Commun 154: 1204–1211.

Attwell D, Laughlin SB (2001) An energy budget for signaling in the grey matter of the brain. J Cereb Blood Flow Metab 21: 1133–1145.

Attwell D, Iadecola C (2002) The neural basis of functional brain imaging signals. Trends Neurosci 25: 621–625.

Aubert A, Costalat R (2002) A model of the coupling between brain electrical activity, metabolism, and hemodynamics: application to the interpretation of functional neuroimaging. NeuroImage 17: 1162–1181.

Axelrod JD, Pilch PF (1983) Unique cytochalasin B binding characteristics of the hepatic glucose carrier. Biochemistry 22: 2222–2227.

Ayata C, Ma J, Meng W, Huang P, Moskowitz MA (1996) L-NA-sensitive rCBF augmentation during vibrissal stimulation in type III nitric oxide synthase mutant mice. J Cereb Blood Flow Metab 16: 539–541.

Azzaro AJ, King J, Kotzuk J, Schoepp DD, Frost J, Schochet S (1985) Guinea pig striatum as a model of human dopamine deamination: the role of monoamine oxidase isozyme ratio, localization, and affinity for substrate in synaptic dopamine metabolism. J Neurochem 45: 949–956.

Bachelard HS, Daniel PM, Love ER, Pratt OE (1973) The transport of glucose into the brain of the rat in vivo. Proc R Soc Lond B 183: 990–993.

Bachelard HS, Brooks KJ, Garofalo O (1991) Studies on the compartmentation of DOG metabolism in the brain. Neurochem Res 16: 1025–1030.

Bagley PR, Tucker SP, Nolan C, Lindsay JG, Davies A, Baldwin SA, Cremer JE, Cunningham VJ (1989) Anatomical mapping of glucose transporter protein and pyruvate dehydrogenase in rat brain: an immunogold study. Brain Res 499: 214–224.

Baker GF, Widdas WF (1973) The asymmetry of the facilitated transfer system for hexose simple kinetics of a two component model. J Physiol 231: 143–165.

Baker PF, Connelly CM (1966) Some properties of the external activation site of the sodium pump in crab nerve. J Physiol (Lond) 185: 270–297.

Balaban RS, Kantor HL, Katz LA, Briggs RW (1986) Relation between work and phosphate metabolites in the in vivo paced mammalian heart. Science 232: 1121–1123.

Baldwin SA, Lienhard GE (1981) Glucose transport across plasma membranes: facilitated diffusion systems. Trends Biochem Sci 6: 208–211.

Baldwin SA, Baldwin JM, Lienhard GE (1982) Monosaccaride transporter of the human erythrocyte. Characterization of an improved preparation. Biochem 21: 3836–3842.

Baldwin SA, Brewster F, Cairns MT, Gardiner RM, Ruggier R (1984) Identification of a D-glucose sensitive cytochalasin B binding component of isolated ovine cerebral microvessels. J Physiol 357: 75P.

Baldwin SA, Cairns MT, Gardiner RM, Ruggier R (1985) A D-glucose sensitive cytochalasin B binding component of cerebral microvessels. J Neurochem 45: 650–652.

Baldwin SA, Lienhard GE (1989) Purification and reconstitution of glucose transporter from human erythrocytes. Meth Enzymol 174: 39–50.

Barnett JEG, Holman RA, Chalkey RA, Munday KA (1975) Evidence for two asymmetric conformational states in the human erythrocyte sugar transport system. Biochem J 145: 417–429.

Berson FG, Spatz M, Klatzo I (1975) Effects of oxygen saturation and $PCO_2$ on brain uptake of glucose analogues in rabbits. Stroke 6: 691–696.

Bassingthwaighte JB (2000) Strategies for the physiome project. Ann Biomed Eng 28(8): 1043–1058.

Bell GI, Kayano T, Buse JB, Burant CF, Takeda J, Lin D, Fukumoto H, Seino S (1990) Molecular biology of mammalian glucose transporters. Diab Care 13: 198–208.

van den Berg CJ, Bruntink R (1983) Glucose oxidation in the brain during seizures: experiments with labeled glucose and deoxyglucose. In: Hertz L, Kvamme E, McGeer EG, Schousboe A (eds) Glutamine, Glutamate and GABA in the Central Nervous System, Alan R Liss, New York, pp. 619–624.

Bergersen LH, Magistretti PJ, Pellerin L (2005) Selective postsynaptic co-localization of MCT2 with AMPA receptor GluR2/3 subunits at excitatory synapses exhibiting AMPA receptor trafficking. Cereb Cortex 15: 361–370.

Berry JJ, Hoffman JM, Steenbergen C, Baker JA, Floyd C, van Trigt P, Hanson MW, Coleman RE (1993) Human pathologic correlation with PET in ischemis and non-ischemic cardiomyopathy. J Nucl Med 34: 39–47.

Betz AL, Gilboe DD, Yudilevich DK, Drewes LR (1973) Kinetics of unidirectional glucose transport into the isolated dog brain. Am J Physiol 225: 586–592.
Betz AL, Gilboe DD (1974) Kinetics of cerebral glucose transport in vivo: inhibition by 3-O-methylglucose. Brain Res 65: 368–372.
Betz AL, Gilboe DD, Drewes LR (1974) Effects of anoxia on net uptake and unidirectional transport of glucose into the isolated dog brain. Brain Res 65: 307–316.
Betz AL, Drewes LR, Gilboe DD (1975) Inhibition of glucose transport into brain by phlorizin, phloretin and glucose analogues. Biochim Biophys Acta 406: 505–515.
Betz AL, Csejtey J, Goldstein GW (1979) Hexose transport and phosphorylation by capillaries isolated from rat brain. Am J Physiol 236: C96–C102.
Betz AL, Goldstein GW (1981) Developmental changes in metabolism and transport properties of capillaries isolated from rat brain. J Physiol Lond 312: 365–376.
Betz AL, Bowman PD, Goldstein GW (1983) Hexose transport in microvascular endothelial cells cultured from bovine retina. Exp Eye Res 36: 269–277.
Betz AL, Iannotti F (1983) Simultaneous determination of regional cerebral blood flow and blood–brain glucose transport kinetics in the gerbil. J Cereb Blood Flow Metab 3: 193–199.
Bidder TG (1968) Hexose translocation across the blood–brain interface: configurational aspects. J Neurochem 15: 867–874.
Bing RJ (1965) Cardiac metabolism. Physiol Rev 45: 171–213.
Birnbaum MJ, Haspel HC, Rosen OM (1986) Cloning and characterization of a cDNA encoding the rat brain glucose-transporter protein. Proc Natl Acad Sci 83: 5784–5788.
Birnbaum MJ (1989) Identification of a novel gene encoding an insulin-responsive glucose transporter protein. Cell 57: 305–315.
Bittar PG, Charnay Y, Pellerin L, Bouras C, Magistretti PJ (1996) Selective distribution of lactate dehydrogenase isoenzymes in neurons and astrocytes of human brain. J Cereb Blood Flow Metab 16: 1079–1089.
Blasberg RG, Patlak CS, Jehle JW, Fenstermacher JD (1978) An autoradiographic technique to measure the permeability of normal and abnormal brain capillaries. Neurology 28: 363.
Blomqvist G (1984) On the construction of functional maps in positron emission tomography. J Cereb Blood Flow Metab 4(4): 629–632.
Blomqvist G, Bergström K, Bergstrom M, Elvin E, Eriksson L, Garmelius B, Lindberg B, Lilja A, Litton JE, Lundmark L, Lundqvist H, Malmberg P, Moström U, Nilsson L, Stone-Elander S, Widén L (1985) Models for $^{11}C$-glucose. In: Greitz T, Ingvar DH, Widen L (eds) The Metabolism of the Brain Studied with Positron Emission Tomography, Raven Press, New York, pp. 185–194.
Blomqvist G, Stone-Elander S, Halldin C, Roland PE, Widén L, Lindqvist M, Swahn CG, Långström, Wiesel FA (1990) Positron emission tomographic measurements of cerebral glucose utilization using D-[1-$^{11}$C]-glucose. J Cereb Blood Flow Metab 10: 467–484.
Blomqvist G, Gjedde A, Gutniak M, Grill V, Widén L, Stone-Elander S, Hellstrand E (1991) Facilitated transport of glucose from blood to brain in man and the effect of moderate hypoglycaemia on cerebral glucose utilization. Eur J Nucl Med 18: 834–837.
Boado RJ, Pardridge WM (1990) The brain-type glucose transporter mRNA is specifically expressed at the blood–brain barrier. Biochem Biophys Res Commun 166: 174–179.
Bohr C (1885) Experimentelle Untersuchungen uber die Sauerstoffaufnahme des Blutfarbstoffes. Copenhagen.
Bohr C (1904) Die Sauerstoffe des genuinen Blutfarbstoffes und des aus dem Blute darstellten Hamoglobins. Zentralbl Physiol 23: 1–7.
Bohr C (1909) Über die spezifische Tätigkeit der Lungen bei der respiratorischen Gasaufnahme und ihr Verhalten zu der durch die Alveolarwand stattfindenden Gasdiffusion. Skand Arch Physiol 22: 221–280.
Boja JW, Cline EJ, Carroll FI, Lewin AH, Philip A, Dannals R, Wong DF, Scheffel U, Kuhar MJ (1992) High potency cocaine analogs – Neurochemical, imaging, and behavioral-studies. Ann NY Acad Sci 654: 282–291.

Borbely K, Brooks RA, Wong DF, Burns RS, Cumming P, Gjedde A, Di Chiro G (1999) NMSP binding to dopamine and serotonin receptors in MPTP-induced parkinsonism: relation to dopa therapy. Acta Neurol Scand 100: 42–52.

Borowsky IW, Collins RC (1989) Metabolic anatomy of brain: a comparison of regional capillary density, glucose metabolism and enzyme activities. J Comp Neurol 288: 401–413.

Bowers C, Zigmond R (1979) Localization of neurons in the rat superior cervical ganglion that project into different postganglionic trunks. J Comp Neurol 185: 381–391.

Bradbury MWB (1985) The blood–brain barrier in vitro. Neurochem Int 7: 27–28.

Brady L (2005) PET and SPECT Consortium: Safety Assessment of Novel, High Specificity, Low Mass Radiotracers for Use in Research and Drug Discovery, National Institute of Mental Health, Bethesda, Maryland.

Brandt RB, Laux JE, Spainhour SE, Kline ES (1987) Lactate dehydrogenase in rat mitochondria. Arch Biochem Biophys 259: 412–422.

Braun LD, Cornford EM, Oldendorf WH (1980) Newbonrn rabbit blood–brain barrier is selectively permeable and differs substantially from the adult. J Neurochem 34: 147–152.

Braun LD, Miller LP, Pardridge WM, Oldendorf WH (1985) Kinetics of regional blood–brain barrier glucose transport and cerebral blood flow determined with the carotid injection technique in conscious rats. J Neurochem 44: 911–915.

Breier A, Su TP, Saunders R, Carson RE, Kolachana BS, de Bartolomeis A, Weinberger DR, Weisenfeld N, Malhotra AK, Eckelman WC, Pickar D (1997) Schizophrenia is associated with elevated amphetamine-induced synaptic dopamine concentrations: evidence from a novel positron emission tomography method. Proc Natl Acad Sci USA 94: 2569–2574.

Brender J, Andersen PE, Rafaelsen OJ (1975) Blood–brain transfer of D- glucose, L-leucine, and L-tryptophan in the rat. Acta Physiol Scand 93: 490–499.

Briggs GE, Haldane JBS (1925) A note on the kinetics of enzyme action. Biochem J 19: 338–339.

Brightman MW, Reese TS (1969) Junctions between intimately apposed cell membranes in the vertebrate brain. J Cell Biol 40: 648–677.

Brodersen P, Jørgensen EO (1974) Cerebral blood flow and oxygen uptake, and cerebrospinal fluid biochemistry in severe coma. J Neurol Neurosurg Psychiat 37: 384–391.

Broer S, Rahman B, Pellegri G, Pellerin L, Martin JL, Verleysdonk S, Hamprecht B, Magistretti PJ (1997) Comparison of lactate transport in astroglial cells and monocarboxylate transporter 1 (MCT 1) expressing Xenopus laevis oocytes. Expression of two different monocarboxylate transporters in astroglial cells and neurons. J Biol Chem 272: 30096–30102.

Brøndsted HE, Gjedde A (1988) Measuring brain glucose phosphorylation with labeled glucose. Am J Physiol 254: E443–E448.

Brooks DJ, Beaney RP, Lammertsma AA, Herold S, Turton DR, Luthra SK, Frackowiak RSJ, Thomas DGT, Marshall J, Jones T (1986a) Glucose transport across the blood–brain barrier in normal human subjects and patients with cerebral tumours studied using [$^{11}$C]3-*O*-methyl-D-glucose and positron emission tomography. J Cereb Blood Flow Metab 6: 230–239.

Brooks DJ, Gibbs JSR, Sharp P, Herold S, Turton DR, Luthra SK, Kohner EM, Bloom SR, Jones T (1986b) Regional cerebral glucose transport in insulin-dependent diabetic patients studied using [$^{11}$C]3-*O*-methyl-D-glucose and positron emission tomography. J Cereb Blood Flow Metab 6: 240–244.

Brooks GA, Brown MA, Butz CE, Sicurello JP, Dubouchaud HJ (1999a) Cardiac and skeletal muscle mitochondria have a monocarboxylate transporter MCT1. Appl Physiol 87: 1713–1718.

Brooks GA, Dubouchaud H, Brown M, Sicurello JP, Butz CE (1999b) Role of mitochondrial lactate dehydrogenase and lactate oxidation in the intracellular lactate shuttle. Proc Natl Acad Sci USA 96: 1129–1134.

Brooks RA, Hatazawa J, DiChiro G, Larson SM, Fishbein DS (1987) Human cerebral glucose metabolism determined by positron emission tomography. J Cereb Blood Flow Metab 7: 427–432.

Brown GC (2001) Regulation of mitochondrial respiration by nitric oxide inhibition of cytochrome c oxide. Biochem Biophys Acta 1504(1): 46–57

Brunken RC, Mody FV, Hawkins RA, Nienaber C, Phelps ME, Schelbert HR (1992) Positron emission tomography detects metabolic viability in myocardium with persistent 24-hour single-photon emission computed tomography 2O1T1 defects. Circulation 86: 1357–1369.

Bryan RM, Hawkins RA, Mans AM, Davis DW, Page RB (1983) Cerebral glucose utilization in awake unstressed rats. Am J Physiol 244: C270–C275.

Brøndsted HE, Gjedde A (1988) Measuring brain glucose phosphorylation with labeled glucose. Am J Physiol 254 (Endocrinol Metab 17): E443–E448.

Buchert R, Thomasius R, Wilke F, Petersen K, Nebeling B, Obrocki J, Schulze O, Schmidt U, Clausen M (2004) A voxel-based PET investigation of the long-term effects of "Ecstasy" consumption on brain serotonin transporters. Am J Psychiatry 161: 1181–1189.

Buschiazzo PM, Terrell EB, Regen DM (1970) Sugar transport across the blood–brain barrier. Am J Physiol 219: 1505–1513.

Buxton RB, Frank LR (1997) A model for the coupling between cerebral blood flow and oxygen metabolism during neural stimulation. J Cereb Blood Flow Metab 17: 64–72.

Caesar K, Akgoren N, Mathiesen C, Lauritzen M (1999) Modification of activity-dependent increases in cerebellar blood flow by extracellular potassium in anaesthetized rats. J Physiol (Lond) 520: 281–292.

Camici P, Ferrannini E, Opic LH (1989) Myocardial metabolism in ischemic heart disease: basic principles and application to imaging by positron emission tomography. Prog Cardiovasc Dis 32: 217–238.

Canfield DR, Spealman RD, Kaufman MJ, Madras BK (1990) Autoradiographic localization of cocaine binding sites by [$^3$H]CFT ([$^3$H]WIN 35,428) in the monkey brain. Synapse 6: 189–195.

Carroll FI, Abraham P, Lewin AH, Parham KA, Boja JW, Kuhar M (1992) Isopropyl and phenyl esters of 3 beta-(4-substituted phenyl)tropan-2beta-carboxylic acids. Potent and selective compounds for the dopamine transporter. J Med Chem 35: 2497–2500.

Carruthers A (1986) ATP regulation of the human red cell sugar transporter. J Biol Chem 261: 11028–11037.

Carruthers A (1990) Facilitated diffusion of glucose. Physiol Rev 70: 1135–1176.

Carter-Su C, Pessin JE, Mora R, Gitomer W, Czech MP (1982) Photoaffinity labeling of the human erythrocyte D-glucose transporter. J Biol Chem 257: 5419–5425.

Chance B (1943) The kinetics of the enzyme-substrate compound of peroxidase. J Biol Chem 151: 553–577.

Charron M, Brosius FC, Alper SL, Lodish HF (1989) A glucose transport protein expressed predominantly in insulin-responsive tissues. Proc Natl Acad Sci 86: 2535–2539.

Chesler M, Kraig RP (1987) Intracellular pH of astrocytes increases rapidly with cortical stimulation. Am J Physiol 253: R666–R670.

Chesler M, Kraig RP (1989) Intracellular pH transients of mammalian astrocytes. J Neurosci 9: 2011–2019.

Chien JB, Wallingford RA, Ewing AG (1990) Estimation of free dopamine in the cytosplasm of the giant dopamine cell of Planorbis corbeus by voltammetry and capillary electrophoresis. J Neurochem 54: 633–638.

Choi IY, Lei H, Gruetter R (2002) Effect of deep pentobarbital anesthesia on neurotransmitter metabolism in vivo: on the correlation of total glucose consumption with glutamatergic action. J Cereb Blood Flow Metab 22: 1343–1351.

Choi TB, Boado RJ, Pardridge WM (1989) Blood–brain barrier glucose transporter mRNA is increased in experimental diabetes mellitus. Biochem Biophys Res Commun 164: 375–380.

Choi Y, Hawkins RA, Huang SC, Gambhir SS, Brunken RC, Phelps ME, Schelbert HR (1991) Parametric images of myocardial metabolic rate of glucose generated from dynamic cardiac PET and 2-[$^{18}$F]fluoro-2-deoxy-D-glucose studies. J Nucl Med 32: 733–738.

Cholet N, Seylaz J, Lacombe P, Bonvento G (1997) Local uncoupling of the cerebrovascular and metabolic responses to somatosensory stimulation after neuronal nitric oxide synthase inhibition. J Cereb Blood Flow Metab 17: 1191–1201.

Christensen TG, Diemer NH, Laursen H, Gjedde A (1981) Starvation accelerates blood–brain glucose transfer. Acta Physiol Scand 112: 221–223.

Christensen O, Andersen HL, Betz AL, Gilboe DD (1982) Transport of glucose across the blood–brain barrier: reevaluation of the accelerative exchange diffusion. Acta Physiol Scand 115: 233–238.

Clapham DE (1995) Calcium signalling. Cell 80: 259–268.

Connett RJ, Gayeski TE, Honig CR (1983) Lactate production in a pure red muscle in absence of anaoxia: mechanisms and significance. Adv Exp Med Biol 159: 327–335.

Connett RJ, Gayeski TE, Honig CR (1984) Lactate accumulation in fully aerobic, working dog gracilis muscle. Am J Physiol 246: H120–H128.

Connett RJ, Gayeski TE, Honig CR (1985) Energy sources in fully aerobic rest-work transitions: a new role for glycolysis. Am J Physiol 248: H922–H929.

Connett RJ (1987) Glycolytic regulation during aerobic rest-to-work transition in dog gracilis muscle. J Appl Physiol 63: 2366–2374.

Cornford EM, Braun LD, Oldendorf WH (1982) Developmental modulations of blood–brain barrier permeability as an indicator of changing nutritional requirements in the brain. Petriatr Res 16: 324–328.

Cornford EM, Cornford ME (1986) Nutrient transport and the blood-brain barrier in developing animals. Federation Proc 45: 2065–2072.

Crane PD, Pardridge WM, Braun LD, Oldendorf WH (1983) Kinetics of transport and phsophorylation of 2-fluoro-2-deoxy-D-glucose in rat brain. J Neurochem 40: 160–167.

Crane PD, Pardridge WM, Braun LD, Oldendorf WH (1985) Two-day starvation does not alter the kinetics of blood–brain barrier transport and phosphorylation of glucose in rat brain. J Cereb Blood Flow Metab 5: 40–46.

Crawley JCW, Crow TI, Johnstone EC, Oldland SRD, Owen F, Owens DGC, Poulter M, Pufter O, Smith T, Veall N, Zanelli GD (1986) Lancet 2: 224–225.

Cremer JE (1976) The influence of liver-bypass on transport and compartmentation in vivo. Adv Exp Med Biol 69: 95–102.

Cremer JE, Braun LD, Oldendorf WH (1976) Changes during development in transport processes of the blood–brain barrier. Biochem Biophys Acta 448: 633–637.

Cremer JE, Cunningham VJ (1979) Effects of some chlorinated sugar derivatives on the hexose transport system of the blood–brain barrier. Biochem J 180: 677–679.

Cremer JE, Cunningham VJ, Pardridge WM, Braun LD, Oldendorf WH (1979) Kinetics of blood–brain transport of pyruvate, lactate and glucose in suckling, weanling and adult rats. J Neurochem 33: 439–446.

Cremer JE, Ray DE, Sarna GS, Cunningham VJ (1981) A study of the kinetic behavior of glucose based on simultaneous estimates of influx and phosphorylation in brain regions of rats in different physiological states. Brain Res 221: 331–342.

Cremer JE, Cunningham VJ, Seville MP (1983) Relationship between extraction and metabolism of glucose, blood flow, and tissue blood volume in regions of rat brain. J Cereb Blood Flow Metab 3: 291–302.

Cremer JE, Seville MP, Cunningham VJ (1988) Tracer deoxyglucose kinetics in brain regions of rats given kainic acid. J Cereb Blood Flow Metab 8: 244–253.

Cressie NAC, Keightley DD (1981) Analysing data from hormone-receptor assays. Biometrics 37: 235–249.

Crone C (1960) The diffusion of some organic non-electrolytes from blood to brain tissue. Acta Physiol Scand 50 (Suppl. 175): 33–34.

Crone C (1963) The permeability of capillaries in various organs as determined by use of the 'indicator diffusion' method. Acta Physiol Scand 58: 292–305.

Crone C (1965) Facilitated transfer of glucose from blood into brain tissue. J Physiol (Lond) 181: 103–113.

Crone C (1975) General properties of the blood–brain barrier with special emphasis on glucose. In: Cserr HF, Fenstermacher JD, Fencl V (eds) Fluid Environment of the Brain. Academic Press, New York, p 33.

Crone C (1985) The blood–brain barrier: a modified tight epithelium. In: Suckling AJ, Rumsby MG, Bradbury MWB (eds) The Blood–Brain Barrier in Health and Disease. Ellis Horwood Limited, Chichester, p 17.

Cumming P, Brown E, Damsma G, Fibiger HC (1992) Formation and clearance of interstitial metabolites of dopamine and serotonin in rat striatum: an in vivo microdialysis study. J Neurochem 59: 1905–1914.

Cumming P, Léger GC, Kuwabara H, Gjedde A (1993) Pharmacokinetics of plasma 6-[$^{18}$F]fluoro-L-3,4-dihydroxy- phenylalanine ([$^{18}$F]FDOPA) in humans. J Cereb Blood Flow Metab 13: 668–675.

Cumming P, Kuwabara H, Ase A, Gjedde A (1995) Regulation of DOPA decarboxylase activity in brain of living rat. J Neurochem 65: 1381–1390.

Cumming P, Ase A, Kuwabara H, Gjedde A (1998) [$^{3}$H]DOPA formed from [$^{3}$H]tyrosine in living rat brain is not committed to dopamine synthesis. J Cereb Blood Flow Metab 18: 491–499.

Cumming P, Hermansen F, Gjedde A (1999) Cerebral dopamine concentrations during levodopa treatment. Neurology 53: 1374–1375.

Cumming P, Wong DF, Gillings N, Hilton J, Scheffel U, Gjedde A (2002) Specific binding of [(11)C]raclopride and *N*-[(3)H]propyl-norapomorphine to dopamine receptors in living mouse striatum: occupancy by endogenous dopamine and guanosine triphosphate-free G protein. J Cereb Blood Flow Metab 22: 596–604.

Cunningham VJ, Sarna GS (1979) Estimation of the kinetic parameters of unidirectional transport across the blood-brain barrier. J Neurochem 33: 433–437.

Cunningham VJ, Cremer JE (1981) A method for the simultaneous estimation of regional rates of glucose influx and phosphorylation in rat brain using radio-labelled 2-deoxy-glucose. Brain Res 221: 319–330.

Cunningham VJ, Hargreaves RJ, Pelling D, Moorhouse SR (1986) Regional blood-brain glucose transfer in the rat: a novel double-membrane kinetic analysis. J Cereb Blood Flow Metab 6: 305–314.

Cunningham VJ, Cremer JE, Hargreaves RJ (1989) Relationships between neuronal activity, energy metabolism and cerebral circulation. In: Battani F et al. (eds) Regulatory mechanisms of neuron to vessel communication in the brain, vol H33. Springer Verlag, Berlin Heidelberg, p 325.

Cunningham VJ, Jones T (1993) Spectral analysis of dynamic PET studies. J Cereb Blood Flow Metab 13: 15–23.

Cutler RWP, Sipe JC (1971) Mediated transport of glucose between blood and brain in the cat. Am J Physiol 220(5): 1182–1186.

Daniel PM, Moorehouse SR, Love ER, Pratt OE (1971) Factors influencing utilization of ketone-bodies by brain in normal rats. Lancet ii: 637–638.

Daniel PM, Donaldson J, Pratt OE (1975a) A method for injecting substances into the circulation to reach rapidly and to maintain a steady level. Med Biol Engng 13: 214–227.

Daniel PM, Love ER, Pratt OE (1975b) Insulin and the way the brain handles glucose. J Neurochem 25: 471–476.

Daniel PM, Love ER, Pratt OE (1978) The effect of age upon the influx of glucose into the brain. J Physiol 274: 141–148.

Daniel PM, Love ER, Pratt OE (1980) Inhibition of glucose transport into the brains of suckling rats by raising the level of galactose in the blood. J Physiol 305: 44–45.

Daniels GM, Amara SG (1999) Regulated trafficking of the human dopamine transporter. Clathrin-mediated internalization and lysosomal degradation in response to phorbol esters. J Biol Chem 274: 35794–35801.

DeFronzo RA, Andres R, Bledsoe TA, Boden G, Faloona GA, Tobin JD (1977) A test of the hypothesis that the rate of fall in glucose concentration triggers counterregulatory hormonal responses in man. Diabetes 26: 445–452.

DeFronzo RA, Tobin JD, Andres R (1979) Glucose clamp technique: a method for quantifying insulin secretion and resistance. Am J Physiol 237: E214–E223.

DeFronzo RA, Hendler R, Christensen N (1980) Stimulation of counterregulatory hormonal responses in diabetic man by a fall in glucose concentration. Diabetes 29: 125–131.

de Graaf RA, Mason GF, Patel AB, Rothman DL, Behar KL (2004) Regional glucose metabolism and glutamatergic neurotransmission in rat brain in vivo. Proc Natl Acad Sci USA 101: 12700–12705.

Denton RM, McCormack JG (1985) Ca transport by mammalian mitochondria and its role in hormone action. Am J Physiol 249: E543–E554.

Desagher S, Glowinski J, Premont JJ (1997) Pyruvate protects neurons against hydrogen peroxide-induced toxicity. Neuroscience 17: 9060–9067.

de Silva R, Yamamoto Y, Rhodes CG, Iida H, Nihoyannopoulos P, Davies GJ, Lammertsma AA, Jones T, Maseri A (1992) Preoperative prediction of the outcome of coronary revascularization using positron emission tomography. Circulation 86: 1738–1742.

Detre JA, Koretsky AP, Williams DS, Ho C (1990a) Absence of pH changes during altered work in the in vivo sheep heart: a 31P-NMR investigation. J Mol Cell Cardiol 22: 543–553.

Detre JA, Williams DS, Koretsky AP (1990b) Nuclear magnetic resonance determination of flow, lactate, and phosphate metabolites during amphetamine stimulation of the rat brain. Nucl Magn Res Biomed 3: 272–278.

Deussen A, Bassingthwaighte JB (1996) Modeling [15O]oxygen tracer data for estimating oxygen consumption. Am J Physiol 270(3 Pt 2): H1115–H1130.

Dichiara G, Imperato A (1988) Drugs abused by humans preferentially increase synaptic dopamine concentrations in the mesolimbic system of freely moving rats. Proc Natl Acad Sci USA 85: 5274–5278.

Dick APK, Harik SI, Klip A, Walker DM (1984) Identification and characterization of the glucose transporter of the glucose transporter of the blood–brain barrier by cytochalasin B binding and immunological reactivity. Proc Natl Acad Sci USA 81: 7233–7237.

Dick APK, Harik SI (1986) Distribution of the glucose transporter in the mammalian brain. J Neurochem 46: 1406–1411.

Diemer NH, Benveniste H, Gjedde A (1985) In vivo cell membrane permeability to deoxyglucose in rat brain. Acta Neurologica Scandinavica 72(1): 87–87.

Dienel GA, Cruz NF, Nakanishi H, Melzer P, Moulis P, Sokoloff L (1992) Comparison of rates of local cerebral glucose utilization determined with deoxy-[1-$^{1}$4C]glucose and deoxy-[6-$^{14}$C]glucose. J Neurochem 59: 1430–1436.

Dienel GA, Cruz NF (2004) Nutrition during brain activation: does cell-to-cell lactate shuttling contribute significantly to sweet and sour food for thought? Neurochem Int 45: 321–351.

Dixon M (1953) The determination of enzyme inhibitor constants. Biochem J 55: 170–171.

Djuricic DM, Kostic VS, Mrsulja BB (1983) Insulin increases entrance of 2-deoxy-D-[$^{3}$H]glucose in isolated rat brain microvessels. Brain Res 275: 186–188.

Drewes LR, Horton RW, Betz AL, Gilboe DD (1977) Cytochalasin B inhibition of brain glucose transport and the influence of blood components on inhibitor concentration. Biochem Biophys Acta 471: 477–486.

Drewes L (1999) Transport of brain fuels, glucose and lactate. In: Paulson OB, Knudsen GM, Moos T (eds) Brain Barrier Systems, Alfred Benzon Symposium 45, Munksgaard, Copenhagen, pp. 285–295.

Drewes LR (2002) Molecular architecture of the brain microvasculature: perspective on blood–brain transport. J Mol Neurosci 16: 93–98.

Duckrow RB, Bryan RM Jr (1987) Regional cerebral glucose utilization during hyperglycemia. J Neurochem 48: 989–993.

Duckrow RB (1988) Glucose transfer into rat brain during acute and chronic hyperglycemia. Metab Brain Dis 3: 201–209.

Duling BR, Berne RM (1970) Longintudinal gradients in periarteriolar oxygen tension: a possible mechanism for the participation of of oxygen in local regulation of blood flow. Circ Res 27: 669–678.

Dyve S, Gjedde A (1991) Glucose metabolism of fetal rat brain in utero measured with labeled deoxyglucose. Acta Neurol Scand 83: 14–19.

Eadie GS (1942) The inhibition of cholinesterase by physostigmine and prostigmine. J Biol Chem 146: 85–93.

Eadie GS (1952) On the evaluation of the constants $V_m$ and $K_M$ in enzyme reactions. Science 116: 688.

Eastman RC, Carson RE, Gordon ME, Berg GW, Lillioja S, Larson SM, Roth J (1990) Brain glucose metabolism in noninsulin-dependent diabetes mellitus: a study in Pima Indians using positron emission tomography during hyperinsulinemia with euglycemic glucose clamp. J Clin Endocrinol Metab 71: 1602–1610.

Ehrlich P (1909) ber den jetzigen Stand der Chemotherapie. Berichte der Deutschen Chemischen Gesellschaft 42: 17–47.

Einstein A (1908) [The elementary theory of the Brownian motion.] Z f Elektrochemie 14: 235–239.

Elhusseiny A, Cohen Z, Olivier A, Stanimirovic DB, Hamel E (1999) Functional acetylcholine muscarinic receptor subtypes in human brain microcirculation: identification and cellular localization. J Cereb Blood Flow Metab 19: 794–802.

Elhusseiny A, Hamel E (2000) Muscarinic–but not nicotinic–acetylcholine receptors mediate a nitric oxide-dependent dilation in brain cortical arterioles: a possible role for the M5 receptor subtype. J Cereb Blood Flow Metab 20: 298–305.

Eisenberg HM, Kadekaro M, Freeman S, Terrell ML (1993) Metabolism in the globus pallidus after fetal implants in rats with nigral lesions. J Neurosurg 78: 83–89.

Erecinska M, Veech RL, Wilson DF (1974) Thermodynamic relationships between the oxidation-reduction reactions and the ATP synthesis in suspensions of isolated pigeon heart mitochondria. Arch Biochem Biophys 160: 412–421.

Erecinska M, Wilson DF (1982) Regulation of cellular energy metabolism. J Membr Biol 70: 1–14.

Erecinska M, Silver I (1989) ATP and brain function. J Cereb Blood Flow Metab 9: 2–19.

Erecinska M, Nelson D, Chance B (1991) Depolarization-induced changes in cellular energy production. Proc Natl Acad Sci USA 88: 7600–7604.

Ernst M, Zametkin AJ, Jons PH, Matochik JA, Pascualvaca D, Cohen RM (1999) High presynaptic dopaminergic activity in children with Tourette's disorder. J Am Acad Child Adol Psychiat 38: 86–94.

Evans AC, Diksic M, Yamamoto YL, Kato A, DAgher A, Redies C, Hakim A (1986) Effect of vascular activity in the determination of rate constants for the uptake of $^{18}F$-labeled 2-fluoro-2-deoxy-D-glucose: error analysis and normal values in older subjects. J Cereb Blood Flow Metab 6: 724–738.

Evans AC (1987) A double integral form of the three-compartmental, four-rate-constant model for faster generation of parameter maps. J Cereb Blood Flow Metab 7 (Suppl 1): S453.

Fabricius M, Lauritzen M (1994) Examination of the role of nitric oxide for the hypercapnic rise of cerebral blood flow in rats. Am J Physiol 266: H1457–H1464.

Fantegrossi WE, Woods JH, Winger G, Kilbourn M, Sherman P, Woolverton WL, Yuan J, Hatzidimitriou G, Ricaurte GA (2004) Behavioral and neurochemical consequences of long-term intravenous self-administration of MDMA and its enantiomers by rhesus monkeys. Neuropsychopharmacology 29: 1270–1281.

Farde L, Ehrin E, Eriksson L, Greitz T, Hall H, Hedstrom CG, Litton JE, Sedvall G (1985) Substituted benzamides as ligands for visualization of dopamine receptor binding in the human brain by positron emission tomography. Proc Natl Acad Sci USA 82: 3863–3867.

Farde L, Hall H, Ehrin E, Sedvall G (1986) Quantitative analysis of $D_2$ dopamine receptor binding in the living human brain by PET. Science 231: 258–261.

Farde L, Halldin C, Stone-Elander S, Sedvall G (1987) PET analysis of human dopamine receptor subtypes using 11C-SCH 23390 and 11C-raclopride. Psychopharmacology (Berl) 92(3): 278–284.

Farde L, Wiesel FA, Halldin C, Sedvall G (1988) Central D2-dopamine receptor occupancy in schizophrenic patients treated with antipsychotic drugs. Arch Gen Psychiatry 45: 71–76.

Farde L, Wiesel FA, Stone-Elander S, Halldin C, Nordstrom AL, Hall H, Sedvall G (1990) Arch Gen Psychiatry 47: 213–219.

Farde L, Hall H, Pauli S, Halldin C (1995) Variability in D2-dopamine receptor density and affinity: a PET study with [11C]raclopride in man. Synapse 20(3): 200–208.

Farde L, Gustavsson JP, Jonsson E (1997) D2 dopamine receptors and personality traits. Nature 385: 590.

Fedosov SN (1994) Creatine-creatine phosphate shuttle modeled as two-compartment system at different levels of creatine kinase activity. Biochim Biophys Acta 1208: 238–246.

Feinendegen LE, Herzog H, Wieler H, Patton DD, Schmid A (1986) Glucose transport and utilization in the human brain: model using carbon-11 methylglucose and positron emission tomography. J Nucl Med 27: 1867–1877.

Fenstermacher JD, Patlak CS, Blasberg RG (1979) New method of estimating plasma to tissue transfer constants. Fed Proc 38(3): 1138–1138.

Fishman RS, Karnovsky ML (1986) Apparent absence of a translocase in the cerebral glucose-6-phosphatase system. J Neurochem 46: 371–378.

Fodor JA (1983) The Modularity of Mind. The MIT Press, Cambridge.

Fodor JA (2000) The Mind Doesn't Work That Way. The MIT Press, Cambridge.

Fox PT, Raichle ME (1986) Focal physiological uncoupling of cerebral blood flow and oxidative metabollism during somatosensory stimulation in human subjects. Proc Natl Acad Sci USA 83: 1140–1144.

Fox PT, Raichle ME, Mintun MA, Dence CE (1988) Nonoxidative glucose consumption during focal physiological activity. Science 241: 462–464.

Frey KA, Hichwa RD, Ehrenkaufer RL, Agranoff BW (1985) Quantitative in vivo receptor binding III: tracer kinetic modeling of muscarinic cholinergic receptor binding. Proc Natl Acad Sci USA 82: 6711–6715.

Frey LD, Locher JT, Hrycaj P, Stratz T, Kovac C, Mennet P, Müller W (1992) Bestimmung der regionalen Glukose-Metabolisierungsrate der Lumbalmuskulatur bei Patienten mit generalisierter Tendomyopathie (GTM) mittels dynamischer $^{18}$F-FDG PET. Z Rheumatol 51: 238–242.

Friedland RP, Budinger TF, Ganz E et al. (1983a) Regional cerebral metabolic alterations in dementia of the Alzheimer type: positron emission tomography with [$^{18}$F]fluorodeoxy- glucose. J Comp Assist Tomogr 7: 590–598.

Friedland RP, Budinger TF, Yano Y, Huesman RH, Knittel B, Derenzo SE, Koss B, Ober BA (1983b) Regional cerebral metabolic alterations in Alzheimer-type dementia: kinetic studies with 18-fluorodeoxyglucose. J Cereb Blood Flow Metab 3(Suppl 1): S510–S511.

Froehner SC, Davis A, Baldwin SA, Leinhard GE (1988) The blood-nerve barrier is rich in glucose transporter. J Neurocytol 17: 173–178.

Frost JJ, Douglass KH, Mayberg HS, Dannals RF, Links JM, Wilson AA, Ravert HT, Crozier WC, Wagner HN Jr. (1989) Multicompartmental analysis of [11C]-carfentanil binding to opiate receptors in humans measured by positron emission tomography. J Cereb Blood Flow Metab 9(3): 398–409.

Fuglsang A, Lomholt M, Gjedde A (1986) Blood–brain transfer of glucose and glucose analogs in newborn rats. J Neurochem 46: 1417–1428.

Fujikawa DG, Dwyer BE, Lake RG, Wasterlain CG (1989) Local cerebral glucose utilization during status epilepticus in newborn primates. Am J Physiol 256 (Cell Physiol 25): C1160–C1167.

Fujita H, Kuwabara H, Reutens DC, Gjedde A (1999) Oxygen consumption of cerebral cortex fails to increase during continued vibrotactile stimulation. J Cereb Blood Flow Metab 19: 266–271.

Fukumoto H, Seino S, Imura H, Seino Y, Bell GI (1988) Characterization and expression of human HepG2/erythrocyte glucose-transporter gene. Diabetes 37: 657–661.

Fukumoto H, Kayano T, Buse JB, Edwards Y, Pilch PF, Bell GI, Seino S (1989) Cloning and characterization of the major insulin-responsive glucose transporter expressed in human skeletal muscle and other insulin-responsive tissues. J Biol Chem 264: 7776–7779.

Furler SM, Jenkins AB, Stolien LH, Kraegen EW (1991) In vivo location of the rate limiting step of hexose uptake in muscle and brain tissue of the rat. Am J Physiol in press.

Gaitonde MK (1965) Rate of utilization of glucose and 'compartmentation' of $\alpha$-oxoglutarate and glutamate in rat brain. Biochem J 95: 803–810.

Gambhir SS, Schwaiger M, Huang SC, Krivokapich J, Schelbert HR, Nienaber CA, Phelps ME (1989) Simple non-invasive quantification method for measuring myocardial glucose utilization in humans employing positron emission tomography and fluorine-18 deoxyglucose. J Nucl Med 30: 359–366.

Garnett ES, Firnau G, Nahmias C (1983) Dopamine visualized in the basal ganglia of living man. Nature 305: 137–138.

Geday J, Gjedde A (2009) Monoaminergic modulation of emotional impact in the inferomedial prefrontal cortex. Synapse 63(2): 160–166.

Gehlen W (1933) Wirkungsstrke intravens verabreichter Arzneimittel als Zeitfunktion. Ein Beitrag zur mathematischen Behandlung pharmakologischer Probleme. Arch Exp Pathol Pharmakol 171: 541.

Gerhart DZ, LeVasseur RJ, Broderuis MA, Drewes LR (1989) Glucose transporter localization in brain using light and electron immunocytochemistry. J Neurosci Res 22: 464–472.

Gerhart DZ, Enerson BE, Zhdankina OY, Leino RL, Drewes LR (1997) Expression of monocarboxylate transporter MCT1 by brain endothelium and glia in adult and suckling rats. Am J Physiol 273: E207–E213.

Gerhart DZ, Enerson BE, Zhdankina OY, Leino RL, Drewes LR (1998) Expression of the monocarboxylate transporter MCT2 by rat brain glia. Glia 22: 272–281.

Germinario RJ, Rockman H, Oliveira M, Manuel S, Taylor M (1982) Regulation of sugar transport in cultured diploid human skin fibroblasts. J Cell Physiol 112: 367–372.

Gibbs EL, Lennox WG, Nims LF, Gibbs FA (1942) Arterial and cerebral venous blood; arterial-venous differences in man. J Biol Chem 144: 325–332.

Ginsberg MD, Chang JY, Kelley RE, Yoshii F, Barker WW, Ingento G, Boothe TE (1988) Increases in both cerebral glucose utilization and blood flow during execution of a somatosensory task. Ann Neurol 23: 152–160.

Gjedde A, Crone C (1975) Induction processes in blood–brain transfer of ketone bodies during starvation. Am J Physiol 229: 1165–1169.

Gjedde A, Siemkowicz E (1978) Post-ischemic coma in rat: effect of glucose and insulin treatment on cerebral metabolic recovery. Trans Am Neurol Assoc 103: 45–47.

Gjedde A (1980) Rapid steady-state analysis of blood–brain glucose transfer in rat. Acta Physiol Scand 108: 331–339.

Gjedde A, Hansen AJ, Siemkowicz E (1980) Rapid simultaneous determination of regional blood flow and blood–brain glucose transfer in brain of rat. Acta Physiol Scand 108: 321–330.

Gjedde A, Rasmussen M (1980a) Blood–brain glucose transport in the conscious rat: comparison of the intravenous and intracarotid methods. J Neurochem 35: 1375–1381.

Gjedde A, Rasmussen M (1980b) Pentobarbital anesthesia reduces blood–brain glucose transfer in the rat. J Neurochem 35: 1382–1387.

Gjedde A (1981a) Regulation and adaptation of substrate transport to the brain. Adv Physiol Sci 7: 307–315.

Gjedde A (1981b) High- and low-affinity transport of D-glucose from blood to brain. J Neurochem 36: 1463–1471.

Gjedde A, Crone C (1981) Blood–brain glucose transfer: repression in chronic hyperglycemia. Science 214: 456–457.

Gjedde A, Hansen AJ, Quistorff B (1981) Blood–brain glucose transfer in spreading depression. J Neurochem 37: 807–812.

Gjedde A (1982) Calculation of cerebral glucose phosphorylation from brain uptake of glucose analogs in vivo: a re-examination. Brain Res 257: 237–274.

Gjedde A (1983) Modulation of substrate transport to the brain. Acta Neurol Scand 67: 3–25.

Gjedde A, Diemer NH (1983) Autoradiographic determination of regional brain glucose content. J Cereb Blood Flow Metab 3: 303–310.

Gjedde A (1984a) On the measurement of glucose in brain. Neurochem Res 9: 1667–1671.

Gjedde A (1984b) Blood–brain transfer of galactose in experimental galactosemia, with special reference to the competitive interaction between galactose and glucose. J Neurochem 43: 1654–1662.

Gjedde A, Christensen O (1984) Estimates of Michaelis-Menten constants for the two membranes of the brain endothelium. J Cereb Blood Flow Metab 4: 241–249.

Gjedde A, Diemer NH (1985a) Relationship between unidirectional and net uptake of glucose and glucose analog into brain: on the variability of transfer and lumped constants. In: Greitz T et al. (eds) The Metabolism of the Human Brain Studied with Positron Emission Tomography. Raven Press, New York, p 207.

Gjedde A, Diemer NH (1985b) Double-tracer study of the fine regional blood–brain glucose transfer in the rat by computer-assisted autoradiography. J Cereb Blood Flow Metab 5: 282–289.

Gjedde A, Lauritzen M (1985) The $C_{O_2}$-reactivity of blood–brain glucose transport in rat. Abstr XII Internat Symp Cereb Blood Flow Metab, Studentlitteratur, Lund.

Gjedde A, Wienhard K, Heiss WD, Kloster G, Diemer NH, Herholz K, Pawlik G (1985) Comparative regional analysis of 2-fluorodeoxyglucose and methylglucose uptake in brain of four stroke patients. With special reference to the regional estimation of the lumped constant. J Cereb Blood Flow Metab 5: 163–178.

Gjedde A, Wong DF, Wagner HN Jr (1986) Transient analysis of irreversible and reversible tracer binding in human brain in vivo. In: Battistin L (ed) PET and NMR: New Perspectives in Neuroimaging and Clinical Neurochemistry, Alan R Liss, New York, pp. 223–235.

Gjedde A, Bodsch W (1987) Facilitated diffusion across the blood–brain barrier: interactions between receptors and transporters. Wissenschaftl Zeitschr Karl Marx University, Leipzig.

Gjedde A, Wong DF (1987) Positron tomographic quantitation of neuroreceptors in human brain in vivo–with special reference to the D2 dopamine receptors in caudate nucleus. Neurosurg Rev 10(1): 9–18.

Gjedde A (1987) Does deoxyglucose uptake in the brain reflect energy metabolism? Biochem Pharmacol 36: 1853–1861.

Gjedde A, Kuwabara H, Hakim A (1990) Reduction of functional capillary density in human brain after stroke. J Cereb Blood Flow Metab 10: 317–316.

Gjedde A, Wong DF (1990) Modeling neuroreceptor binding of radioligands in vivo. In: Frost J, Wagner HN Jr (eds) Quantitative Imaging of Neuroreceptors, Raven Press, New York, pp 51–79.

Gjedde A, Reith J, Dyve S, Léger G, Guttman M, Diksic M, Evans AC, Kuwabara H (1991a) Dopa decarboxylase activity of the living human brain. Proc Nat Acad Sci USA 88: 2721–2725.

Gjedde A, Ohta S, Kuwabara H, Meyer E (1991b) Is oxygen diffusion limiting for blood–brain transfer of oxygen? In: Lassen NA, Ingvar DH, Raichle ME, Friberg L (eds) Brain Work and Mental Activity, Alfred Benzon Symposium 31, Munksgaard, Copenhagen, pp. 177–184.

Gjedde A (1992) Blood–brain glucose transfer. In: Bradbury MBW (ed) Handbook of Experimental Pharmacology, Vol 103. Physiology and Pharmacology of the Blood–Brain Barrier, Springer Verlag, Berlin-Heidelberg, pp. 65–114.

Gjedde A, Léger G, Cumming P, Yasuhara Y, Evans AC, Guttman M, Kuwabara H (1993) Striatal L-DOPA decarboxylase activity in Parkinson's disease in vivo: implications for the regulation of dopamine synthesis. J Neurochem 61: 1538–1541.

Gjedde A (1993a) The energy cost of neuronal depolarization. In: Gulyas B, Ottoson D, Roland PE (eds) Functional Organization of the Human Visual Cortex, Pergamon Press, Oxford, pp. 291–306.

Gjedde A (1993b) Interpreting physiology maps of the living human brain. In: Uemura K, Lassen NA, Jones T, Kanno I (eds) Quantification of Brain Function. Tracer Kinetics and Image Analysis in Brain PET, Elsevier, Amsterdam, pp. 187–196.

Gjedde A (1995a) Compartmental analysis. In: Wagner HN Jr, Szabo Z, Buchanan JW (eds) Principles of Nuclear Medicine, 2nd edn, Saunders, Philadelphia, pp. 451–461.

Gjedde A (1995b) Glucose metabolism. In: Wagner HN Jr, Szabo Z, Buchanan JW (eds) Principles of Nuclear Medicine, 2nd edn, Saunders, Philadelphia, pp. 54–71.

Gjedde A (1996a) PET criteria of cerebral tissue viability in ischemia. Acta Neurol Scand [Suppl] 166: 3–5.

Gjedde A (1996b) The relation between brain function and cerebral blood flow and metabolism. In: Batjer HH (ed) Cerebrovascular Disease, Lippincott-Raven, New York, pp. 23–40.

Gjedde A, Poulsen PH, Ostergaard L (1999) On the oxygenation of hemoglobin in the human brain. Adv Exp Med Biol 471: 67–81.

Gjedde A, Gee AD, Smith DF (2000) Basic CNS drug transport and binding kinetics in vivo. In: Begley DJ, Bradbury MW, Kreuter J (eds) The Blood–Brain Barrier and Drug Delivery to the CNS, Marcel Dekker, New York, pp. 225–242.

Gjedde A (2001) Brain energy metabolism and the physiological basis of the hemodynamic response. In: Jezzard P, Matthews PM, Smith S (eds) Functional Magnetic Resonance Imaging: Methods for Neuroscience. Oxford University Press, Oxford, pp. 37–65.

Gjedde A, Marrett S (2001) Glycolysis in neurons, not astrocytes, delays oxidative metabolism of human visual cortex during sustained checkerboard stimulation in vivo. J Cereb Blood Flow Metab 21: 1384–1392.

Gjedde A, Wong DF (2001) Quantification of neuroreceptors in living human brain. V. Endogenous neurotransmitter inhibition of haloperidol binding in psychosis. J Cereb Blood Flow Metab 21: 982–994.

Gjedde A (2002) Cerebral blood flow change in arterial hypoxemia is consistent with negligible oxygen tension in brain mitochondria. NeuroImage 17: 1876–1881.

Gjedde A, Marrett S, Vafaee M (2002) Oxidative and nonoxidative metabolism of excited neurons and astrocytes. J Cereb Blood Flow Metab 22: 1–14.

Gjedde A (2003) Modeling metabolite and tracer kinetics. Chap. 7 In: Feinendegen LE, Shreeve WW, Eckelman WC, Bahk Y-W, Wagner HN Jr (eds) Molecular Nuclear Medicine, Springer-Verlag, Berlin-Heidelberg, pp. 121–169.

Gjedde A, Johannsen P, Cold GE, Ostergaard L (2005a) Cerebral metabolic response to low blood flow: possible role of cytochrome oxidase inhibition. J Cereb Blood Flow Metab 25(9): 1183–1196.

Gjedde A, Wong DF, Rosa-Neto P, Cumming P (2005b) Mapping neuroreceptors at work: on the definition and interpretation of binding potentials after 20 years of progress. Int Rev Neurobiol 63: 1–20.

Gjedde A (2005) Blood–brain transfer and metabolism of oxygen. In: Dermietzel RS, Spray DC, Nedergaard M (eds) Blood–Brain Barriers: From Ontogeny to Artificial Interfaces, volume 2, Chap. 22, Wiley, Hoboken, pp. 523–549.

Gjedde A (2007) Brain energetics. Integration of molecular and cellular processes. In: Lajtha AG, Gray E, Dienel GA (eds.) Handbook of Neurochemistry and Molecular Neurobiology, 3rd ed.

Gjedde A, Geday J (2009) Deep brain stimulation reveals emotional impact processing in ventromedial prefrontal cortex. PLoS One 4(12): e8120.

Gjedde A, Kumakura Y, Cumming P, Linnet J, Møller A (2010a) Inverted-U-shaped correlation between dopamine receptor availability in striatum and sensation seeking. Proc Natl Acad Sci USA 107(8): 3870–3875.

Gjedde A, Keiding S, Vilstrup H, Iversen P (2010b) No oxygen delivery limitation in hepatic encephalopathy. Metab Brain Dis 25(1): 57–63.

Gnaiger E (1993) Homeostatic and microxic regulation of respiration in transitions to anaerobic metabolism. In: Bicudo JEPW (ed) The Vertebrate Gas Transport Cascade: Adaptations to Environment and Mode of Life, CRC Press, Boca Raton, pp. 358–370.

Gnaiger E, Steinlechner-Maran R, Mendez G, Eberl T, Margreiter R (1995) Control of mitochondrial and cellular respiration by oxygen. J Bioenerg Biomembr 27: 583–596.

Gnaiger E, Lassnig B, Kuznetsov A, Rieger G, Margreiter R (1998) Mitochondrial oxygen affinity, respiratory flux control and excess capacity of cytochrome c oxidase. J Exp Biol 201: 1129–1139.

Goldman DE (1943) Potential, impedance, and rectification in membranes. J Gen Physiol 27: 37–60.

Goldberg MA, Lee MJ, Fischman AJ, Mueller PR, Alpert NM, Thrall JH (1993) Fluorodeoxyglucose PET of abdominal and pelvic neoplasms: potential role in oncologic imaging. Radiographics 13(5): 1047–1062.

Gonzalez-Lima F, Jones D (1994) Quantitative mapping of cytochrome oxidase activity in the central auditory system of the gerbil: a study with calibrated activity standards and metal-intensified histochemistry. Brain Res 660: 34–49.

Goodwin RFW (1956) Distribution of sugar between red cells and plasma: variations associated with age and species. J Physiol 134: 88–101.

Graham MM, Muzi M, Spence AM, O'Sullivan F, Lewellen TK, Link JM, Krohn KA (2002) The FDG lumped constant in normal human brain. J Nucl Med 43(9): 1157–1166.

Growdon WA, Bratton TS, Houston MC, Tarpley HL, Regen DM (1971) Brain glucose metabolism in the intact mouse. Am J Physiol 221: 1738–1745.

Grunder G, Carlsson A, Wong DF (2003) Mechanism of new antipsychotic medications: occupancy is not just antagonism. Arch Gen Psychiatry 60: 974–977.

Guilivi C (1998) Functional implications of nitric oxide produced by mitochondria in mitochondrial metabolism. Biochem J 332: 673–679.

Gunn RN, Gunn SR, Cunningham VJ (2001) Positron emission tomography compartmental models. J Cereb Blood Flow Metab 21: 635–652.

Gunn RN, Gunn SR, Turkheimer FE, Aston JAD, Cunningham VJ (2002) Positron emission tomography compartmental models: a basis pursuit strategy for kinetic modelling. J Cereb Blood Flow Metab 22: 1425–1439.

Gutniak M, Blomqvist G, Stone-Elander S, Widén L, Hamberger B, Grill V (1990) Cerebral blood flow and substrate utilization in insulin-treated diabetic subjects. Am J Physiol 258: E805–E812.

Hakim AM, Moss G (1972) Direct in vivo quantitation of the pentose phosphate pathway in brain. Trans N Y Acad Sci 34: 473–490.

Hakim AM, Moss G, Scuderi D (1980) The pentose phosphate pathway in brain during development. Biol Neonate 37: 15–21.

Haldane JBS (1957) Untitled. Nature 179: 832.

Halestrap AP (1975) The mitochondrial pyruvate carrier. Biochem J 148: 85–96.

Halestrap AP (1978) Stimulation of pyruvate transport in metabolizing mitochondria through changes in the transmembrane pH gradient induced by glucagon treatment of rat. Biochem J 172: 389–398.

Halestrap AP, Armston AE (1984) A re-evaluation of the role of mitochondrial pyruvate transport in the hormonal control of rat liver mitochondrial pyruvate metabolism. Biochem J 223: 677–685.

Halestrap AP, Price NT (1999) The proton-linked monocarboxylate transporter (MCT) family: structure, function and regulation. Biochem J 343: 281–299.

Hansen AJ (1984) Ion and membrane changes in the brain during anoxia. Behav Brain Res 14: 93–98.

Hansen AJ (1985) Effect of anoxia on ion distribution in the brain. Physiol Rev 65: 101–148.

Hargreaves RJ, Planas AM, Cremer JE, Cunningham VJ (1986) Studies on the relationship between cerebral glucose transport and phosphorylation using 2-deoxyglucose. J Cereb Blood Flow Metab 6: 708–716.

Harik SI, LaManna JC (1988) Vascular perfusion and blood–brain glucose transport in acute and chronic hyperglycemia. J Neurochem 51: 1924–1929.

Harik SI, Gravina SA, Kalaria RN (1988) Glucose transporter of the blood–brain barrier and brain in chronic hyperglycemia. J Neurochem 51: 1930–1934.

Harik SI, Kalaria RN, Andersson L, Lundahl P, Perry G (1990) Immunocytochemical localization of the erythroid glucose transporter: abundance in tissues with barrier functions. J Neurosci 10: 3862–3872.

Harik SI, Behmand RA, Arafah BM (1991a) Chronic hyperglycemia increases the density of glucose transporters in human erythrocyte membranes. J Clin Endocrinol Metab 72: 814–818.

Harik SI, Behmand RE, LaManna JC (1991b) Chronic hypobaric hypoxia increases the density of cerebral capillaries and their glucose transporter protein. J Cereb Blood Flow Metab 11 (Suppl. 2): S496.

Harris M, Prout BJ (1970) Relative hypoglycemia. Lancet ii: 317.

Hassel B, Bråthe A (2000) Cerebral metabolism of lactate in vivo: evidence for neuronal pyruvate carboxylation. J Cereb Blood Flow Metab 20: 327–336.

Hasselbalch SG, Moos-Knudsen G, Jakobsen J, Holm S, Høgh P, Paulson O (1991) The effect of insulin on brain FDG transport and metabolism in man studied by PET. J Cereb Blood Flow Metab 11 (Suppl. 2): S465.

Hasselbalch SG, Knudsen GM, Holm S, Hageman LP, Capaldo B, Paulson OB (1996) Transport of D-glucose and 2-fluorodeoxyglucose across the blood–brain barrier in humans. J Cereb Blood Flow Metab 16(4): 659–666.

Hasselbalch SG, Madsen PL, Knudsen GM, Holm S, Paulson OB (1998) Calculation of the FDG lumped constant by simultaneous measurements of global glucose and FDG metabolism in humans. J Cereb Blood Flow Metab 18(2): 154–160.

Hasselbalch SG, Holm S, Pedersen HS, Svarer C, Knudsen GM, Madsen PL, Paulson OB (2001). The (18)F-fluorodeoxyglucose lumped constant determined in human brain from extraction fractions of (18)F-fluorodeoxyglucose and glucose. J Cereb Blood Flow Metab 21: 995–1002.

Haubold-Reuter B, Haldemann R, Amann F, Turina J, Reist W, Leenders KL, von Schulthess GK (1992) Die Positronenemissionstomographie (PET) zur Beurteilung des metabolischen Zustandes des Herzmuskels. Schweiz Med Wochenschr 122: 14–21.

Hawkins R, Miller AL, Cremer JE, Veech RL (1974) Measurements of the rate of glucose utilization by rat brain in vivo. J Neurochem 23: 917–923.

Hawkins R, Hass K, Ransohoff J (1979) Measurement of regional brain glucose utilization in vivo using [2-$^{14}$C]glucose. Stroke 10: 690–703.

Hawkins RA, Mazziotta JC, Phelps ME, Huang SC, Kuhl DE, Carson RE, Mettler EJ, Riege WH (1983a) Cerebral glucose metabolism as a function of age in man: influence of the rate constants in the fluorodeoxyglucose model. J Cereb Blood Flow Metab 3: 250–253.

Hawkins RA, Mans AM, Davis DW, Hibbard LS, Lu DM (1983b) Glucose availability to individual cerebral structures is correlated to glucose metabolism. J Neurochem 40: 1013–1018.

Hawkins RA, Phelps ME, Huang SC (1984) Effects of temporal sampling, glucose metabolic rates, and disruptions of the blood–brain barrier on the FDG model with and without a vascular compartment: studies in human brain tumor with PET. J Cereb Blood Flow Metab 6: 170–183.

Hawkins R, Mans AM, Davis DW, Vina JR, Hubbard LS (1985) Cerebral glucose use measured with [$^{14}$C]glucose labeled in the 1, 2, or 6 position. Am J Physiol 248: C170–C176.

Hawkins RA, Phelps ME, Huang SC (1986) Effects of temporal sampling, glucose metabolic rates, and disruptions of the blood–brain barrier on the FDG model with and without a vascular compartment: studies in human brain tumor with PET. J Cereb Blood Flow Metab 6: 170–183.

Helgerson AL, Hebert DN, Naderi S, Carruthers A (1989) Characterization of two independent modes of action of ATP on human erythrocyte sugar transport. Biochemistry 28: 6410–6417.

Helgerson AL, Carruthers A (1989) Analysis of protein-mediated 3-O-methylglucose transport in rat erythrocytes: rejection of the alternating conformation carrier model for sugar transport. Biochemistry 28: 4580–4594.

Hemphill JC, III, Wade S. Smith, Christian Sonneb D, Diane Morabitoc and Geoffrey T. Manley (2005) Relationship between brain tissue oxygen tension and CT perfusion: feasibility and initial result. American Journal of Neuroradiology 26: 10951100.

Henry PG, Lebon V, Vaufrey F, Brouillet E, Hantraye P, Bloch G (2002) Decreased TCA cycle rate in the rat brain after acute 3-NP treatment measured by in vivo 1H-[13C] NMR spectroscopy. J Neurochem 82: 857–66.

Herman NL, Kostreva DR (1986) Alterations in cardiac [$^{14}$C]deoxyglucose uptake induced by two renal afferent stimuli. Am J Physiol 251: R867–R877.

Hertz L, Schousboe A (1975) Ion and energy metabolism of the brain at the cellular level. Int Rev Neurobiol 18: 141–211.

Hertz L, Swanson RA, Newman GC, Marrif H, Juurlink BH, Peng L (1998) Can experimental conditions explain the discrepancy over glutamate stimulation of aerobic glycolysis? Dev Neurosci 20: 339–347.

Hertz MM, Paulson OB, Barry DI, Christensen JS, Svendsen PA (1981) Insulin increases glucose transfer across the blood–brain barrier in man. J Clin Invest 67: 597–604.

Hertz MM, Paulson OB (1982) Transfer across the human blood–brain barrier: evidence for capillary recruitment and for a paradox glucose permeability increase in hypocapnia. Microvasc Res 24: 364–376.

Hertz MM, Paulson OB (1983) Glucose transfer across the blood–brain barrier. In: Szabo AJ (ed) Advances in metabolic disorders, vol. 10. CNS regulation of carbohydrate metabolism. Academic Press, , New York, p. 177.

Hertz MM (1986) Thesis: Glucose Transfer Across the Blood-Brain Barrier (In Danish). FADL's Forlag, Copenhagen.

Hevner RF, Liu S, Wong-Riley MT (1995) A metabolic map of cytochrome oxidase in the rat brain: histochemical, densitometric and biochemical studies. Neuroscience 65: 313–342.

Hill AV (1910) The possible effect of aggregation of the molecules of haemoglobin on its dissociation curve. J Physiol 40: iv–vii.

Hill AV (1913) The combination of haemoglobin with oxygen and with carbon monoxide. Biochem J 7: 471-480.

Himwich HE, Fazekas JF (1937) Effect of hypoglycemia on metabolism of brain. Endocrinology 21: 800–807.

Himwich WA, Himwich HE (1946) Pyruvic acid exchange of brain. J Neurophysiol 9: 133–136.

Hjorth JSU (1994) Computer Intensive Statistical Methods Validation, Model Selection and Bootstrap. Chapman & Hall, London.

Hodgkin AL, Katz B (1949) The effect of sodium ions on the electrical activity of the giant axon of the squid. J Physiol (Lond) 108: 37–77.

Hodgkin AL, Huxley (1952) A quantitative description of membrane current and its application to conductance and excitation in nerve. J Physiol (Lond) 117: 500–544.

Hodgkin AL, Keynes RD (1956) Experiments on the injection of substances into squid giant axons by means of a microsyringe. J Physiol (Lond) 131: 592–616.

Hofstee BHJ (1952) On the evaluation of the constants $V_m$ and $K_M$ in enzyme reactions. Science 116: 329–331.

Hofstee BHJ (1954–1956) Graphical analysis of single enzyme systems. Enzymologia 17: 273–278.

Hoge RD, Atkinson J, Gill B, Crelier GR, Marrett S, Pike GB (1999) Linear coupling between cerebral blood flow and oxygen consumption in activated human cortex. Proc Natl Acad Sci USA 96: 9403–9408.

Holden JE, Jivan S, Ruth TJ, Doudet DJ (2002) In vivo receptor assay with multiple ligand concentrations: an equilibrium approach. J Cereb Blood Flow Metab 22: 1132–1141.

Honig CR, Connett RJ, Gayeski TE (1992) $O_2$ transport and its interaction with metabollism; a systems view of aerobic capacity. Med Sci Sports Exerc 24: 47–53.

Horton RW (1973) The kinetics of glucose influx into the mouse brain in vivo and the effects of pentobarbitone anaesthesia and hypothermia. 54th Meeting Trans Biochem Soc 1: 127–128.

Hoshi H, Kuwabara H, Léger G, Cumming P, Guttman M, Gjedde A (1993) 6-[$^{18}$F]fluoro-L-DOPA metabolism in living human brain: a comparison of six analytical methods. J Cereb Blood Flow Metab 13: 57–69.

Huang SC, Phelps ME (1986) Principles of tracer kinetic modeling in positron emission tomography and autoradiography. In: Phelps ME, Mazziotta J, Schelbert H (eds) Positron Emission Tomography and Autoradiography: Principles and Applications for the Brain and Heart, Raven Press, New York, pp. 287–346.

Huang SC, Bahn MM, Barrio JR, Hawkins RA, Hoffmann JM, Satyamurthy N, Bida GT, Mazziotta JC, Phelps ME (1987) J Cereb Blood Flow Metab 7(suppl I ): S359.

Hunziker O, Abdel'al S, Schulz U (1979) The aging human cerebral cortex: a stereological characterization of changes in the capillary net. J Gerontol 34: 345–350.

Hyder F, Patel AB, Gjedde A, Rothman DL, Behar KL, Shulman RG (2005) Neuronal-glia glucose oxidation and glutamatergic-GABAergic function (in press).

Iadecola C (1992) Does nitric oxide mediate the increases in cerebral blood flow elicited by hypercapnia? Proc Natl Acad Sci USA 89: 3913–3916.

Iadecola C, Pelligrino DA, Moskowitz MA, Lassen NA (1994) Nitric oxide synthase inhibition and cerebrovascular regulation. J Cereb Blood Flow Metab 14: 175–192.

Iadecola C (2004) Neurovascular regulation in the normal brain and in Alzheimer's disease. Nat Rev Neurosci 5: 347–360.

Ide K, Schmalbruch IK, Quistorff B, Horn A, Secher NH (2000) Lactate, glucose and O2 uptake in human brain during recovery from maximal exercise. J Physiol 522: 159–164.

Iida H, Jones T, Miura S. (1993) Modeling approach to eliminate the need to separate arterial plasma in oxygen-15 inhalation positron emission tomography. J Nucl Med 34(8): 1333-1340

Ikezaki K, Black KL, Conklin SG, Becker DP (1992) Histochemical evaluation of energy metabolism in rat glioma. Neurol Res 14: 289–293.

Innis RB, Cunningham VJ, Delforge J, Fujita M, Gjedde A, Gunn RN, Holden J, Houle S, Huang S-C, Ichise M, Lida H, Ito H, Kimura Y, Koeppe RA, Knudsen GM, Knuuti J, Lammertsma AA, Laruelle M, Logan J, Maguire RP, Mintun MA, Morris ED, Parsey R, Price JC, Slifstein M, Sossi V, Suhara T, Votaw JR, Wong DF and Carson RE (2007) Concensus nomenclature for in vivo imaging of reversibly binding radioligands. J Cereb Blood Flow Metab 27(9): 1533–1539.

Itoh Y, Esaki T, Shimoji K, Cook M, Law MJ, Kaufman E, Sokoloff L (2003) Dichloroacetate effects on glucose and lactate oxidation by neurons and astroglia in vitro and on glucose utilization by brain in vivo. Proc Natl Acad Sci USA 100: 4879–4884.

Jacquez JA (1984) Red blood cell as glucose carrier: significance for placental and cerebral glucose transfer. Am J Physiol 246: R289–R298.

Jagust WJ, Seab JP, Huesman RH, Valk PE, Mathis CA, Reed BR, Coxson PG, Budinger TF (1991) Diminished glucose transport in Alzheimer's disease: dynamic PET studies. J Cereb Blood Flow Metab 11: 323–330.

James DE, Brown R, Navarro J, Pilch PF (1988) Insulin-regulatable tissues express a unique insulin-sensitive glucose transport protein. Nature 333: 183–185.

James DE, Strube M, Mueckler M (1989) Molecular cloning and characterization of an insulin-regulatable glucose transporter. Nature 338: 83–87.

Johnson JA, Wilson TA (1966) A model for capillary exchange. Am J Physiol 210: 1299–1303.

Johnson RG, Scarpa A (1979) Protonmotive force and catecholamine transport in isolated chromaffin granules. J Biol Chem 254: 3750–3760.

Joliot M, Ribary U, Llinas R (1995) Human oscillatory brain activity near 40 Hz coexists with cognitive temporal binding. Proc Natl Acad Sci USA 91: 11748–11751.

Joo F (1985) The blood–brain barrier in vitro: ten years of research on microvessels isolated from the brain. Neurochem Int 7: 1–25.

Kadekaro M, Crane AM, Sokoloff L (1985) Differential effects of electrical stimulation of sciatic nerve on metabolic activity in spinal cord and dorsal root ganglion in the rat. Proc Natl Acad Sci USA 82: 6010–6013.

Kalaria RN, Gravina SA, Schmidley JW, Perry G, Harik SI (1988) The glucose transporter of the human brain and blood–brain barrier. Ann Neurol 24: 757–764.

Kalaria RN, Harik SI (1989) Reduced glucose transporter at the blood–brain barrier and in cerebral cortex in Alzheimer's disease. J Neurochem 53: 1083–1088.

Kaplan NO, Everse J (1972) Regulatory characteristics of lactate dehydrogenases. Adv Enzyme Regul 10: 323–336.

Kapoor R, Spence AM, Muzi M, Graham MM, Abbott GL, Krohn K (1989) Determination of the deoxyglucose and glucose phosphorylation ratio and the lumped constant in rat brain and a transplantable rat glioma. J Neurochem 53: 37–44.

Kasahara M, Hinkle PC (1976) Reconstitution of D-glucose transport catalyzed by a protein fraction from human erythrocytes in sonicated liposomes. Proc Natl Acad Sci USA 73: 396–400.

Kasahara M, Hinkle PC (1977) Reconstitution and purification of the D-glucose transporter from human erythrocytes. J Biol Chem 252: 7384–7390.

Kasanicki MA, Cairns MT, Davis A, Gardiner RM, Baldwin SA (1987) Identification and characterization of the glucose-transport fof the bovine blood/brain barrier. Biochem J 247: 101–108.

Kasischke KA, Vishwasrao HD, Fisher PJ, Zipfel WR, Webb WW (2004) Neural activity triggers neuronal oxidative metabolism followed by astrocytic glycolysis. Science 305(5680): 99–103.

Kassissia IG, Goresky CA, Rose CP, Schwab AJ, Simard A, Huet PM, Bach GG (1995) Tracer oxygen distribution is barrier-limited in the cerebral microcirculation. Circ Res 77: 1201–1211.

Kastrup A, Krüger G, Neumann-Haefelin T, Glover GH, Moseley ME (2002) Changes of cerebral blood flow, oxygenation, and oxidative metabolism during graded motor activation. NeuroImage 15: 74–82.

Katayama Y, Tsubokawa T, Hirayama T, Kido G, Tsukiyama T, Iio M (1986) Response of regional cerebral blood flow and oxygen metabolism to thalamic stimulation in humans as revealed by positron emission tomography. J Cereb Blood Flow Metab 6: 637–634.

Katayama Y, Fukuchi T, Mc Kee A, Terashi A (1998) Effect of hyperglycemia on pyruvate dehydrogenase activity and energy metabolites during ischemia and reperfusion in gerbil brain. Brain Res 788: 302–304.

Kato A, Diksic M, Yamamamoto YL, Strother SC, Feindel W (1984) An improved approach for measurement of regional cerebral rate constants in the deoxyglucose method with positron emission tomography. J Cereb Blood Flow Metab 4: 555–563.

Kayano T, Fukumoto H, Eddy RL, Fan Y-S, Byers MG, Shows TB, Bell GI (1988) Evidence for a family of human glucose transporter-like proteins. J Biol Chem 263: 15245–15248.

Kayano T, Burant CF, Fukumoto G, Gould W, Fan Y-S, Eddy RL, Byers MG, Shows TB, Scine S, Bell GI (1990) Human facilitative glucose transporters. J Biol Chem 265: 13276–13282.

Kety SS (1949) The physiology of the human cerebral circulation. Anesthesiology 10: 610–614.

Kety SS (1951) The theory and application of the exchange of inert gas at the lungs and tissues Pharmacol Rev 3: 1–41.

Kety SS (1960a) Theory of blood-tissue exchange and its application to measurement of blood flow. In: Bruner HD (ed) Methods in Medical Research, Vol. 8, The Year Book Publishers, Inc, New York, p. 223.

Kety SS (1960b) Measurements of local blood flow by the exchange of an inert diffusible substance. In: Methods in Medical Research, Vol. 8, Year Book Publ. Chicago, IL, pp. 223–236.

Kintner D, Costello DJ, Levin AB, Gilboe DD (1980) Brain metabolism after 30 minutes of hypoxic or anoxic perfusion or ischemia. Am J Physiol 239: E501–E509.

Kiss JP, Vizi ES (2001) Nitric oxide: a novel link between synaptic and nonsynaptic transmission. Trends Neurosci 24(4): 211–215. Review.

Kiukka JT, Bassingthwaighte JB, Henrich MM, Feinendegen LE (1991) Mathematical modelling in nuclear medicine. Eur J Nucl Med 18(5): 351–362.

Kuikka JT et al. (1991) Mathematical modeling in Nuclear Medicine. Eur J Nucl Med 18(5): 351–362.

Kjaer TW, Bertelsen C, Piccini P, Brooks D, Alving J, Lou HC (2002) Increased dopamine tone during meditation-induced change of consciousness. Brain Res Cogn Brain Res 13: 255–259.

Kung HF, Kim HJ, Kung MP, Meegalla SK, Plossl K, Lee HK (1996) Imaging of dopamine transporters in humans with technetium-99m TRODAT-1. Eur J Nucl Med 23: 1527–1530.

Kletzien RF, Perdue JC, Springer A (1972) Inhibition of sugar uptake in cultured cells. J Biol Chem 247: 2964–2966.

Kletzien RF, Perdue JF (1985) Induction of sugar transport in chick embryo fibroblasts by hexose starvation: evidence for transcriptional regulation of transport. J Biol Chem 250: 593–600.

Klip A, Paquet MR (1990) Glucose transport and glucose transporters in muscle and their metabolic regulation. Diab Care 13: 228–243.

Klip A, Dimitrakoudis D, Ramlal T, Burdett E, Marette A, Bilan P, Mitsumoto Y, Kojvisto U-M, Sarabia V (1991) Regulation of glucose transport and transporters in muscle: acute and chronic effects of insulin and glucose. Symposium on Advances in Regulation of Carbohydrate Metabolism, Jerusalem, Israel.

Knudsen GM, Pettigrew KD, Paulson OB, Hertz MM, Patlak CS (1990) Kinetic analysis of blood-brain barrier transport of D-glucose in man: quantitative evaluation in the presence of tracer backflux and capillary heterogeneity. Microvasc Res 39: 28–49.

Knudsen GM, Hertz M, Paulson OB (1991) Does capillary recruitment exist in the human brain? J Cereb Blood Flow Metab 11 (Suppl. 2): S442.

Knuuti MJ, Nuurtila P, Ruotsalainen U, Saraste M, Härkönen R, Ahonen A, Teräs M, Haaparanta M, Wegelius U, Hartiala J, Voipio-Pulkki LM (1992) Euglycemic hyperinsulinemic clamp and oral glucose load in stimulating myocardial glucose tomography. J Nucl Med 33: 1255–1262.

Kojima S, Nakamura T, Nidaira T, Nakamura K, Ooashi N, Ito E, Watase K, Tanaka K, Wada K, Kudo Y, Miyakawa H (1999) Optical detection of synaptically induced glutamate transport in hippocampal slices. J Neurosci 19: 2580–2588.

Kolber AR, Bagnell CR, Krigman MR, Hayward J, Morell P (1979) Transport of sugars into microvessels isolated from rat brain: a model for the blood-brain barrier. J Neurochem 33: 419–432.

Kolter T, Uphues I, Wichelhaus A, Reinauer H, Eckel J (1992) Contraction-induced translocation of the glucose transporter Glut4 in isolated ventricular cardiomyocytes. Biochem Biophys Res Commun 189: 1207–1214.

Kraig RP (1990) Astrocytic acid-base homeostasis in cerebral ischemia. In: Schurr A, Rigor BM (eds) Cerebral Ischemia and Resuscitation, CRC Press, Boca Raton, pp. 89–99.

Krivokapich J, Huang SC, Phelps ME, Barrio JR, Watanabe CR, Selin CE, Shine KI (1982) Estimation of rabbit myocardial metabolic rate for glucose using fluorodeoxyglucose. Am J Physiol 243: H884–H895.

Krivokapich J, Huang SC, Selin CE, Phelps ME (1987) Fluorodeoxyglucose rate constants, lumped constant, and glucose metabolic rate in rabbit heart. Am J Physiol 252: H777–H787.

Krogh A (1919) The number and distribution of capillaries in muscles with calculations of the oxygen pressure head necessary for supplying the tissue. J Physiol 52: 405–415.

Krogh A (1946) The active and passive exchanges of inorganic ions through the surface of living cells and through living membranes generally. Proc R Soc (Lond) (B) 133: 140–200.

Kuhar MJ (2002) Social rank and vulnerability to drug abuse. Nat Neurosci 5: 88–90.

Kuhl DE, Barrio JR, Huang SC, Selin C, Ackermann RF, Lear JL, Wu JL, Lin TH, Phelps ME (1982) Quantifying local cerebral blood flow by *N*-isopropyl-p-[123I]iodoamphetamine (IMP) tomography. J Nucl Med 23: 196–203.

Kumakura Y, Vernaleken I, Gründer G, Bartenstein P, Gjedde A, Cumming P (2005) PET studies of net blood–brain clearance of FDOPA to human brain: age-dependent decline of [18F]fluorodopamine storage capacity. J Cereb Blood Flow Metab 25: 807–819.

Kumakura Y, Gjedde A, Danielsen EH, Christensen S, Cumming P (2006) Dopamine storage capacity in caudate and putamen of patients with early Parkinson's disease: correlation with asymmetry of motor symptoms. J Cereb Blood Flow Metab 26: 358–370.

Kuschinsky W, Paulson OB (1992) Capillary circulation in the brain. Cerebrovasc Brain Metab Rev 4: 261–286.

Kuwabara H, Evans AC, Gjedde A (1990) Michaelis-Menten constraints improved cerebral glucose metabolism and regional lumped constant measurements with [$^{18}$F]fluroodeoxyglucose. J Cereb Blood Flow Metab 10: 180–189.

Kuwabara H, Gjedde A (1991) Measurements of glucose phosphorylation with FDG and PET are not reduced by dephosphorylation of FDG-6-phosphate. J Nucl Med 32: 692–698.

Kuwabara H, Ohta S, Brust P, Meyer E, Gjedde A (1992) Density of perfused capillaries in living human brain during functional activation. Prog Brain Res 91: 209–215.

Kwong KK, Belliveau JW, Chesler DA, Goldberg IE, Weisskoff RM, Poncelet BP, Kennedy DN, Hoppel BE, Cohen MS, Turner R et al. (1992) Dynamic magnetic resonance imaging of human brain activity during primary sensory stimulation. Proc Natl Acad Sci USA 89: 5675–5679.

LaManna JC, Harik SI (1985) Regional comparisons of brain glucose influx. Brain Res 326: 299–305.

LaManna JC, Harik SI (1986) Regional studies of blood–brain transport of glucose and leucine in awak and anesthetized rats. J Cereb Blood Flow Metab 6: 717–723.

Lammertsma AA, Brooks DJ, Frackowiak SJ, Beaney RP, Herold S, Heather JD, Palmer AJ, Jones T (1987) Measurement of glucose utilization with 2-[$^{18}$F]fluoro-2-deoxy-D-glucose: a comparison of different analytical methods. J Cereb Blood Flow Metab 7: 161–172.

Langley JN (1878) On the physiology of the salivary secretion. J Physiol (Lond) 1: 339–369.

Langmuir I (1916) The constitution and fundamental properties of solids and liquids. J Am Chem Soc 37: 2221–2295.

LaNoue KF, Schoolwerth AC (1979) Metabolite transport in mitochondria. Ann Rev Biochem 48: 871–922.

LaNoue KF, Jeffries FM, Radda GK (1986) Kinetic control of mitochondrial ATP synthesis. Biochemistry 25: 7667–7675.

Laptook AR, Peterson J, Porter AM (1988) Effects of lactic acid infusions and pH on cerebral blood flow and metabolism. J Cereb Blood Flow Metab 8: 193–200.

Laruelle M, Abi-Dargham A, van Dyck CH, Gil R, D'Souza CD, Erdos J, McCance E, Rosenblatt W, Fingado C, Zoghbi SS, Baldwin RM, Seibyl JP, Krystal JH, Charney DS, Innis RB (1996) Single photon emission computerized tomography imaging of amphetamine-induced dopamine release in drug-free schizophrenic subjects. Proc Natl Acad Sci USA 93: 9235–9240.

Lassen NA (1959) Cerebral blood flow and oxygen consumption in man. Physiol Rev 39: 183–238.

Lassen NA, Trap-Jensen AJ, Alexander SC, Olesen J, Paulson OB (1971) Blood–brain barrier studies in man using the double-indicator method. Am J Physiol 220: 1627–1633.

Lassen NA, Perl W (1979) Tracer Kinetic Methods in Medical Physiology, Raven Press, New York.

Lassen NA, Gjedde A (1983) Kinetic analysis of the uptake of glucose and some of its analogs in the brain, using the single capillary model: comments on some points of controversy. In: Lambrecht RM, Rescigno A (eds) Lecture Notes in Biomathematics 48: Tracer Kinetics and Physiologic Modeling, Springer Verlag, New York, pp. 387–410.

Laughton JD, Charnay Y, Belloir B, Pellerin L, Magistretti PJ, Bouras C (2000) Differential messenger RNA distribution of lactate dehydrogenase LDH-1 and LDH-5 isoforms in the rat brain. Neuroscience 96: 619–625.

Lauritzen M (2005) Lauritzen M. Reading vascular changes in brain imaging: is dendritic calcium the key? Nat Rev Neurosci 6: 77–85.

Lear JL, Ackermann RF (1988) Comparison of cerebral glucose metabolic rates measured with fluorodeoxyglucose and glucose labeled in the 1, 2, 3–4, and 6 positions using double label quantitative digital autoradiography. J Cereb Blood Flow Metab 8: 575–585.

LeFauconnier JM (1992) Transport of amino acids. In: Bradbury MWB (ed) Physiology and Pharmacology of the Blood–Brain Barrier, Springer-Verlag, Berlin, pp. 117–150.

LeFèvre PG (1948) Evidence of active transfer of certain non- electrolytes across the human red cell membrane. J Gen Physiol 31: 505–527.

LeFèvre PG (1961) Sugar transport in the red blood cell: structure activity relationships in substrates and antagonists. Pharmacol Rev 16: 39–70.

Levine R, Goldstein M, Klein S, Huddlestun B (1949) The action of insulin on the distribution of galactose in eviscerated nephrectomized dogs. J Biol Chem 179: 985–986.

Levy DE, Sidtis JJ, Rottenberg DA, Jarden JO, Strother SC, Dhawan V, Ginos JZ, Tramo MJ, Evans AC, Plum F (1987) Differences in cerebral blood flow and glucose utilization in vegetative versus locked-in patients. Ann Neurol 22: 673–682.

Li Z, Yipintsoi T, Bassingthwaighte JB (1997) Nonlinear model for capillary-tissue oxygen transport and metabolism. Ann Biomed Eng 25(4): 604–619.

Lilavivathane U, Brodows RG, Woolf PD, Campbell RG (1979) Counterregulatory hormonal responses to rapid glucose lowering in diabetic man. Diabetes 28: 873–877.

Lin S, Spudich JA (1974) Biochemical studies on the mode of action of cytochalasin B. J Biol Chem 249: 5778–5783.

Lindauer U, Megow D, Matsuda H, Dirnagl U (1999) Nitric oxide: a modulator, but not a mediator, of neurovascular coupling in rat somatosensory cortex. Am J Physiol 277: H799–H811.

Logan J, Wolf AP, Shiue CY, Fowler JS (1987) J Neurochem 48: 73–83.

Logan J, Fowler JS, Volkow ND, Wolf AP, Dewey SL, Schlyer DJ, MacGregor RR, Hitzemann R, Bendriem B, Gatley SJ, Christman DR (1990) Graphical analysis of reversible radioligand binding from time-activity measurements applied to [N-11C-methyl]-(-)-cocaine PET studies in human subjects. J Cereb Blood Flow Metab 10: 740–747.

Logan J, Fowler JS, Volkow ND, Ding YS, Wang GJ, Alexoff DL (2001) A strategy for removing the bias in the graphical analysis method. J Cereb Blood Flow Metab 21: 307–320.

Lou HC, Rosa P, Pryds O, Karrebaek H, Lunding J, Cumming P, Gjedde A (2004) ADHD: increased dopamine receptor availability linked to attention deficit and low neonatal cerebral blood flow. Dev Med Child Neurol 46: 179–183.

Lowe AG, Walmsley AR (1986) The kinetics of glucose transport in human red blood cells. Biochim Biophys Acta 857: 146–154.

Lucignani G, Paolini G, Landoni C, Zuccari M, Paganelli G, Galli L, di Credico G, Vanoli G, Rossetti C, Mariani MA, Gilardi MC, Colombo F, Grossi A, Fazio F (1992) Presurgical identification of hibernating myocardium by combined use of technetium-99m hexakis 2-methoxybutylisonitrile single photonemission tomography and fluorine-18 fluoro-2-deoxy-D-glucose positron emission tomography in patients with coronary artery disease. Eur J Nucl Med 19: 874–881.

Lund-Andersen H, Kjeldsen CS (1976) Uptake of glucose analogues by rat brain cortex slices: a kinetic analysis based upon a model. J Neurochem 27: 361–368.

Lund-Andersen H, Kjeldsen CS (1977) Uptake of glucose analogues by rat brain cortex slices: membrane transport versus metabolism of 2-deoxy-D-glucose. J Neurochem 29: 205–211.

Lund-Andersen H (1979) Transport of glucose from blood to brain. Physiol Rev 59: 305–352.

Lundsgaard E (1939) On the mode of action of insulin. Uppsala Läk Fören Förh 45: 1–4.

Ma J, Ayata C, Huang PL, Fishman MC, Moskowitz MA (1996) Regional cerebral blood flow response to vibrissal stimulation in mice lacking type I NOS gene expression. Am J Physiol 270: H1085–H1090.

Madras BK, Spealman RD, Fahey MA, Neumeyer JL, Saha JK, Milius RA (1989) Cocaine receptors labeled by [3H]2 beta-carbomethoxy-3 beta-(4-fluorophenyl)tropane. Mol Pharmacol 36: 518–524.

Magistretti PJ, Pellerin L, Rothman DL, Shulman RG (1999) Energy on demand. Science 283: 496–497.

Mans AM, Davis DW, Biebuyck JF, Hawkins RA (1986) Failure of glucose and branched-chain amino acids to normalize brain glucose use in portacaval shunted rats. J Neurochem 47: 1434–1443.

Markram H (2006) The blue brain project. Nat Rev Neurosci 7: 153–160.

Marrett S, Gjedde A (1997) Changes of blood flow and oxygen consumption in visual cortex of living humans. Adv Exp Med Biol 413: 205–208.

Marshall RC, Tillisch J, Phelps ME, Huang SC, Carson R, Henze E, Schelbert HR (1983a) Identification and differentiation of resting myocardial ischemia and infarction in man with positron emission tomography, $^{18}$F-labeled fluorodeoxyglucose and N-13 ammonia. Circulation 67: 766–778.

Marshall RC, Huang SC, Nash WW, Phelps ME (1983b) Assessment of the [$^{18}$F]fluoro deoxyglucose kinetic model in calculations of myocardial glucose metabolism during ischemia. J Nucl Med 24: 1060–1064.

Mata M, Fink DG, Gainer H, Smith CB, Davidsen L, Savaki HE, Schwarts WJ, Sokoloff L (1980) Activity-dependent energy metabolism in rat posterior pituitary primarily reflects sodium pump activity. J Neurochem 34: 213–215.

Matthaei S, Horuk R, Olefsy JM (1986) Blood–brain glucose transfer in diabetes mellitus. Decreased number of glucose transporters at blood-brain barrier. Diabetes 35: 1181–1184.

Matthaei S, Garvey WT, Horuk R, Hueckstadt TP, Olefsky JM (1987a) Human adipocyte glucose transport system: biochemical and functional heterogeneity of hexose carriers. J Clin Invest 79: 703–709.

Matthaei S, Olefsky JM, Horuk R (1987b) Biochemical characterization and subcellular distribution of the glucose transporter from rat brain microvessels. Biochim Biophys Acta 905: 417–425.

Mathiesen C, Caesar K, Akgoren N, Lauritzen M (1998) Modification of activity-dependent increases of cerebral blood flow by excitatory synaptic activity and spikes in rat cerebellar cortex. J Physiol (Lond) 512: 555–566.

Matthews PM, Bland JL, Gadian DG, Radda GK (1981) The steady-state rate of ATP synthesis in the perfused rat heart measured by 31P NMR saturation transfer. Biochem Biophys Res Commun 103: 1052–1059.

Maxwell K, Berliner JA, Cancilla PA (1989) Stimulation of glucose analogue uptake by cerebral microvessel endothelial cells by a product released by astrocytes. J Neuropathol Exp Neurol 48: 69–80.

Mayman CI, Gatfield PD, Breckenbridge BM (1964) The glucose content of brain in anaesthesia. J Neurochem 11: 483–487.

McCall AL, Millington WR, Wurtman RJ (1982) Metabolic fuel and aminoacid transport into the brain in experimental diabetes mellitus. Proc Natl Acad Sci USA 79: 5406–5410.

McCall AL, Fixman LB, Fleming N, Tornheim K, Chick W, Ruderman NB (1986) Chronic hypoglycemia incrases brain glucose transport. Am J Physiol 251: E442–E447.

McCann UD, Szabo Z, Seckin E, Rosenblatt P, Mathews WB, Ravert HT, Dannals RF, Ricaurte GA (2005) Quantitative PET studies of the serotonin transporter in MDMA users and controls using [$^{11}$C]McN5652 and [$^{11}$C]DASB. Neuropsychopharmacology 30: 1741–1750.

McCann UD, Wong DF, Yokoi F, Villemagne V, Dannals RF, Ricaurte GA (1998) Reduced striatal dopamine transporter density in abstinent methamphetamine and methcathinone users: evidence from positron emission tomography studies with [$^{11}$C]WIN-35,428. J Neurosci 18: 8417–8422.

McCormick DA (1990) Membrane properties and neurotransmitter actions. In Shepherd G (ed) The Synaptic Organization of the Brain, 3rd edn., Oxford University Press, New York, pp. 32–66.

McCully KK, Kakihira H, van den Borne K, Kent-Braun J (1991) Non-invasive measurements of activity-induced changes in muscle metabolism. J Biomechan 24: Suppl 1: 153–161.

McIlwain H (1951) Metabolic response in vitro to electrical stimulation of section. Biochem J 49: 382–393.

Melikian HE, Buckley KM (1999) Membrane trafficking regulates the activity of the human dopamine transporter. J Neurosci 19: 7699–7710.

Messa C, Choi Y, Hoh CK, Jacobs EL, Glaspy JA, Rege S, Nitzsche E, Huang SC, Phelps ME, Hawkins RA (1992) Quantification of glucose utilization in liver metastases: parametric imaging of FDG uptake with PET. J Comput Assist Tomogr 16: 684–689.

Michaelis L, Menten ML (1913) Die Kinetik der Invertinwirkung. Biochem Z 49: 333–369.

Mintun MA, Raichle ME, Kilbourn MR, Wooten GF, Welch MJ (1984) A quantitative model for the in vivo assessment of drug binding sites with positron emission tomography. Ann Neurol 15: 217–227.

Møller M, Andersen G, Gjedde A (2007a) Serotonin 5HT1A receptor availability and pathological crying after stroke. Acta Neurol Scand 116(2): 83–90.

Møller M, Jakobsen S, Gjedde A (2007b) Parametric and regional maps of free serotonin 5HT1A receptor sites in human brain as function of age in healthy humans. Neuropsychopharmacology 32(8): 1707–1714.

Mooradian AD, Morin AM (1991) Brain uptake of glucose in diabetes mellitus: the role of glucose transporters. Am J Med Sci 301: 173–177.

Moore TJ, Lione AP, Sugden MSM, Regen DM (1976) $\beta$-hydroxybutyrate transport in rat brain: developmental and dietary modulations. Am J Physiol 230: 619–630.

Mora BN, Narasimhan PT, Ross BD (1992) $^{31}$P magnetization transfer studies in the monkey brain. Magn Reson Med 26: 100–115.

Morgan D, Grant KA, Gage HD, Mach RH, Kaplan JR, Prioleau O, Nader SH, Buchheimer N, Ehrenkaufer RL, Nader MA (2002) Social dominance in monkeys: dopamine D2 receptors and cocaine self-administration. Nature Neurosci 5: 169–174.

Mori K, Schmidt K, Jay T, Palombo E, Lucignani G, Pettigrew K, Kennedy C, Sokoloff L (1990) Optimal duration of experimental period in measurement of local cerebral glucose utilization with the deoxyglucose method. J Neurochem 54: 307–319.

Morin AM, Dwyer BE, Fujikawa DG, Wasterlain CG (1988) Low [3H]cytochalasin B binding in the cerebral cortex of newborn rat. J Neurochem 51: 206–211.

Morowitz HJ (1978) Foundations of Bioenergetics. Academic Press, New York.

Mukherjee J, Christian BT, Dunigan KA, Shi B, Narayanan TK, Satter M, Mantil J (2002) Brain imaging of $^{18}$F-fallypride in normal volunteers: blood analysis, distribution, test-retest studies, and preliminary assessment of sensitivity to aging effects on dopamine D-2/D-3 receptors. Synapse 46: 170–188.

Mukherjee J, Christian BT, Narayanan TK, Shi B, Collins D (2005) Measurement of d-amphetamine-induced effects on the binding of dopamine D-2/D-3 receptor radioligand, $^{18}$F-fallypride in extrastriatal brain regions in non-human primates using PET. Brain Res 1032: 77–84.

Mukherjee J, Shi B, Christian BT, Chattopadhyay S, Narayanan TK (2004) $^{11}$C-Fallypride: radiosynthesis and preliminary evaluation of a novel dopamine D2/D3 receptor PET radiotracer in non-human primate brain. Bioorg Med Chem 12: 95–102.

Mueckler MC, Caruso C, Baldwin SA, Panico M, Blench I, Morris HR, Allard WJ, Lienhard GE, Lodish HF (1985) Sequence and structure of a human glucose transporter. Science 229: 941–945.

Muir D, Berl S, Clarke DD (1986) Acetate and fluoroacetate as possible markers for glial metabolism in vivo. Brain Res 380: 336–340.

Naftalin RJ (1988) Pre-steady-state uptake of D-glucose by the human erythrocyte is inconsistent with a circulating carrier mechanism. Biochim Biophys Acta 946: 431–438.

Nalecz KA, Kaminska J, Nalecz MJ, Azzi A (1992) The activity of pyruvate carrier in a reconstituted system: substrate specificity and inhibitor sensitivity. Arch Biochem Biophys 297: 162–168.

Namba H, Lucignani G, Nehlig A, Patlak C, Pettigrew K, Kennedy C, Sokoloff L (1987) Effects of insulin on hexose transport across blood–brain barrier in normoglycemia. Am J Physiol 252: E299–E303.

Nehlig A, Wittendorp-Rechenmann E, Lam CD (2004) Selective uptake of [14C]2-deoxyglucose by neurons and astrocytes: high-resolution microautoradiographic imaging by cellular 14C-trajectography combined with immunohistochemistry. J Cereb Blood Flow Metab 24(9): 1004–1014.

Nemoto EM, Stezoski SW, MacMurdo D (1978) Glucose transport across the rat blood–brain barrier during anesthesia. Anesthesiology 49: 170–176.

Newsholme EA, Crabtree B (1979) Theoretical principles in the approaches to control of metabolic pathways and their application to glycolysis in muscle. J Mol Cell Cardiol 11: 839–856.

Ng CK, Holden JE, DelGrado TR, Raffel DM, Kornguth ML, Gatley SJ (1991) Sensitivity of myocardial fluorodeoxyglucose lumped constant to glucose and insulin. Am J Physiol 260: H593–H603.

Nirenberg MJ, Chan J, Pohorille A, Vaughan RA, Uhl GR, Kuhar MJ, Pickel VM (1997) The dopamine transporter: comparative ultrastructure of dopaminergic axons in limbic and motor compartments of the nucleus accumbens. J Neurosci 17: 6899–6907.

Norenberg MD (1994) Astrocyte responses to CNS injury. J Neuropathol Exp Neurol 53(3): 213–220.

Ogawa M, Magata Y, Ouchi Y, Fukuyama H, Yamauchi H, Kimura J, Yonekura Y, Konishi J (1994) Scopolamine abolishes cerebral blood flow response to somatosensory stimulation in anesthetized cats: PET study. Brain Res 650: 249–252.

Ogawa S, Lee TM, Nayak AS et al. (1990a) Oxygenation-sensitive contrast in magnetic resonance imaging of rodent brain at high magnetic fields. Magn Reson Med 14: 68–78.

Ogawa S, Lee TM, Kay AR, Tank DW (1990b) Brain magnetic resonance imaging with contrast dependent on blood oxygenation. Proc Natl Acad Sci USA 87: 9868–9872.

Ogawa S, Menon RS, Tank DW, Kim SG, Merkle H, Ellermann JM, Ugurbil K (1993) Functional brain mapping by blood oxygenation level-dependent contrast magnetic resonance imaging. A comparison of signal characteristics with a biophysical model. Biophys J 64: 803–812.

Ohira Y, Tabata I (1992) Muscle metabolism during exercise: anaerobic threshold does not exist. Ann Physiol Anthropol 11: 319–323.

Ohta S, Meyer E, Thompson CJ, Gjedde A (1992) Oxygen consumption of the living human brain measured after a single inhalation of positron emitting oxygen. J Cereb Blood Flow Metab 12(2): 179–192.

Ohta S, Meyer E, Gjedde A (1996) Cerebral [$^{15}$O]water clearance in humans determined by PET. I. Theory and normal values. J Cereb Blood Flow Metab 16: 765–780.

Ohta S, Reutens DC, Gjedde A (1999) Brief vibrotactile stimulation does not increase cortical oxygen consumption when measured by single inhalation of positron emitting oxygen. J Cereb Blood Flow Metab 19: 260–265.

Oka Y, Asano T, Shibasaki Y, Kasuga M, Kanazawa Y, Takaku F (1989) Studies with antipeptide antibody suggest the presence of at least two types of glucose transporter in rat brain and adipocyte. J Biol Chem 263: 13432–13439.

Oka Y, Asano T, Shibasaki Y, Lin J-L, Tsukuda K, Katagiri H, Akanuma Y, Takaku F (1990) C-terminal truncated glucose transporter is locked into an inward-facing form without transport activity. Nature 345: 550–553.

Okazumi S, Isono K, Enomoto K, Kikuchi T, Ozaki M, Yamamoto H, Hayashi H, Asano T, Ryu M (1992) Evaluation of liver tumors using fluorine-18-fluorodeoxyglucose PET: characterization of tumor and assessment of effect of treatment. J Nucl Med 33: 333–339.

Oldendorf WH (1970) Measurement of brain uptake of radiolabeled substances using a tritiated water internal standard. Brain Res 24: 372–376.

Oldendorf WH (1971) Brain uptake of labeled amino acids, amines, and hexoses after arterial injection. Am J Physiol 221: 1629–1639.

Oldendorf WH (1973) Carrier-mediated blood–brain barrier transport of short-chain monocarboxylic organic acids. Am J Physiol 224: 1450–1453.

Oldendorf WH, Braun LD (1976) [$^3$H]tryptamine and $^3$H-water as diffusible internal standards for measuring brain extraction of radiolabeled substances following carotid injection. Brain Res 113: 219–224.

Olesen J (1970) Total $CO_2$, lactate, and pyruvate in brain biopsies taken after freezing the tissue in situ. Acta Neurol Scand 46: 141–148.

Ostergaard L, Johannsen P, Host-Poulsen P, Vestergaard-Poulsen P, Asboe H, Gee AD, Hansen SB, Cold GE, Gjedde A, Gyldensted C (1998) Cerebral blood flow measurements by magnetic resonance imaging bolus tracking: comparison with [(15)O]H2O positron emission tomography in humans. J Cereb Blood Flow Metab 18: 935–940.

Oz G, Berkich DA, Henry PG, Xu Y, LaNoue K, Hutson SM, Gruetter R (2004) Neuroglial metabolism in the awake rat brain: $CO_2$ fixation increases with brain activity. J Neurosci 24: 11273–11279.

Pappenheimer JR, Setchell BP (1973) Cerebral glucose transport and oxygen consumption in sheep and rabbits. J Physiol 233: 529–551.

Pardridge WM, Oldendorf WH (1975) Kinetics of blood–brain barrier transport of hexoses. Bioclin Biophys Acta 382: 377–392.

Pardridge WM, Oldendorf WH (1977) Transport of metabolic substrates through the blood–brain barrier. J Neurochem 28: 5–12.

Pardridge WM (1981) Transport of nutrients and hormones through the blood–brain barrier. Diabetol$\gamma$a 20: 246–254.

Pardridge WM, Crane PD, Mietus LJ, Oldendorf WH (1982) Kinetics of regional blood–brain barrier transport and brain phosphorylation of glucose and 2-deoxyglucose in the barbiturate-anesthetized rat. J Neurochem 38: 560–568.

Pardridge WM (1983) Brain metabolism: a perspective from the blood–brain barrier. Physiol Rev 63: 1481–1535.

Pardridge WM, Landaw EM, Miller LP, Braun LD, Oldendorf WH (1985) Carotid artery injection technique: bounds for bolus mixing by plasma and by brain. J Cereb Blood Flow Metab 5: 576–583.

Pardridge WM, Boado RJ, Farrell C (1990a) Downregulation of blood-brain barrier glucose transporter in experimental diabetes. Diabetes 39: 1040–1044.

Pardridge WM, Boado RJ, Farrell C (1990b) Brain-type glucose transporter (GLUT-1) is selectively localized to the blood-brain barrier. J Biol Chem 265: 18035–18040.

Patel AB, de Graaf RA, Mason GF, Kanamatsu T, Rothman DL, Shulman RG, Behar KL (2004) Glutamatergic neurotransmission and neuronal glucose oxidation are coupled during intense neuronal activation. J Cereb Blood Flow Metab 24: 972–985.

Patlak C, Blasberg RG, Fenstermacher JD (1983) Graphical evaluation of blood-to-brain transfer constants from multiple-time uptake data. J Cereb Blood Flow Metab 3: 1–7.

Paul R, Johannson R, Kellokumpu-Lehtinen PL, Söderström KO, Kangas L (1985) Tumor localization with $^{18}F$-2-fluoro-2-deoxy-D-glucose: comparative autoradiography, glucose 6-phosphate histochemistry, and histology of renally implanted sarcoma of the rat. Res Exp Med (Berl) 185: 87–94.

Paulson OB, Newman EA (1987) Does the release of potassium from astrocyte endfeet regulate cerebral blood flow? Science 237: 896–898.

Pearlson GD, Wong DF, Tune LE, Ross CA, Chase GA, Links JM, Dannals RF, Wilson AA, Ravert HT, Wagner HN Jr et al. (1995) In vivo D2 dopamine receptor density in psychotic and nonpsychotic patients with bipolar disorder. Arch Gen Psychiatry 52(6): 471–477.

Pedroarena C, Llinas R (1997) Dendritic calcium conductances generate high-frequency oscillation in thalamocortical neurons. Proc Natl Acad Sci USA 94: 724–728.

Pellerin L, Magistretti PJ (1994) Glutamate uptake into astrocytes stimulates aerobic glycolysis: a mechanism coupling neuronal activity to glucose utilization. Proc Natl Acad Sci USA 91: 10625–10629.

Pelligrino DA, Segil LJ, Albrecht RF (1990a) Brain glucose utilization and transport and cortical function in chronic vs. acute hypoglycemia. Am J Physiol 259: E729–E735.

Pelligrino DA, Lipa MD, Albrecht RF (1990b) Regional blood–brain glucose transfer and glucose utilization in chronically hyperglycemic, diabetic rats following acute glycemic normalization. J Cereb Blood Flow Metab 10: 775–780.

Perlmutter JS, Larson KB, Raichle ME, Markham J, Mintun MA, Kilbourn MR, Welch MJ (1986) J Cereb Blood Flow Metab 6: 154–169.

Perlmutter JS, Kilbourn MR, Raichle ME, Welch MJ (1987) MPTP induced up-regulation of in vivo dopaminergic radioligand-receptor binding in man. Neurology 37: 1575–1579.

Pertsch M, Duncan GE, Stumpf WE, Pilgrim C (1988) A histochemical study of the regional distribution in the rat brain of enzymatic activity hydrolyzing glucose- and 2-deoxyglucose-6-phosphate. Histochemistry 88: 257–262.

Pette D (1985) Metabolic heterogeneity of muscle fibres. J Exp Biol 115: 179–189.

Pessin JE, Tillotson LG, Yamada K, Gitomer W, Carter-Su C, Mora R, Isselbacher KJ, Czech M (1982) Identification of the stereospecific hexose transporter from starved and fed chicken embryo fibroblasts. Proc Natl Acad Sci (USA) 79: 2286–2290.

Phelps ME, Huang SC, Hoffman EJ, Selin C, Sokoloff L, Kuhl DE (1979) Tomographic measurement of local cerebral glucose metabolic rate in humans with 2-[$^{18}$F]fluoro-2-deoxy-D-glucose: validation of method. Ann Neurol 6: 371–388.

Phelps ME, Schelbert HR, Hoffman EJ, Huang SC, Kuhl DE (1980) Positron tomography of the heart. Prog Nucl Med 6: 183–209.

Pierre K, Pellerin L (2005) Monocarboxylate transporters in the central nervous system: distribution, regulation and function. J Neurochem 94: 1–14.

Pilch PF (1990) Glucose transporters: what's in a name? Endocrinology 126: 3–5.

Pirker W, Asenbaum S, Bencsits G, Prayer D, Gerschlager W, Deecke L, Brucke T (2000) [$^{123}$I]beta-CIT SPECT in multiple system atrophy, progressive supranuclear palsy, and corticobasal degeneration. Mov Disord 15: 1158–1167.

Planas AM, Cunningham VJ (1987) Uncoupling of cerebral glucose supply and utilization after hexane-2,5-dione intoxication in the rat. J Neurochem 48: 816–823.

Pohjalainen T, Rinne JO, Någren K, Syvlahti E, Hietala J. (1998) Sex differences in the striatal dopamine D2 receptor binding characteristics in vivo. Am J Psychiatry 155(6): 768–773.

Poitry-Yamate CL, Tsacopoulos M (1992) Glucose metabolism in freshly isolated Muller glial cells from a mammalian retina. J Comp Neurol 320: 257–266.

Poitry-Yamate CL, Poitry S, Tsacopoulos M (1995) Lactate released by Muller glial cells is metabolized by photoreceptors from mammalian retina. J Neurosci 15: 5179–5191.

Pollay M, Stevens FA (1979) Starvation-induced changes in transport of ketone bodies across the blood–brain barrier. J Neurosci Res 5: 163–172.

Poole RC, Halestrap AP (1993) Transport of lactate and other monocarboxylates across mammalian plasma membranes. Am J Physiol 264: C761–C782.

Prichard J, Rothman D, Novotny E, Petroff O, Kuwabara T, Avison M, Howseman A, Hanstock C, Shulman R (1991) Lactate rise detected by 1H NMR in human visual cortex during physiologic stimulation. Proc Natl Acad Sci USA 88: 5829–5831.

Pristupa ZB, McConkey F, Liu F, Man HY, Lee FJ, Wang YT, Niznik HB (1998) Protein kinase-mediated bidirectional trafficking and functional regulation of the human dopamine transporter. Synapse 30(1): 79–87.

Pullman ME, Penefsky HS, Datta A, Racker E (1960) Partial resolution of the enzymes catalyzing oxidative phosphorylation. I. Purification and properties of soluble dinitrophenol-stimulated adenosine triphosphatase. J Biol Chem 235: 3322–3329.

Qian H, Bassingthwaighte JB (2000) A class of flow bifurcation models with lognormal distribution and fractal dispersion. J Theor Biol 205(2): 261–268.

Raichle ME, Larson KB, Phelps ME, Grubb RL Jr, Welch MJ, Ter-Pogossian MM (1975) In vivo measurement of brain glucose transport and metabolism employing glucose-$^{11}$C. Am J Physiol 228: 1936–1948.

Raichle ME, Grubb RL Jr, Gado MH, Eichling JO, Ter-Pogossian MM (1976) Correlation between regional cerebral blood flow and oxidative metabolism. In vivo studies in man. Arch Neurol 33: 523–526.

Raichle ME, MacLeod AM, Snyder AZ, Powers WJ, Gusnard DA, Shulman GL (2001) A default mode of brain function. Proc Natl Acad Sci USA 98(2): 676–682.

Ramlal T, Rastogi S, Vranic M, Klip A (1989) Decrease in glucose transporter number in skeletal muscle of mildly diabetic (streptozotocin-treated) rats. Endocrinology 125: 890–897.

Rasmussen M, Poulsen PH, Treiber A, Delahaye S, Tankisi A, Cold GE, Therkelsen K, Gjedde A, Astrup J (2003) No influence of the endothelin receptor antagonist bosentan on basal and indomethacin-induced reduction of cerebral blood flow in pigs. Acta Anaesthesiol Scand 47: 200–207.

Ratib O, Phelps ME, Huang SC, Henze E, Selin CE, Schelbert HR (1982) Positron tomography with deoxyglucose for estimating local myocardial glucose metabolism. J Nucl Med 23: 577–586.

Redies C, Gjedde A (1989) Double-label and conventional deoxyglucose methods: a practical guide for the user. Cereb Vasc Brain Metab Rev 1: 319–367.

Redies C, Hoffer LJ, Beil C, Marliss EB, Evans AC, Lariviere F, Marrett S, Meyer E, Diksic M, Gjedde A, Hakim AM (1989) Generalized decrease in brain glucose metabolism during fasting in humans studied by PET. Am J Physiol 256: E805–E810.

Reese TS, Feder N, Brightman MW (1971) Electron microscopic study of the blood–brain and blood-cerebrospinal fluid barriers with microperoxidase. J Neuropathol Exp Neurol 30: 137–138.

Reeves JT, Wolfel EE, Green HJ, Mazzeo RS, Young AJ, Sutton JR, Brooks GA (1992) Oxygen transport during exercise at altitude and the lactate paradox: lessons from Operation Everest II and Pike's Peak. Exerc Sport Sci Rev 20: 275–296.

Reith J, Ermisch A, Diemer NH, Gjedde A (1987) Saturable retention of vasopressin in vivo, associated with inhibition of blood–brain transfer of large neurtral amino acids. J Neurochem 49: 1471–1480.

Reith J, Dyve S, Kuwabara H, Guttman M, Diksic M, Gjedde A (1990) Blood–brain transfer and metabolism of 6-[$^{18}$F]fluoro-L-DOPA. J Cereb Blood Flow Metab 10: 707–719.

Reith J, Benkelfat C, Sherwin A, Yasuhara Y, Kuwabara H, Andermann F, Bachneff S, Cumming P, Diksic M, Dyve SE, Etienne P, Evans AC, Lal S, Shevell M, Savard G, Wong DF, Chouinard G, Gjedde A (1994) Elevated dopa decarboxylase activity in living brain of patients with psychosis. Proc Natl Acad Sci USA 91: 11651–11654.

Reivich M, Kuhl D, Wolf A, Greenberg J, Phelps M, Ido T, Casella V, Fowler J, Hoffman E, Alavi A, Som P, Sokoloff L (1979) The [18F]fluorodeoxyglucose method for the measurement of local cerebral glucose utilization in man. Circ Res 44: 127–137.

Reivich M, Alavi A, Wolf A, Fowler J, Russell J, Arnett C, MacGregor RR, Shiue CY, Atkins H, Anand A, Dann R, Greenberg JH (1985) Glucose metabolic rate kinetic model parameter determination in humans: the lumped constants and rate constants for [$^{18}$F]fluorodeoxy- and [$^{11}$C]deoxyglucose. J Cereb Blood Flow Metab 5: 179–192.

Rescigno A, Beck JS (1972a) In: Rosen R (ed) Foundations of Mathematical Biology, Vol 2, Academic Press, New York, pp. 255–322.

Rescigno A, Beck JS (1972b) Compartments. In: Rosen R (ed) Foundations of Mathematical Biology, Vol 2, Academic Press, New York, pp. 255–322.

Rescigno A, Beck JS, Goren HJ (1983a) Determination of dependence of binding parameters on receptor occupancy. Bull Math Biol 44: 477–489.

Rescigno A, Lambrecht RM, Duncan CC (1983b) Mathematical methods in the formulation of pharmacokinetic models. In: Levin (ed) Lecture Notes in Biomathematics. Man; Tracer Kinetics and Physiologic Modeling: Proceedings, St. Louis 1983, Lambrecht, Rescigno (eds); Springer-Verlag, New York, p. 59.

Reutens DC, McHugh MD, Toussaint PJ, Evans AC, Gjedde A, Meyer E, Stewart DJ (1997) L-arginine infusion increases basal but not activated cerebral blood flow in humans. J Cereb Blood Flow Metab 17: 309–315.

Ribak CE (1981) The histochemical localization of cytochrome oxidase in the dentate gyrus of the rat hippocampal. Brain Res 212: 169–174.

Ribeiro L, Bercovic SF, Kuwabara H, Andermann F, Gjedde A (1991) Functional capillary density is reduced in brain of patients with MERFF (myoclonus epilepsy, ragged-red fiber syndrome). The 21st Annual Meeting for Society for Neuroscience, November 10–15, 1991, New Orleans, Louisiana.

Ribeiro L, Kuwabara H, Meyer E, Fujita H, Marrett S, Evans A, Gjedde A (1993) Cerebral blood flow and metabolism during nonspecific bilateral visual stimulation in normal subjects. In: Uemura K, Lassen NA, Jones T, Kanno I (eds) Quantification of Brain Function. Tracer Kinetics and Image Analysis in Brain PET, Elsevier, Amsterdam, pp. 217–224.

Rinne JO, Laihinen A, Rinne UK, Nagren K, Bergman J, Ruotsalainen U (1993) PET study on striatal dopamine D2 receptor changes during the progression of early Parkinson's disease. Mov Disord 8: 134–138.

Ritchie JM (1967) The oxygen consumption of mammalian non-myelinated fibres at rest and during activity. J Physiol (Lond) 188: 309–329.

Robin ED, Murphy BJ, Theodore J (1984) Coordinate regulation of glycolysis by hypoxia in mammalian cells. J Cell Physiol 118: 287–290.

Roland PE, Eriksson L, Stone-Elander S, Widen L (1987) Does mental activity change the oxidative metabolism of the brain? J Neurosci 8: 2373–2389.

Roland PE, Eriksson L, Widen L, Stone-Elander S (1989) Changes in regional cerebral oxidative metabolism induced by tactile learning and recognition in man. Eur J Neurosci 7: 2373–2389.

Rosa-Neto P, Lou HC, Cumming P, Pryds O, Karrebaek H, Lunding J, Gjedde A (2005) Methylphenidate-evoked changes in striatal dopamine correlate with inattention and impulsivity in adolescents with attention deficit. Neuroimage 25: 868–876.

Rose SP (1975) Cellular compartmentation of brain metabolism and its functional significance. J Neurosci Res 1: 19–30.

Rosenthal HE (1967) A graphic method for the determination and presentation of binding parameters in a complex system. Anal Biochem 20(3): 525–532.

Roth K, Weiner MW (1991) Determination of cytosolic ADP and AMP concentrations and the free energy of ATP hydrolysis in huan muscle and brain tissues with $^{31}$P NMR spectroscopy. Magn Reson Med 22: 505–511.

Rothstein JD, Dykes Hoberg M, Pardo CA, Bristol LA, Jin L, Kuncl RW, Kanai Y, Hediger MA, Wang Y, Schielke JP, Welty DF (1996) Knockout of glutamate transporters reveals a major role for astroglial transport in excitotoxicity and clearance of glutamate. Neuron 16: 675–686.

Rousset OG, Deep P, Kuwabara H, Evans AC, Gjedde A, Cumming P (2000) Effect of partial volume correction on estimates of the influx and cerebral metabolism of 6-[(18)F]fluoro-L-dopa studied with PET in normal control and Parkinson's disease subjects. Synapse 37(2): 81–89.

Rousset OG, Ma Y, Kamber M, Evans AC (1993) 3D simulations of radiotracer uptake in deep nuclei of human brain. Comput Med Imag Graph 17: 373–379.

Roy CS, Sherrington CS (1890) On the regulation of the blood supply of the brain. J Physiol 11: 85–108.

Rubart M, Breull W, Hahn N (1991) Regional metabolic rate of exogenous glucose in the isoprenaline and dolbutamine stimulated canine myocardium as estimated by the 2-deoxy-D[1-$^{14}$C]glucose method. Int J Rad Appl Instrumen B 18: 157–166.

Russell RR 3rd, Mrus JM, Mommessin JI, Taegtmeyer H (1992) Compartments of hexokinase in rat heart. A critical marker for tracer kinetic analysis of myocardial glucose metabolism. J Clin Invest 90: 1972–1977.

Sabol KE, Lew R, Richards JB, Vosmer GL, Seiden LS (1996) Methylenedioxy-methamphetamine-induced serotonin deficits are followed by partial recovery over a 52-week period. Part I: Synaptosomal uptake and tissue concentrations. J Pharmacol Exp Therapeut 276: 846–854.

Sakoh M, Gjedde A (2003) Neuroprotection in hypothermia linked to redistribution of oxygen in brain. Am J Physiol 285: H17–H25.

Salceda R, Vilchis C, Coffe V, Hernandez-Munoz R (1998) Changes in the redox state in the retina and brain during the onset of diabetes in rats. Neurochem Res 23: 893–897.

Saperstein LA (1958) Regional blood flow by fractional distribution of indicators. Am J Physiol 193: 161–165.

Sappey-Marinier D, Calabrese G, Fein G, Hugg JW, Biggins C, Weiner MW (1992) Effect of photic stimulation on human visual cortex lactate and phosphates using 1H and 31p magnetic resonance spectroscopy. J Cereb Blood Flow Metab 12: 584–592.

Sargent PA, Kjaer KH, Bench CJ, Rabiner EA, Messa C, Meyer J et al. (2000) Brain serotonin1A receptor binding measured by positron emission tomography with [11C]WAY-100635: effects of depression and antidepressant treatment. Arch Gen Psychiatry 57: 174–180.

Sarna GS, Bradbury MW, Cavanagh J (1977) Permeability of the blood–brain barrier after portocaval anastomosis in the rat. Brain Res 138: 550–554.

Sarna GS, Bradbury MWB, Cremer JE, Lai JCK, Teal HM (1979) Brain metabolism and specific transport at the blood–brain barrier after portocaval anastomosis in the rat. Brain Res 160: 69–83.

Sawada Y, Patlak CS, Blasberg RG (1989) Kinetic analysis of cerebrovascular transport based on indicator diffusion technique. Am J Physiol 256: 794–812.

Scatchard G (1949) The attractions of proteins for small molecules and ions. Ann NY Acad Sci 51: 660–672.

Schaefer JA, Gjedde A, Plum F (1976) Regional cerebral blood flow using n-(14C)butanol. Neurology 26: 394.

Scheufler KM, Rohrborn HJ, Zentner J (2002) Does tissue oxygen-tension reliably reflect cerebral oxygen delivery and consumption? Anesth Analg 95: 1042–1048.

Scheffel U, Pogun S, Stathis M, Boja JW, Kuhar MJ (1991) In vivo labeling of cocaine binding sites on dopamine transporters with [$^{3}$H]WIN 35,428. J Pharmacol Exp Therapeut 257: 954–958.

Schlaepfer TE, Pearlson GD, Wong DF, Marenco S, Dannals RF (1997) PET study of competition between intravenous cocaine and [$^{11}$C]raclopride at dopamine receptors in human subjects. Am J Psychiatry 154: 1209–1213.
Schmidt K, Lucignani G, Mori K, Jay T, Palombo E, Nelson T, Pettigrew K, Holden JE, Sokoloff L (1989) Refinement of the kinetic model of the 2-[$^{14}$C]deoxyglucose method to incorporate effects of intracellular compartmentation in brain. J Cereb Blood Flow Metab 9: 290–303.
Schneier FR, Liebowitz MR, Abi-Dargham A, Zea-Ponce Y, Lin S, Laruelle M (2000) Low dopamine D2 receptor binding potential in social phobia. Am J Psychiatry 157: 457–459.
Seeman P, Lee T, Chau-Wong M, Wong K (1976) Antipsychotic drug doses and neuroleptic/dopamine receptor. Nature 262: 717–719.
Seibyl JP (2003) Imaging studies in movement disorders. Seminars Nucl Med 33: 105–113.
Seitz RJ, Roland PE (1992) Vibratory stimulation increases and decreases the regional cerebral blood flow and oxidative metabolism: a positron emission tomography (PET) study. Acta Neurol Scand 86: 60–67.
Shalit MN, Beller AJ, Feinsod M, Drapkin AJ, Cotev S (1970) The blood flow and oxygen consumption of the dying brain. Neurology 20: 740–748.
Shalit MN, Beller AJ, Feinsod M (1972) Clinical equivalents of cerebral oxygen consumption in coma. Neurology 22: 155–160.
Shao J (1993) Linear model selection by cross-validation. J Am Statist Assoc 88: 486–494.
Shapiro ET, Cooper M, Chen C-T, Given BDF, Polonsky KS (1990) Change in hexose distribution volume and fractional utilization of [18F]-2-deoxy-2-fluoro-D-glucose in brain during acute hypoglycemia in humans. Diabetes 39: 175–180.
Shearman MS, Halestrap AP (1984) The concentration of the mitochondrial pyruvate carrier in rat liver and heart mitochondria determined with alpha-cyano-beta-(1-phenylindol-3-yl)acrylate. Biochem J 223: 673–676.
Sheppard CW (1948) The theory of the study of transfers within a multi-compartmental system using isotopic tracers. J Appl Phys 19: 70–76.
Shows TB, Eddy RL, Byers MG, Fukushima Y, Dehaven CR, Murray JC, Bell GI (1987) Polymorphic human glucose transport gene (GLUT) is on chromosome 1p31.3→p35. Diabetes 36: 546–549.
Shulman RG, Rothman DL (1998) Interpreting functional imaging studies in terms of neurotransmitter cycling. Proc Natl Acad Sci USA 95: 11993 11998.
Shulman RG, Rothman DL, Behar KL, Hyder F (2004) Energetic basis of brain activity: implications for neuroimaging. Trends Neurosci. 27: 489–495.
Sibson NR, Dhankhar A, Mason GF, Rothman DL, Behar KL, Shulman RG (1998) Stoichiometric coupling of brain glucose metabolism and glutamatergic neuronal activity. Proc Natl Acad Sci U S A 95: 316–321.
Silver IA, Erecinska M (1994) Extracellular glucose concentration in mammalian brain: continous monitoring of changes during increased neuronal activity and upon limitation in oxygen supply in normo-, hypo-, and hyperglycemic animals. J Neurosci 14: 5068–5076.
Silver IA, Erecinska M (1997) Energetic demands of the Na+/K+ ATPase in mammalian astrocytes. Glia 21: 35–45.
Singer HS, Szymanski S, Giuliano J, Yokoi F, Dogan AS, Brasic JR et al. (2002) Elevated intrasynaptic dopamine release in Tourette's syndrome measured by PET. Am J Psychiatry 159: 1329–1336.
Skou JC (1960) Further investigations on a $Mg^{++}$—$Na^{+}$-activated adenosine-triphospha- tase, possibly related to the active, linked transport of $Na^{+}$ and $K^{+}$ across the nerve membrane. Biochim Biophys Acta 42: 6–23.
Slot JW, Moxley R, Geuze HJ, James DE (1990) No evidence for expression of the insulin-regulatable glucose transporter in endothelial cells. Nature 346: 369–371.
von Smoluchowsky M (1914) [The Kinetic Theory of Matter and Electricity], Leipzig and Berlin.
Snyder CD, Wilson JE (1983) Relative levels of hexokinase in isolated neuronal, astrocytic, and oligodendroglial fractions from rat brain. J Neurochem 40: 1178–1181.

Sokoloff L (1960) Metabolism of the central nervous system in vivo. In: Field J, Magoun HW, Hale VE (eds) Handbook of Physiology, Vol 3, American Physiological Society, Washington, pp. 1843–1864.
Sokoloff L, Reivich M, Kennedy C, DesRosiers MH, Patlak CS, Pettigrew KD, Sakurada O, Shinohara M (1977) The [$^{14}$C]deoxyglucose method for the measurement of local cerebral glucose utilization: Theory, procedure, and normal values in the conscious and anesthetized albine rat. J Neurochem 28: 897–916.
Sokoloff L, Takahashi S, Gotoh J, Driscoll BF, Law MJ (1996) Contribution of astroglia to functionally activated energy metabolism. Dev Neurosci 18: 344–352.
Sokoloff L (1999) Energetics of functional activation in neural tissues. Neurochem Res 24: 321–329.
Spence AM, Graham MM, Muzi M, Abbott GL, Krohn K, Kapoor R, Woods SD (1990) Deoxyglucose lumped constant estimated in a transplanted rat astrocytic glioma by the hexose utilization index. J Cereb Blood Flow Metab 10: 190–198.
Springett R, Wylezinska M, Cady EB, Cope M, Delpy DT (2000) Oxygen dependency of cerebral oxidative phosphorylation in newborn piglets. J Cereb Blood Flow Metab 20: 280–289.
Stein WD (1986) Transport and Diffusion Across Cell Membranes. Academic Press, New York.
Stern L, Gautier R (1921) Recherches sur le liquide céphalo-rachidien. I. Les rapports entre le liquide céphalo-rachidien et la circulation sanguine. Arch Internat Physiol 17: 138–192.
Suolinna EM, Haaparanta M, Paul R, Harkonen P, Solin O, Sipila H (1986) Metabolism of 2-[$^{18}$F]fluoro-2-deoxyglucose in tumor-bearing rats: chromatographic and enzymatic studies. Int J Rad Appl Instrumen B 13: 577–581.
Takasato Y, Rapoport SI, Smith QR (1984) An in situ brain perfusion technique to study cerebrovascular transport in the rat. Am J Physiol 247: H484–H493.
Tanaka K, Watase K, Manabe T, Yamada K, Watanabe M, Takahashi K, Iwama H, Nishikawa T, Ichihara N, Kikuchi T, Okuyama S, Kawashima N, Hori S, Takimoto M, Wada K (1997) Epilepsy and exacerbation of brain injury in mice lacking the glutamate transporter GLT-1. Science 276: 1699–1702.
Taverna RD, Langdon RG (1973) Reversible association of cytochalasin B with the human erythrocyte membrane: inhibition of glucose transport and the stoichiometry of cytochalasin binding. Biochim Biophys Acta 323: 207–219.
Tholey G, Roth-Schechter BF, Mandel P (1981) Activity and isoenzyme pattern of lactate dehydrogenase in neurons and astroblasts cultured from brains of chick embryos. J Neurochem 36: 77–81.
Thompson JK, Peterson MR, Freeman RD (2003) Single-neuron activity and tissue oxygenation in the cerebral cortex. Science 299: 1070–1072.
Thorens B, Sarkar HK, Kaback HR, Lodish HF (1988) Cloning and functional expression in bacteria of a novel glucose transporter present in liver, intestine, kidney, and beta-pancreatic islet cells. Cell 55: 281–290.
Tsukada H, Nishiyama S, Fukumoto D, Ohba H, Sato K, Kakiuchi T (2004) Effects of acute acetylcholinesterase inhibition on the cerebral cholinergic neuronal system and cognitive function: functional imaging of the conscious monkey brain using animal PET in combination with microdialysis. Synapse 52: 1–10.
Tune LE, Wong DF, Pearlson G, Strauss M, Young T, Shaya EK et al. (1993) Dopamine D2 receptor density estimates in schizophrenia: a positron emission tomography study with 11C-N-methylspiperone. Psychiatry Res 49: 219–237.
Tucker SP, Cunningham VJ (1988) Autoradiography of [$^{3}$H]cytochalasin B binding in rat brain. Brain Res 450: 131–136.
Ullrey DB, Gammon BMT, Kalckar HM (1975) Uptake patterns and transport enhancements in cultures of hamster cells deprived of carbohydrates. Arch Biochem Biophys 167: 410–416.
Ullrey DB, Kalckar HM (1981) The nature of regulation of hexose transport in cultured mammalian fibroblasts: aerobic 'repressive' control by D-glucosamine. Arch Biochem Biophys 209: 168–174.

Vafaee MS, Meyer E, Marrett S, Evans AC, Gjedde A (1998) Increased oxygen consumption in human visual cortex: respond to visual stimulation. Acta Neurol Scand 98: 85–89.

Vafaee MS, Meyer E, Marrett S, Paus T, Evans AC, Gjedde A (1999) Frequency-dependent changes in cerebral metabolic rate of oxygen during activation of human visual cortex. J Cereb Blood Flow Metab 19: 272–277.

Vafaee MS, Gjedde A (2000) Model of blood–brain transfer of oxygen explains nonlinear flow-metabolism coupling during stimulation of visual cortex. J Cereb Blood Flow Metab 20: 747–754.

Vafaee MS, Meyer E, Gjedde A (2000) Impaired activation of oxygen consumption and blood flow in visual cortex of patients with mitochondrial encephalomyopathy. Ann Neurol 48(4): 676–679.

Vafaee MS, Gjedde A (2004) Spatially dissociated flow-metabolism coupling in brain activation. Neuroimage 21: 507–515.

Vandenberg RJ (1998) Molecular pharmacology and physiology of glutamate transporters in the central nervous system. Clin Exp Pharmacol Physiol 25: 393–400.

Van Hall G (2000) Acta Physiol Scand 168: 643 Lactate as a fuel for mitochondrial respiration. Van Hall G Acta Physiol Scand 168(4): 643–656.

Van Uitert RL, Levy DE (1978) Regional brain blood flow in the conscious gerbil. Stroke 9: 67.

Varrone A, Marek KL, Jennings D, Innis RB, Seibyl JP (2001) [(123)I]beta-CIT SPECT imaging demonstrates reduced density of striatal dopamine transporters in Parkinson's disease and multiple system atrophy. Mov Disord 16: 1023–1032.

Villringer A, Dirnagl U (1995) Coupling of brain activity and cerebral blood flow: basis of functional neuroimaging. Cerebrovasc Brain Metab Rev 7: 240–276.

Villemagne V, Yuan J, Wong DF, Dannals RF, Hatzidimitriou G, Mathews WB, Ravert HT, Musachio J, McCann UD, Ricaurte GA (1998) Brain dopamine neurotoxicity in baboons treated with doses of methamphetamine comparable to those recreationally abused by humans: evidence from [11C]WIN-35,428 positron emission tomography studies and direct in vitro determinations. J Neurosci 18: 419–427.

Volkow ND, Wang GJ, Fowler JS, Logan J, Gatley SJ, Hitzemann R, Chen AD, Dewey SL, Pappas N (1997) Decreased striatal dopaminergic responsiveness in detoxified cocaine-dependent subjects. Nature 386: 830–833.

Volkow ND, Fowler JS, Wang GJ (2003) Positron emission tomography and single-photon emission computed tomography in substance abuse research. Semin Nucl Med 33: 114–128.

Volkow ND, Wang GJ, Fowler JS, et al. (2005) Imaging the effects of methylphenidate on brain dopamine: new model on its therapeutic actions for attention-deficit/hyperactivity disorder. Biol Psychiatry 57: 1410–1415.

Volkow ND, Wang GJ, Newcorn J, Fowler JS, Telang F, Solanto MV, Logan J, Wong C, Ma Y, Swanson JM, Schulz K, Pradhan K (2007) Brain dopamine transporter levels in treatment and drug naive adults with ADHD. Neuroimage 34: 1182–1190. Epub 2006 Nov 27.

Wagner HN Jr, Burns HD, Dannals RF, Wong DF, Langstrom B, Duelfer T, Frost JJ, Ravert HT, Links JM, Rosenbloom SB, Lukas SE, Kramer AV, Kuhar MJ (1983) Imaging dopamine receptors in the human brain by positron tomography. Science 221: 1264–1266.

Walker PS, Ramial T, Donovan JA, Doering TP, Sandra A, Klip A, Pessin JE (1989) Insulin and glucose-dependent regulaton of the glucose transport system in the rate L6 skeletal muscle cell line. J Biol Chem 264: 6587–6595.

Wang CY, Bassingthwaighte J (2001) Capillary supply regions. Math Biosci 173: 103–114.

Wang H, Oster G (1998) Energy transduction in the F1 motor of ATP synthase. Nature 396: 279–282.

Wang SZ, Zhu SZ, el-Fakahany EE (1994) Efficient coupling of m5 muscarinic acetylcholine receptors to activation of nitric oxide synthase. J Pharmacol Exp Ther 268: 552–557.

Weibel ER (1984) The Pathway For Oxygen, Harvard University Press, Cambridge.

Werner H, Adamo M, Lowe WL, Roberts CT, LeRoith D (1989) Developmental regulation of rat brain/HepG2 glucose transporter gene expression. Mol Endocrinol 3: 273–279.

Wertheimer E, Sasson S, Cerasi E, and Ben-Neriah Y (1991) The ubiquitous glucose transporter GLUT-1 belongs to the glucose to the glucose-regulated protein family of stress-inducible proteins, PNAS 15, 88(6): 2525–2529.

Whittam R (1962) The dependence of the respiration of brain cortex on active cation transport. Biochem J 82: 205–212.

Widdas WF (1952) Inability of diffusion to account for placental glucose transfer in the sheep and consideration of the kinetics of a possible carrier transfer. J Physiol Lond 118: 23–39.

Widdas WF (1955) Hexose permeability of foetal erythrocytes. J Physiol 127: 318–327.

Widmark EMP (1919) Studies in the concentration of indifferent narcotics in blood and tissues. Acta Med Scand 52: 87–164.

Widmark EMP, Tandberg J (1924) ber die Bedinungen fr die Akkumulation indifferenter narkotiken: theoretische Bereckerunger. Biochem Z 147: 358–369.

Wong DF, Wagner HN Jr, Dannals RF, Links JM, Frost JJ, Ravert HT, Wilson AA, Rosenbaum AE, Gjedde A, Douglass KH, Petronis JD, Folstein MF, Toung JKT, Burns HD, Kuhar MJ (1984) Effects of age on dopamine and serotonin receptors measured by positron tomography in the living human brain. Science 226(4681): 1393–1396.

Wong DF, Gjedde A, Wagner HN Jr (1986a) Quantification of neuroreceptors in the living human brain. I. Irreversible binding of ligands. J Cereb Blood Flow Metab 6: 137–146.

Wong DF, Gjedde A, Wagner HN Jr, Dannals RF, Douglass KH, Links JM, Kuhar MJ (1986b) Quantification of neuroreceptors in the living human brain. II. Inhibition studies of receptor density and affinity. J Cereb Blood Flow Metab 6: 147–153.

Wong DF, Wagner HN Jr, Tune LE, Dannals RF, Pearlson GD, Links JM, Tamminga CA, Broussolle EP, Ravert HT, Wilson AA, Toung JKT, Malat JN, Williams JA, O'Tuama LA, Snyder SH, Kuhar MJ, Gjedde A (1986c). Positron emission tomography reveals elevated D2 dopamine receptors in drug naive schizophrenics. Science 234(4783): 1558–1563. Erratum in: Science 235(4789): 623.

Wong DF, Gjedde A, Dannals RF, Lever JR, Hartig P, Ravert H, Wilson A, Links J, Villemagne V, Braestrup C, Harris J, Wagner HN Jr (1987) J Cereb Blood Flow Metab 7(Suppl I ): S356.

Wong DF, Yung B, Dannals RF, Shaya EK, Ravert HT, Chen CA, Chan B, Folio T, Scheffel U, Ricuarte GA, Neumeyer J, Wagner HN Jr, Kuhar MJ (1993) In vivo imaging of baboon and human dopamine transporters by positron emission tomography using [$^{11}$C]WIN 35,428. Synapse 15: 130–142.

Wong DF, Gjedde A (1996) Compartments and reaction volumes of brain fluid spaces: shaken, not stirred. J Nucl Med 37: 126–127.

Wong DF, Young D, Wilson PD, Meltzer CC, Chan B, Dannals RF, Ravert HT, Kuhar MJ, Gjedde A (1997a) Quantification of neuroreceptors in the living human brain: III. D2-like dopamine receptors: theory, validation and changes during normal aging. J Cereb Blood Flow Metab 17(3): 316–330.

Wong DF, Pearlson GD, Tune LE, Young LT, Meltzer CC, Dannals RF, Ravert HT, Reith J, Kuhar MJ, Gjedde A (1997b) Quantification in living human brain. IV. Effects of aging and density elevations of D2-like receptors in schizophrenia and bipolar illness. J Cereb Blood Flow Metab 17: 331–342.

Wong DF, Singer HS, Brandt J, Shaya E, Chen C, Brown J, Kimball AW, Gjedde A, Dannals RF, Ravert HT, Wilson PD, Wagner HN Jr (1997c) D2-like dopamine receptor density in Tourette syndrome measured by PET. J Nuc Med 38: 1243–1247.

Wong DF, Sølling T, Yokoi F, Gjedde A (1998a) Quantification of extracellular dopamine release in schizophrenics and cocaine use by means of TREMBLE. In: Carson RE, Daube-Witherspoon ME, Herscovitch P (eds) Quantitative Functional Brain Imaging and Positron Emission Tomography, Academic Press, San Diego, pp. 463–468.

Wong DF, Ricaurte G, Grunder G, Rothman R, Naidu S, Singer H, Harris J, Yokoi F, Villemagne V, Szymanski S, Gjedde A, Kuhar M (1998b) Dopamine transporter changes in neuropsychiatric disorders. Adv Pharmacol 42: 219–223.

Wong DF, Potter WZ, Brasic J (2000) Proof of concept: functional models for drug development in humans. In: Charney D, Coyle J, Davis K, Nemeroff, C (eds) Psychopharmacology. The Fifth Generation of Progress, Lippincott Williams & Wilkins, Baltimore, Maryland.

Wong DF, Gjedde A (2009) Monoamines: Human brain imaging. In: Squire LR (ed) Encyclopedia of Neuroscience, Vol. 5, Oxford, Academic Press, pp. 939–952.

Wong DF, Tauscher J, Gründer G (2009) The role of imaging in proof of concept for CNS drug discovery and development. Neuropsychopharmacology. 34(1): 187–203. Epub 2008 Oct 8.

Wong DF, Pomper MG (2003) Predicting the success of a radiopharmaceutical for in vivo imaging of central nervous system neuroreceptor systems. Mol Imag Biol 5: 350–362.

Wong DF, Kuwabara H, Schretlen DJ, Bonson KR, Zhou Y, Nandi A, Brasic JR, Kimes AS, Maris MA, Kumar A, Contoreggi C, Links J, Ernst M, Rousset O, Zukin S, Grace AA, Lee JS, Rohde C, Jasinski DR, Gjedde A, London ED (2006) Increased occupancy of dopamine receptors in human striatum during cue-elicited cocaine craving. Neuropsychopharmacology 31: 2716–2727.

Wong-Riley MT (1989) Cytochrome oxidase: an endogenous metabolic marker for neuronal activity. Trends Neurosci 12: 94–101.

Woolf B (1951) Computation and interpretation of multiple regressions. J R Stat Soc. Ser B (Methodol) 13: 100–119.

Wyke BD (1959) Electroencephalographic studies in the syndrome of relative cerebral hypoglycemia. Electroenceph Clin Neurophysiol 11: 602.

Wyss M, Smeitink J, Wevers RA, Wallimann T (1992) Mitochondrial creatine kinase: a key enzyme of aerobic energy metabolism. Biochim Biophys Acta 1102: 119–166.

Yarowsky PJ, Ingvar DH (1981) Neuronal activity and energy metabolism. Fed Proc 40: 2353–2362.

Ye JM, Colquhoun EQ, Hettiarachchi M, Clark MG (1990) Flow-induced oxygen uptake by the perfused rat hindlimb is inhibited by vasodilators and augmented by norepinephrine: a possible role for the microvasculature in hindlimb thermogenesis. Can J Physiol Pharmacol 68: 119–125.

Young SW, Kuhar MJ (1980) Eur J Pharmacol 62: 237–239.

Yudelevich DL, DeRose N (1971) Blood–brain barrier transfer of glucose and other molecules measured by rapid indicator dilution. Am J Physiol 220: 841–846.

Zahniser NR, Sorkin A (2004) Rapid regulation of the dopamine transporter: role in stimulant addiction? Neuropharmacology 47 (Suppl 1): 80–91.

Zauner A, Daugherty WP, Bullock MR, Warner DS (2002) Brain oxygenation and energy metabolism: Part I – Biological function and pathophysiology. Neurosurgery 51: 289–302.

Zierler K (2000) Indicator dilution methods for measuring blood flow, volume, and other properties of biological systems: a brief history and memoir. Ann Biomed Eng 28: 836–848.

Zlokovic BV, Begley DJ, Djuricic BM, Mitrovic DM (1986) Measurement of solute transport transport across the blood–brain barrier in the perfused guinea pig brain: method and application to N-methyl-$\alpha$-aminoisobutyric acid. J Neurochem 46: 1444–1451.

Zonta M, Angulo MC, Gobbol S, Rosengarten B, Hossmann KA, Pozzan T, Carmignoto G (2003) Neuron-to-astrocyte signaling is central to the dynamic control of brain microcirculation. Nature Neurosci 6: 43–50.

Zucker M, Weizman A, Rehavi M (2005) Repeated swim stress leads to down-regulation of vesicular monoamine transporter 2 in rat brain nucleus accumbens and striatum. Eur Neuropsychopharmacol 15: 199–201.

# Index

## N



## O



## P

## T



## V



## W